D1293052

Semiconductor Measurements and Instrumentation

TEXAS INSTRUMENTS ELECTRONICS SERIES

Carr and Mize ▪ MOS/LSI DESIGN AND APPLICATION

Crawford ▪ MOSFET IN CIRCUIT DESIGN

Delhom ▪ DESIGN AND APPLICATION OF TRANSISTOR SWITCHING CIRCUITS

The Engineering Staff of Texas Instruments Incorporated ▪ CIRCUIT DESIGN FOR AUDIO, AM/FM, AND TV

The Engineering Staff of Texas Instruments Incorporated ▪ SOLID-STATE COMMUNICATIONS

The Engineering Staff of Texas Instruments Incorporated ▪ TRANSISTOR CIRCUIT DESIGN

The IC Applications Staff of Texas Instruments Incorporated ▪ DESIGNING WITH TTL INTEGRATED CIRCUITS

Hibberd ▪ INTEGRATED CIRCUITS

Hibberd ▪ SOLID-STATE ELECTRONICS

Kane and Larrabee ▪ CHARACTERIZATION OF SEMICONDUCTOR MATERIALS

Luecke, Mize, and Carr ▪ SEMICONDUCTOR MEMORY DESIGN AND APPLICATION

Runyan ▪ SEMICONDUCTOR DESIGN AND INSTRUMENTATION

Runyan ▪ SILICON SEMICONDUCTOR TECHNOLOGY

Sevin ▪ FIELD-EFFECT TRANSISTORS

Semiconductor Measurements and Instrumentation

W. R. RUNYAN

Special Circuits Department
Texas Instruments Incorporated

McGRAW-HILL BOOK COMPANY

New York St. Louis San Francisco Auckland Düsseldorf Johannesburg Kuala Lumpur London Mexico Montreal New Delhi Panama Paris São Paulo Singapore Sydney Tokyo Toronto

Library of Congress Cataloging in Publication Data

Runyan, W R

Semiconductor measurements and instrumentation.

(Texas Instruments electronics series)

Bibliography: p.

1. Semiconductors. I. Title.

QC611.24.R86 537.6′22 75-19035

ISBN 0-07-054273-2

1234567890 HDBP 784321098765

The editors for this book were Tyler G. Hicks and Lester Strong, and the production supervisor was George Oechsner.
It was set in Times Roman by York Graphic Services, Inc.
It was printed by Halliday Lithograph Corporation and bound by The Book Press.

Contents

Preface

In the course of twenty years, since the introduction of the first commercial silicon transistor, semiconductor evaluation has progressed from some rather simple lifetime, Hall, and resistivity measurements and quite insensitive chemical analyses to very elegant analytical procedures and extremely sophisticated equipment for measuring the various electrical properties. In addition, the ability to detect and to appreciate the myriads of crystallographic defects which may appear in semiconductors has improved manyfold in that same twenty years. Keeping pace with this has been the literature, which regularly reports new methods or refinements of older ones and has by now reached major proportions.

My hope in preparing this book is that it will provide a guide through that maze of literature and methods that have accumulated so that the most appropriate technique for a particular problem may be chosen, and that the directions will be precise enough for most bulk measurements to be made without difficulty. It was written principally for semiconductor process engineers, failure analysis laboratory personnel, and university students faced with measuring the properties of semiconductors. However, many of the procedures are applicable to other materials as well. The depth of treatment varies from topic to topic, depending not only on its importance to the business of semiconductors, but also on the availability of background reading and how dependent some particular method is on techniques that develop only with usage, but which are seldom written down. In addition, there are hopefully enough references provided about the more important device-oriented measurements to direct the reader to them as well.

I wish to express my appreciation to the many people who helped in the preparation of this book. Mr. Stacy Watelski (Texas Instruments) coauthored Chap. 7, supplied various photographs, and generally provided useful comments. Dr. Murray Bullis (NSB), Dr. Lyndon Taylor (Univ. of Texas), Dr. W. F. Keenan, and G. P. Pollack, and Messrs A. N. Akridge, Andreas Niewold, and A. F. Polack (all of Texas Instruments) critically read various chapters. Mrs. Hettie Smith typed the manuscript as it went through its many iterations, and my daughter Kay assisted with the proofreading.

Dallas, Texas

W. R. Runyan

1

Crystal Orientation

The values obtained when the various measurements discussed in later chapters are made may depend on the orientation of the crystallographic face on which measurements are being made and the direction in which some stress (voltage, force, temperature, etc.) is applied. Because of these dependencies, and because many of the semiconductor processes are dependent on orientation, considerable emphasis must be placed on crystal orientation. Some of the more common properties and their behavior are summarized in Table 1.1. Table 1.2 summarizes the orientation-dependent processes that might be expected. In general, if a property is anything other than a scalar, it will be direction-sensitive in most crystalline materials.[1] Those properties which are described by a second-rank tensor, and this includes almost all the properties of common interest in the semiconductor industry (resistivity, thermal conductivity, diffusion coefficients), are independent of direction in cubic crystals (which encompasses the majority of present commercially important semiconductors).

There are, however, some problems in trying to predict behavior a priori. They arise when it is difficult to determine the defining equation of the property (e.g., hardness and etch rate) or because an apparently simple measurement of one property may in some subtle way involve additional phenomena. Resistivity should be independent of the direction of the electric field for cubic crystals, but for the specific case of spreading-resistance measurements, different values are obtained with the same equipment when measurements are made on different faces of silicon. Also, even though the diffusion coefficient should be independent of direction, it is observed that under the same condition of surface ambient, temperature, and time, the diffusion-depth of phosphorus can be different in the [111] and the [100] directions. Further, there may be physical constraints, e.g., very thin layers, in which the crystal is no longer three-dimensional. Under such circumstances, properties normally isotropic may become directionally dependent. It is this phenomenon which causes orientation differences in the carrier mobility of silicon inversion layers.

1.1 CRYSTALLOGRAPHY

Before the various means for determining crystallographic orientation are discussed, a brief survey of the appropriate crystallographic nomenclature will be given.

Plane Indices. The various planes that pass through a crystal may be described in terms of the reciprocal of the intercepts of that plane with the crystallographic axes. These reciprocals are usually expressed as the smallest possible integers having the same ratio and, for those crystal systems with three axes, are written as (hkl). For the hexagonal crystal system, in which there are three coplanar axes as well as one perpendicular to the plane of the first three, the indices are $hkil$. h, k, and

Table 1.1. Partial Listing of Directional Properties of Crystals

Tensor rank	Property	Symbol	Relates	Form of relation	Isotropic in
0	Density Heat capacity	δ C	Mass to volume Heat transferred to temperature changes	A scalar to a scalar	All classes
1	Pyroelectric coefficient	P_i	Electrical polarization to temperature change	A vector to a scalar	None
2	Electrical conductivity Electrical mobility Thermal Diffusion coefficient	σ_{ik} μ_{ik} k_{ij} D_{ij}	Current density to applied field Current density to applied field and number of carriers Heat transferred to temperature gradient Current density to concentration gradient	A vector to a vector	All cubic
	Thermal expansion	α_{ij}	Elongation to temperature change	A scalar and 2d-rank tensor	
3	Piezoelectric coefficient	d_{ijk}	Polarization to applied stress	A vector to a 2d-rank tensor	None
4	Elastic constants Piezoresistance	C_{ijkl} π_{ijkl}	Stress to elongation Change of resistivity to applied stress	Two 2d-rank tensors	None

Adapted from J. F. Nye, "Physical Properties of Crystals," Oxford University Press, New York, 1960.

Table 1.2. Orientation-Dependent Semiconductor Processes and Parameters

Wet etch (will depend on etchant)	Important aspect of dielectrically isolated Si ICs
Diffusion depth	Not expected. See, for example, L. E. Katz, "Orientation Dependent Diffusion Phenomena," Paper 23, National Bureau of Standards, *Spec. Pub.* 337, 1970, for a discussion
Depth of ion implant	
Buried-layer pattern wash out	Effect minimized when growth surface slightly misoriented from low-index plane
Vapor-phase epitaxy	Best growth when surface slightly misoriented from low-index plane. Growth surface rate may depend on orientation
Scribe and break	Scribe lines for good scribe-break operation must be accurately aligned with predetermined orientation
Mechanical polish	Also reflected in depth of damage after mechanical polish or abrasion. See, for example, A. W. Fisher and J. A. Amick, *J. Electrochem. Soc.,* **113**:1054–1060 (1966)
Channel mobility in Si MOS devices	
Surface-charge density of Si MOS devices	

i are the reciprocals of the intercepts of the plane in question with the three coplanar axes, and from the geometry it can be readily shown that $h + k = -i$. Since *h* and *k* together completely determine *i*, it is not necessary to write all four indices, but some method of indicating the hexagonal system is required. One common notation is *hk*:*l*. Examples of various planes are shown in Fig. 1.1. A complete family, composed of all possible planes resulting from permuting a given set of indices (including negative values), is denoted by braces; e.g., {111} represents the eight planes (111), $(\bar{1}11)$, $(1\bar{1}1)$, $(11\bar{1})$, $(\bar{1}\bar{1}1)$, $(1\bar{1}\bar{1})$, $(\bar{1}1\bar{1})$, $(\bar{1}\bar{1}\bar{1})$. The bar over an index number indicates a negative intercept.

The indices of a direction through a crystal are written as [*hkl*]. A complete set of equivalent directions is written as ⟨*hkl*⟩. For a cubic system, the direction indices will always be perpendicular to a plane with the same indices, but in other systems this is not generally true.

Angles between Planes. Expressions for calculating the angle between any two planes (*hkl*) and (*h'k'l'*) are available.[2] Except for cubic crystals, dimensions of the unit cell are involved and, since these numbers will be different for different materials, the angles between given planes can be determined only independent of the material when the structure is cubic. Table 1.3 lists angles between low-index cubic-crystal planes; it is very useful in the identification of facets.

Relative Position of Planes. Probably the easiest way to visualize the positions of the various planes is by means of a model. Figure 1.2*a* is a photograph and plans of a very convenient one for cubic crystals. The pattern has been reduced for publication, and unless one is very dexterous, it will probably need to be enlarged before construction. If a small magnet is glued to the inside of one face, the model

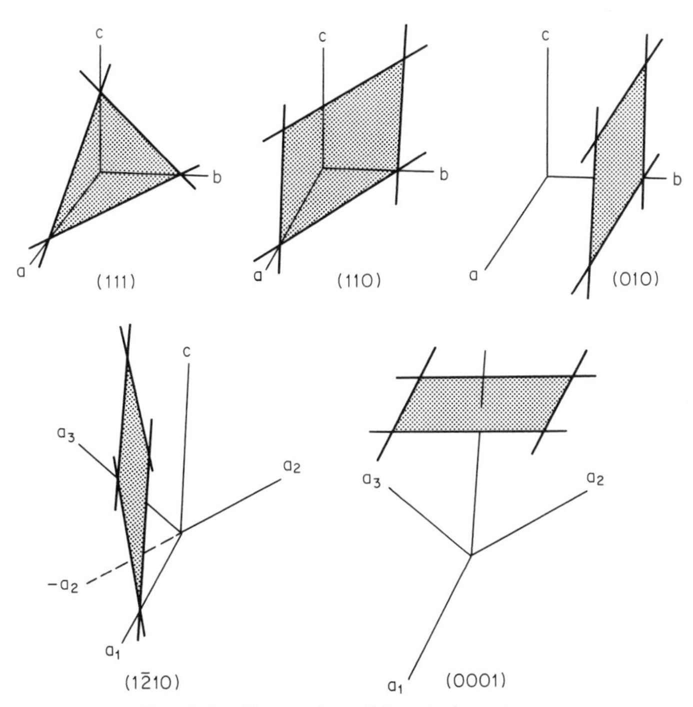

Fig. 1.1. Examples of low-index planes.

Table 1.3. Angles between Crystallographic Planes (and between Crystallographic Directions) in Crystals of the Cubic System

{*HKL*}	{*hkl*}	Values of angles between *HKL* and *hkl* planes (or directions)					
100	100	0.00	90.00				
	110	45.00	90.00				
	111	54.74					
	210	26.56	63.43	90.00			
	211	35.26	65.90				
	221	48.19	70.53				
	310	18.43	71.56	90.00			
	311	25.24	72.45				
110	110	0.00	60.00	90.00			
	111	35.26	90.00				
	210	18.43	50.77	71.56			
	211	30.00	54.74	73.22	90.00		
	221	19.47	45.00	76.37	90.00		
	310	26.56	47.87	63.43	77.08		
	311	31.48	64.76	90.00			
111	111	0.00	70.53				
	210	39.23	75.04				
	211	19.47	61.87	90.00			
	221	15.79	54.74	78.90			
	310	43.09	68.58				
	311	29.50	58.52	79.98			
210	210	0.00	36.87	53.13	66.42	78.46	90.00
	211	24.09	43.09	56.79	79.48	90.00	
	221	26.56	41.81	53.40	63.43	72.65	90.00
	310	8.13	31.95	45.00	64.90	73.57	81.87
	311	19.29	47.61	66.14	82.25		
211	211	0.00	33.56	48.19	60.00	70.53	80.40
	221	17.72	35.26	47.12	65.90	74.21	82.18
	310	25.35	40.21	58.91	75.04	82.58	
	311	10.02	42.39	60.50	75.75	90.00	
221	221	0.00	27.27	38.94	63.61	83.62	90.00
	310	32.51	42.45	58.19	65.06	83.95	
	311	25.24	45.29	59.83	72.45	84.23	
310	310	0.00	25.84	36.87	53.13	72.54	84.26
	311	17.55	40.29	55.10	67.58	79.01	90.00
311	311	0.00	35.10	50.48	62.96	84.78	

A more complete listing may be found in R. J. Peavler and J. L. Lenusky, "Angles between Planes in Cubic Crystals," *Spec. Rept.* 8, American Institute of Mining, Metallurgical and Petroleum Engineers.

can be stored by sticking to metal wall paneling. The shape is not a standard crystallographic form but rather is comprised solely of the complete sets of {100}, {110}, and {111} planes. Crystal models such as those shown in Fig. 1.2*b* are also helpful, and paper patterns for a great number of them are available.*

The relative positions of the various planes may also be indicated by stereographic-projection charts.[3],† These projections are made by surrounding the crystal with an imaginary sphere and drawing lines from its center to the spherical surface

*Arthur J. Gude, "Three Dimension Models of the Basic Crystal Forms," cut-out kit available from Polycrystal Book Service, Pittsburgh, Pa.

†Because the angles between planes for noncubic crystals vary with cell dimensions, projections are normally used only for the cubic system, and the following discussion assumes cubic crystals.

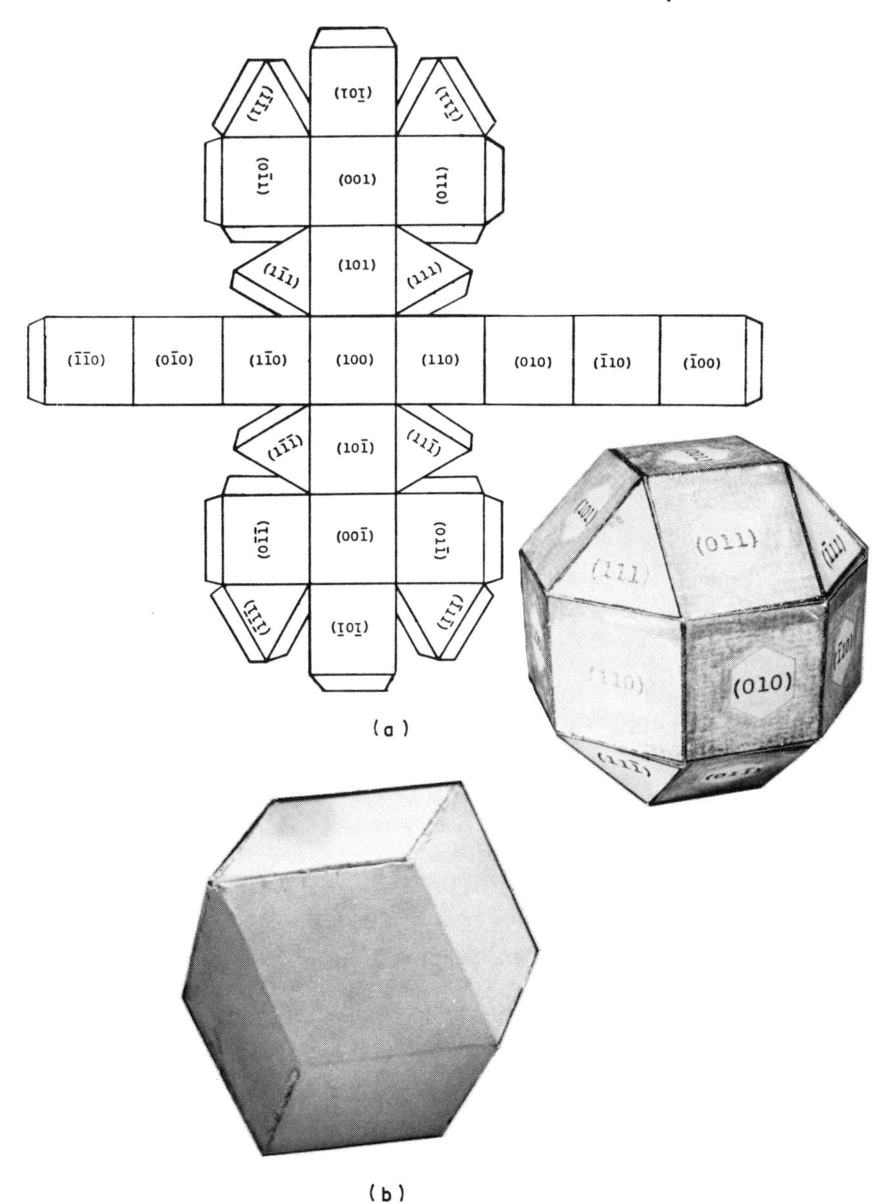

Fig. 1.2. Paper models. (*a*) Pattern and photograph of a figure comprised of {111}, {110}, and {100} faces. (*b*) Dodecahedron.

in such a manner that they are normal to the planes of interest. If an *hkl* projection is desired, the crystal is oriented so that its (*hkl*) plane is normal to the "north pole" of the imaginary sphere and the equatorial (projection) plane which divides the sphere into an upper and lower hemisphere is parallel to the (*hkl*) plane. Lines are then drawn connecting the "south pole" with the intersection of the various plane normals and the northern hemisphere.

The points where the lines pass through the projection plane represent the planes in question. This is illustrated in Fig. 1.3 for the specific case of a (111) plane

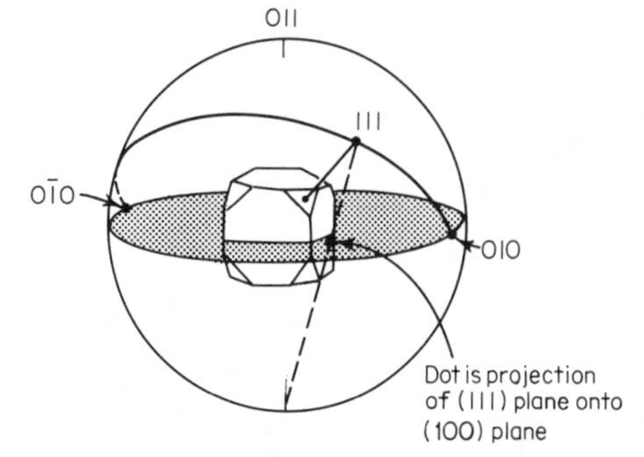

Fig. 1.3. Development of a stereographic projection.

projection onto a (100) plane. Note that all dots on the periphery of the projection plane represent planes perpendicular to the plane of projection and their angles relative to each may be obtained graphically by drawing tangents to the circle as shown in Fig. 1.4*a*.

The traces* of other planes (i.e., those represented by the interior dots) on the projection plane may be found by drawing a line through the appropriate dot and perpendicular to the line connecting the dot with the center of the projection as in Fig. 1.4*b*. However, angles between the planes themselves can be found only by superimposing a nonlinear gridwork over the projection. Because of this, it is more convenient to use tables (e.g., Table 1.3) for angles between planes,† but angles between traces are most easily obtained from the projections just described. Projections for several of the more common planes are shown in Figs. 1.5 to 1.7. It is very helpful to be able to mark on the projection; so it is suggested that a sheet of thin, clear plastic (e.g., a viewgraph jacket) be kept handy for laying on top of the page. A grease pencil can then be used for nondestructive sketching.

In the event that it is desired to calculate trace angles rather than obtain them from the projections (or for planes not on the projections), the direction cosines u, v, and w of the line of intersection of a plane $h'k'l'$ with plane hkl are given by

$$u = \begin{vmatrix} kl \\ k'l' \end{vmatrix} \qquad v = \begin{vmatrix} lh \\ l'h' \end{vmatrix} \qquad w = \begin{vmatrix} hk \\ h'k' \end{vmatrix}$$

*The line formed by the intersection of two planes.

†Similar gridworks are used in interpreting x-ray diffraction patterns and for that purpose are most useful.

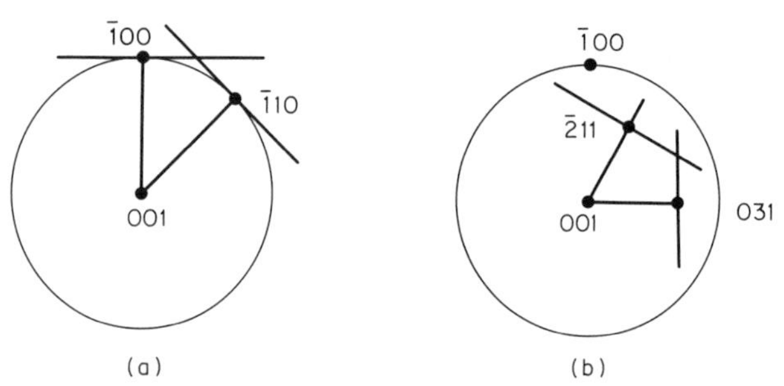

Fig. 1.4. Standard projection interpretation. (*a*) The ($\bar{1}$00) and ($\bar{1}$10) planes are both perpendicular to the (001) plane and intersect at a 45° angle. (*b*) The traces of the (211) and (031) planes on the (001) reference surface make an angle of 63.5° with each other.

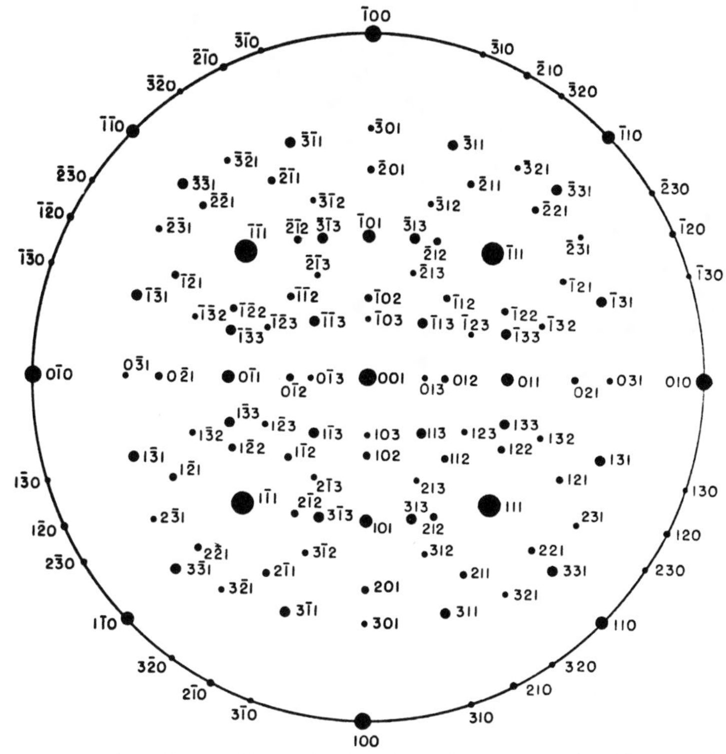

Fig. 1.5. Standard (001) projection for a face-centered cubic crystal.

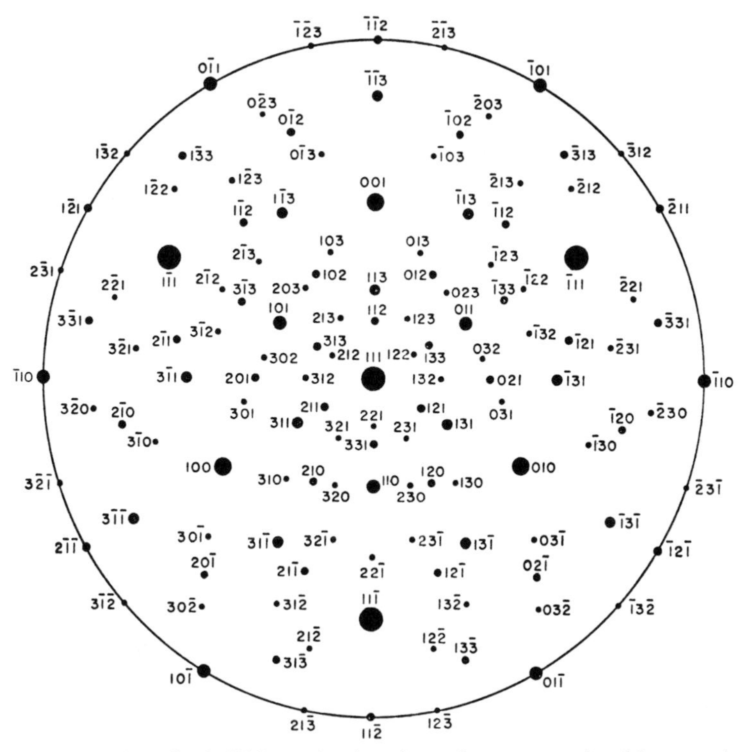

Fig. 1.6. Standard (111) projection for a face-centered cubic crystal.

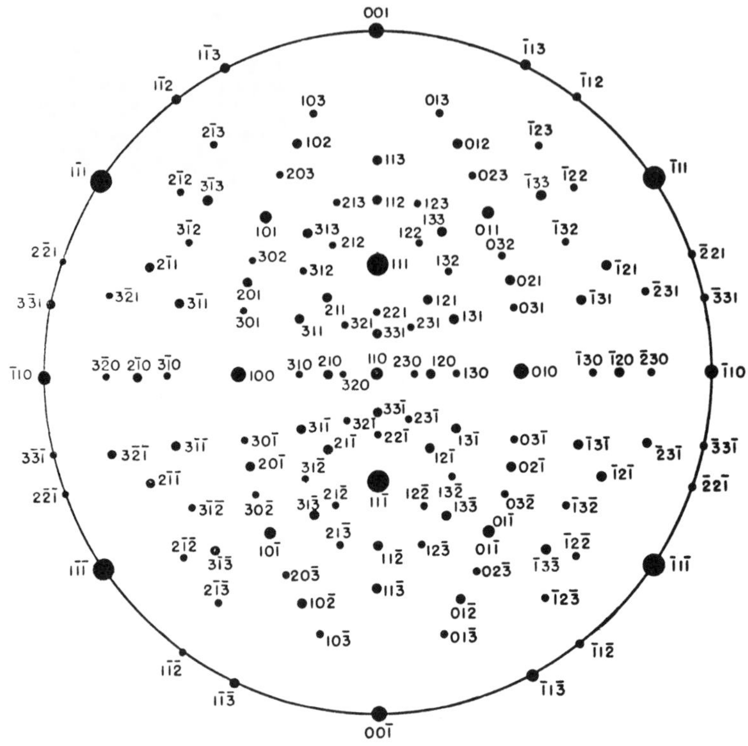

Fig. 1.7. Standard (110) projection for a face-centered cubic crystal.

The angle between the two lines is then given by the same equation used to compute the angle between planes.

1.2 ORIENTATION

There are numerous ways in which the orientation of crystals can be determined (see Table 1.4), but they can be broken down into the two broad categories of visual observation of distinguishing features such as growth facets or etch pits, and x-ray diffraction. All the visual methods require prior knowledge of growth, etch, or

Table 1.4. Orientation Methods

Method	Comments
Visual observation of gross features	Must show natural growth faces, poor accuracy
Microscopic observation of minute growth facets	Accuracy medium, usually restricted to low-index planes
Examination of fracture characteristics	May be destructive, poor accuracy
Visual or microscopic observation of etch pits	Selective etches must be available—accuracy medium, most appropriate to low-angle planes
Optical reflectograms	Accuracy quite good when orienting low-index planes
Laue x-ray	Applicable to any orientation
X-ray diffractometer	Highest accuracy, but no ability to orient azimuthally

fracture characteristics of the particular material being studied. The use of x-ray orientation relieves the need for this additional information but requires specialized equipment. In any event, it is assumed that the crystal structure of the semiconductor being examined has already been determined, since those procedures are considered outside the scope of this book.

Gross Features. When grown from an essentially pure melt, silicon, germanium, GaAs, InSb, and many other materials have as their most slowly growing planes the $\{111\}$ family. Even though thermal gradients may offer severe growth constraints, it is virtually impossible to grow a crystal by Czochralski or float zone without some trace of $\{111\}$ planes being visible. Zone leveling and Bridgeman growth, however, usually completely obliterate such features. Vapor-phase growth of these materials also usually has (111) planes as the most slowly growing ones, but this can change with both the choice of feedstock and the concentrations used. For example, Te-transported Si has (100) faces as the slow ones. Further, most semiconductor vapor-phase crystal-growth conditions (e.g., for silicon epitaxy) are chosen to give replicas of the seeding surface and to produce as few facets as possible.

When pulled crystals (Si, Ge, etc.) are grown in the [111] direction, and if the top is quite flat, the (111) plane perpendicular to the growth direction can develop and large "flats" will be visible. If the crystal diameter is increased more slowly as it is being grown so that the crystal grows out at an angle near 70° [the angle between the (111) and the $(1\bar{1}1)$, $(\bar{1}11)$, $(11\bar{1})$ planes], flats will occur where the $(\bar{1}11)$, $(1\bar{1}1)$, and $(11\bar{1})$ planes are tangent to the growing crystal. An example of this is shown in Fig. 1.8. If the crystal starts growing in, i.e., reducing in diameter, and if the angle is again near 70°, $(\bar{1}\bar{1}1)$, $(\bar{1}1\bar{1})$, and $(1\bar{1}\bar{1})$ facets will develop and will be rotated 60° on the crystal from the previous set.

Ordinarily, regardless of the angle at which the top grows out, six ridges will be visible at the point where facets should be found. When a straight-sided crystal is grown, depending on specific growth conditions, it may have either three or six marks extending down the sides. Three of these will be relatively flat; they are due to the $(\bar{1}11)$, $(1\bar{1}1)$, and $(11\bar{1})$ planes and will always show. The other three will be raised and correspond to the $(\bar{1}\bar{1}1)$, $(1\bar{1}\bar{1})$, and $(\bar{1}1\bar{1})$ planes.

When other orientations are grown, the same general features occur. That is, there are marks where the various (111) planes intersect the growing periphery of the

Fig. 1.8. Silicon crystal grown in [111] direction and showing pronounced (111) faceting.

crystal. Figure 1.9 shows typical tops for (111), (100), and (110) grown crystals and also indicates the various crystallographic directions with respect to the ridges and facets.

In looking for natural features, caution must be exercised, since they may be confused with artificially induced growth constraints. For example, a zone-leveled crystal grown in a boat with a smooth, flat surface and then subsequently broken into several chunks could show an apparent facet on some of the pieces. If the history of those pieces were unknown, an erroneous orientation might be assumed. Likewise, if a crystal were observed which had a regular cross section but was long compared with its cross-sectional dimension, it might be assumed that the crystal was hexagonal or perhaps rhombohedral. However, if there were some unexpected growth condi-

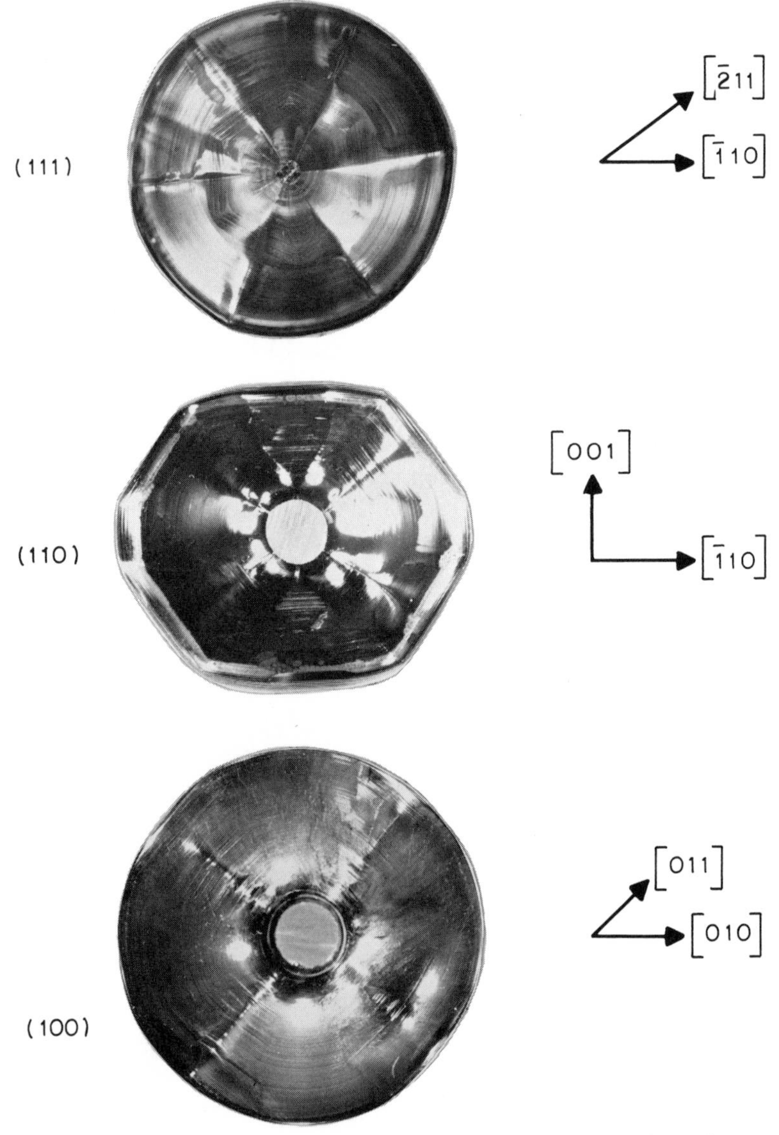

Fig. 1.9. Shape of silicon-crystal tops for various growth orientations.

tions, the crystal could, for example, be cubic, and growth in the (111) direction then be enhanced by a vapor-liquid-solid (VLS) process. When several facets are visible, a measurement of their relative angles will often allow the orientation to be determined. If the crystal is sizable, a protractor may suffice, but for small samples or small facets on a large sample, a vertically illuminated microscope coupled with a precision goniometer[4] ("microgoniometer") is required.

Cleavage or Fracture. Most brittle semiconductors exhibit cleavage planes, and if these planes are known, they can be used for orientation. Silicon, germanium, and diamond cleave most easily and most often between (111) planes, whereas for the III–V compounds, (110) planes are most likely to separate.

If the semiconductor surface is sandblasted, it will usually chip out pieces in such a manner that the cleavage planes are exposed. In this case optical orientation techniques to be described later can be used without the requirement for first etching the surface. Thin samples such as silicon can be broken by placing them on a resilient backing such as a stack of absorbent paper and pressing down in the center with a blunt probe (e.g., a pencil eraser) until they shatter.

Typical fracture patterns for the low-index planes (111) and (100) are shown in Fig. 1.10. If azimuthal orientation within the pieces is desired, it can be determined by remembering that the traces of the cleavage planes for materials which break along (111) planes are identical with those of the exposed planes of a (111) etch pit, so that the directions of Fig. 1.11* can be used. While it might at first be considered that fracture for orientation is destructive and very wasteful, it can sometimes be used with little or no additional loss or damage over that which normally occurs. For example, the seeding tip of Bridgeman crystals can be broken off to determine the approximate orientation, and a slight increase in the chipping of the edges of silicon slices is often observed where the traces of (111) planes are tangent to (100)-oriented slices. Further, under some circumstances, it may be less expensive to break a slice than to use some of the more elegant methods to be mentioned later.

Twin Planes. If the material in question has only a single twinning plane, and if that plane is known, visible twins can be used to establish directions. Si, Ge, diamond, and most III–V compounds twin only on a (111) plane.

Etch Pits.[5] The etch-pit shape can be observed in a microscope and used to make

*See the discussion of etch pits below.

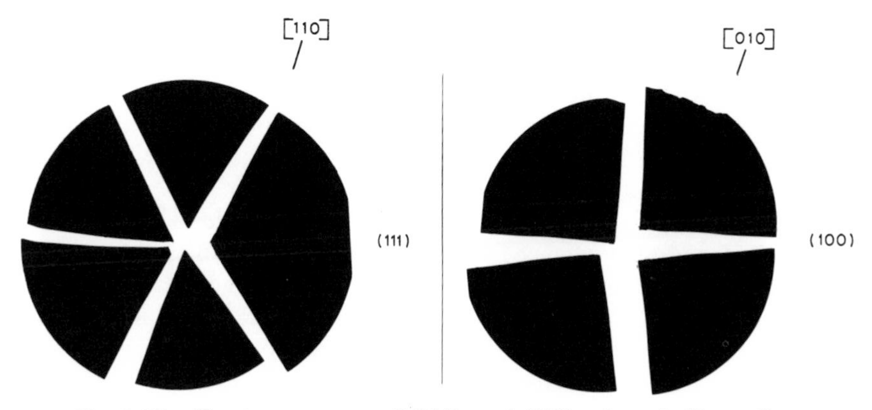

Fig. 1.10. Fracture patterns of (111)- and (100)-oriented silicon slices.

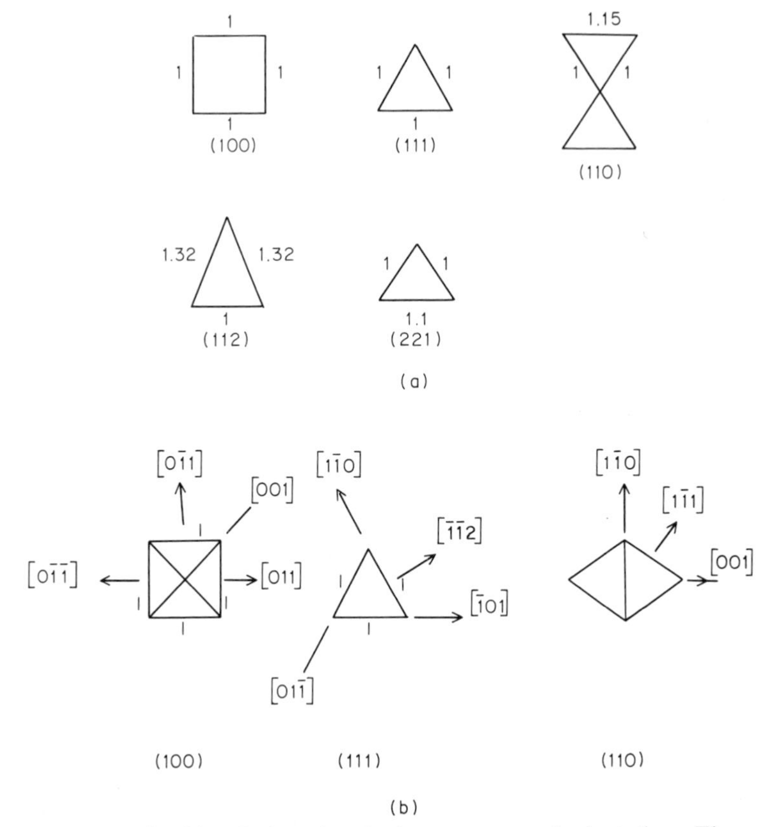

Fig. 1.11. Stacking-fault and etch-pit geometry and orientation. These drawings are applicable to cubic crystals (*a*) having a (111) stacking-fault system and (*b*) when using a {111} selective etch. Well-formed (110) etch pits are hard to generate in Si. Overetching may change their character and produce serious error.

a rough determination of plane, but the most useful aspect is that it allows directions in the plane of a thin slice to be rather easily determined (a thing not readily done otherwise). The use of etch pits depends on the availability of a very selective etch which preferentially exposes a given orientation. For diamond and zinc blende structures, etchants specifically for (111) planes are readily available and commonly used. For this class of etchant, if the surface were originally near a (111) plane, there would be little three-sided inverted pyramids (tetrahedra) etched in the surface. For a (100) plane, four-sided pyramids will result. In the case of (110), a diamond-shaped aperture is formed. The surface patterns are shown in Fig. 1.11 along with azimuthal directions and relative dimensions.[6] Etchants that are suitable for dislocation etch-pit studies are also applicable to orientation work, since well-developed pits (whatever their source) are the prime requirement. Numerous specific etch formulations are tabulated in Chap. 2, and the reader is referred to them as well as to the etchants used specifically for optical orientation. In addition to the more usual aqueous etchants, many molten metals produce sharp, well-defined pits. If contacts are removed from devices, it is often found that enough microalloying occurred to allow the original slice orientation to be determined. Slow evaporation (thermal etching) and various high-temperature vapor-phase reactions

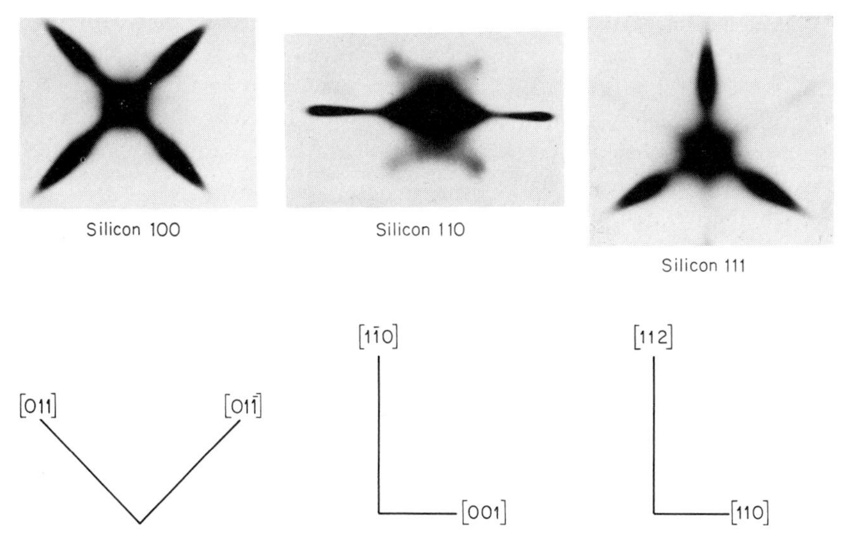

Fig. 1.12. Optical-orientation patterns for materials which develop (111)-bounded etch pits. (*Micromech Manufacturing Co.*)

also can give etch figures useful for orientation. In any case, care must be taken to ensure that only known planes are exposed. Most etchants, after prolonged time, either round off corners or show higher-order planes. In some circumstances the original shape of the pit may be so changed that the initial planes are no longer recognizable. Germanium (110) surfaces in particular are quite susceptible to a change in apparent orientation because of overetching. In some materials, the etch-pit geometry can be radically changed by the etchant. For example, diamond may develop either (111) or (100) pit faces, depending on etch composition. Therefore, when using etch pits whose history is unknown, considerable caution in interpretation is indicated.

If stacking faults are available, they too can be used for orientation purposes. In the diamond and zinc blende structures such faults occur along (111) planes, and hence their intersection with the surface produces the same pattern as that of an etch pit. Stacking faults are prevalent in epitaxial layers and may sometimes be seen directly with the aid of a phase- or interference-contrast microscope. Otherwise etching is required (see Chap. 2 for details). Even if etching is required, they have the advantage of not rounding off as etch pits often do. Thus, accurate measurements of the side lengths can be made and deviations from symmetry used for determining a small amount of misorientation. For very small misorientations from the (111) plane, Ref. 7 gives directions for converting relative lengths into the amount of misorientation.

If the etch-pit geometry is unknown, the angles between the exposed planes can be determined, and by assuming that they are low-index, an assignment can often be made by comparing the measured angle with those in Table 1.3. For some cases, a graduated rotatable stage combined with a metallurgical microscope will suffice for making the measurements, but usually a microgoniometer is required.

Optical Reflectograms.[8-18] More accurate optical orientation can be accomplished by observing the pattern which etch or fracture pits reflect back from a beam of collimated light impinging normal to the surface. The little facets act as mirrors and return the light in well-defined patterns. Typical reflectograms for

{111}-bounded etch pits in cubic crystals are shown in Fig. 1.12 along with directions in the plane of the crystal. Two slightly different optical systems as shown in Fig. 1.13 may be used for producing the light figures. When constructing the shape of expected reflections from geometrical considerations, it should be remembered that because of the sidewall angle of most pits, a beam must suffer an additional reflection, as shown in Fig. 1.13, before emerging. If the physical surface is slightly misoriented from the crystallographic plane, the image will be asymmetrical. By measuring the amount of tilt necessary to produce a symmetrical pattern, the amount of misorientation of the crystal lattice from the physical surface is directly determined. The azimuthal position of the spots relative to the sample also allows directions in the plane to be uniquely determined.

The success and accuracy of the reflectogram method depend on both the equipment used and the generation of sharp, well-defined pits. Table 1.5 lists both aqueous and molten-metal etches which are suitable. The former are easier to use than the others but provide less accuracy than the metals. The pits may also be formed by abrasive cleaving, i.e., lapping or sandblasting, but if the cleavage planes are different from slow etching planes, a different interpretation from Fig. 1.12 is required.[16] In addition, the mechanism used for holding the slice or crystal and changing its orientation relative to the screen must be sturdy and have precision-measuring capability if any but the very crudest orientation is to be done.

Polarity Differentiation. For crystals of the diamond family, the (111) and $(\bar{1}\bar{1}\bar{1})$ faces appear identical, but for crystals with zinc blende structure, this is not true, since each double layer of the stacking sequence consists of one sheet of component A atoms and one sheet of component B atoms. Thus, depending on whether the crystal is traversed in the [111] or a $[\bar{1}\bar{1}\bar{1}]$ direction, A atoms or B atoms will first be encountered. A similar disparity exists between the [0001] and the $[000\bar{1}]$ directions of wurtzite crystals; so in either case differences in such characteristics as etching behavior and x-ray scattering are to be expected.

For an original determination, anomalous x-ray dispersion is most applicable.[20-22] This method makes use of the fact that x-ray reflections from (111) and $(\bar{1}\bar{1}\bar{1})$ or (0001) and $(000\bar{1})$ faces of noncentrosymmetric crystals will have slightly different

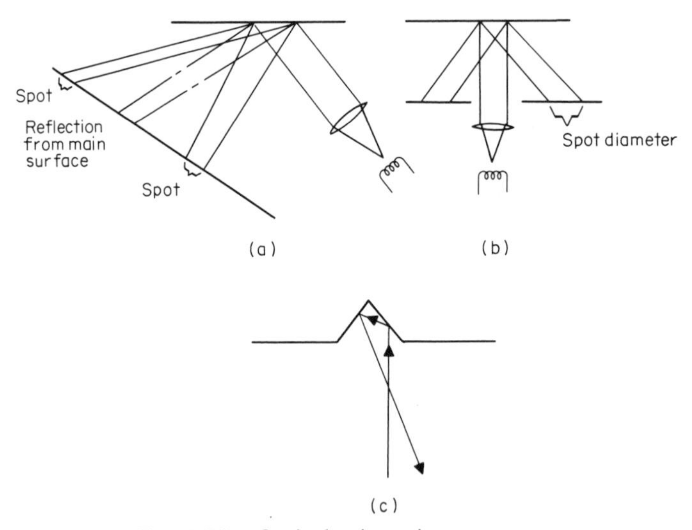

Fig. 1.13. Optical-orientation geometry.

Table 1.5. Etches for Optical Reflectograms

Semiconductor	Etch	Reference
Silicon	5% sodium hydroxide in water at 85°C for 5 min 50% sodium hydroxide (by weight) solution or 50% potassium hydroxide (by weight) solution at 65°C for 5 min	19
	Molten indium at 600°C, remove indium with HCl	12
	Molten Ga at 500°C, remove Ga with HCl	12
Germanium	1 part (volume) hydrofluoric acid (49%) 1 part (volume) hydrogen peroxide (30%) 4 parts (volume) water at 25°C for 1 min	19
	Molten indium at 450°C, remove indium with HCl	12
	Molten Ga at 137°C, remove Ga with HCl	12
GaAs	1 part (volume) HNO_3 (60%), 1 part water	18
	Molten Ga at 800°, remove Ga with HCl	12
InSb	Molten indium at 350°, remove indium with HCl	12
GaSb	Molten Ga at 500°, remove Ga with HCl	12
InAs	Molten indium at 600°, remove indium with H_2SO	12
AlSb	Molten aluminum at 800°, remove aluminum with NaOH	12

intensities which can be calculated from scattering factors. If the crystal is electrically active, the sign of the piezoelectric voltage can be used for face determination.[23] After a unique determination is once made, differences in etching characteristics may be used.[23-27] Table 1.6 lists some appropriate etches, but most of them are somewhat technique-oriented; so care should be exercised in their use and interpretation.

For some materials, e.g., GaAs, one set of (111) planes may grow more slowly than the other. When this occurs, a crystal grown in the [111] direction will have a shape distinctly different from one grown in the $[\bar{1}\bar{1}\bar{1}]$ direction. If the shape is not completely symmetrical, optical reflectograms can also be used for face determination if the original growth direction is known.[16] This is because etch or fracture pits developed on the two faces will be rotated 60° (for cubic crystals) from each other in the plane of the surface and thus the reflection pattern and the sample outline will bear a different relation to each other depending on which face is reflecting.

X-Ray Goniometer. If the wavelength of the x-rays used and the spacing of the desired plane are known, the angle at which coherent scattering, or "reflection," is expected can be calculated. The source and detector may be set at the proper angles and the surface of the crystal rocked until a maximum occurs. It follows that the desired crystallographic plane makes an angle θ with the beam and thus lies along the reference plane of Fig. 1.14. If the surface of the crystal does not also lie in this plane, it is misoriented by the difference between the two. The angle θ for a given plane is found from

$$\sin\theta = \frac{n\lambda}{2d}$$

where λ is the x-ray wavelength being used and d the spacing between the planes

Table 1.6. Etches for Polarity Determination

Material	Etch*	Comments	References
AlSb	1 H_2O_2 + 1 conc. HF + 1 water for 1 min, followed by 1 conc. HCl + 1 conc. HNO_3 for 2 s		26
CdS {0001}	6 fuming HNO_3 + 6 glacial acetic acid + 1 H_2O for 2 min	Sulfur film forms on Cd surface. Hexagonal pits on sulfur surface	24
{0001}	1 conc. HCl + 1 conc. HNO_3	Cd surface grainy, sulfur surface develops conical etch pits	24
CdSe {111}	30 conc. HNO_3 + 0.1 conc. HCl + 10 glacial acetic acid + 20 of 18*N* H_2SO_4 for 8 s at 40°C	Cd surface develops hexagonal-shaped pits, both surfaces develop selenium films	24
CdTe {111}	3 HF + 2 H_2O_2 + 1 H_2O for 2 min	Triangular etch pits on tellurium surface	24
GaAs {111}	2 HCl + 1 HNO_3 + 2 H_2O	Ga face has etch pits, As face none	25
	1 HNO_3 + 2 H_2O	Ga face has pits, As face none	23
GaSb	2 conc. HNO_3 + 1 conc. HF + 1 glacial acetic acid for 15 s	Etch pits on Ga face	26
HgSe {111}	6 conc. HCl + 2 conc. HNO_3 + 3 H_2O, 2–5 min intervals at 25°C, remove film in 50 HNO_3, 10 acetic, 1 HCl, 20 of 18*N* H_2SO_4	Se surface develops triangular etch figures, Hg surface develops craterlike structure	24
HgTe {111}	1 conc. HCl + 1 conc. HNO_3 1-min intervals after chemical polishing in 6 conc. HNO_3, 1 conc. HCl, 1 H_2O	Mercury surface develops triangular etch pits, Te surface has flat, grainy appearance	24
InAs	0.4*N* Fe^{3+} in conc. HCl for 30 min	Etch pits on In face	26
InP	0.4*N* Fe^{3+} in conc. HCl for 1.5 min		26
InSb {111}	2 conc. HNO_3 + 1 conc. HF + 1 glacial acetic acid, 4 s	Etch pits on In (111) face	26
	0.2*N* Fe^{3+} in 6*N* HCl	Hexagonal etch figures on In face triangular figures on Sb face	29
SiC {111}	6% Cl_2 + 26% O_2 in argon at 850–900°C, fused 75% NaOH, 25% Na_2O_2 at 700°C	Etch pits on {111} Si face	27
{0001}	Molten Na_2O_2 + $NaNO_2$	Smooth Si surface, rough carbon surface	28
ZnS {111}	0.5m $K_2Cr_2O_7$ in 16*N* H_2SO_4 for 10 min at 95°C	Sulfur surface irregular, triangular etch figures on zinc surface	24
ZnTe {111}	3 HF + 2 H_2O_2 + 1 H_2O for 2 min	Film on zinc surface, triangular etch figures on tellurium surface	24

*All parts by volume. Conc. HNO_3 = 70%. Conc. HF = 48%.

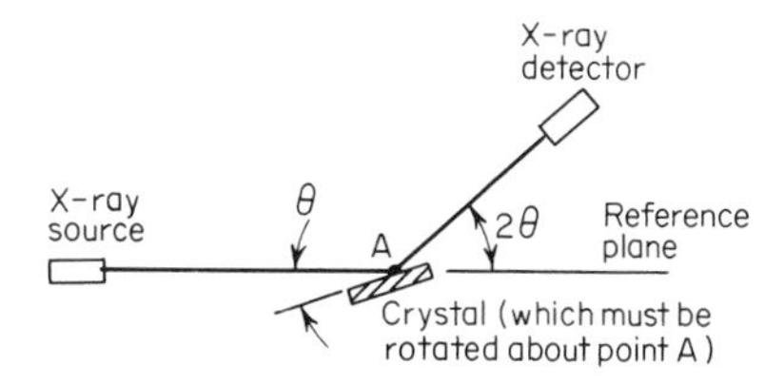

Fig. 1.14. X-ray orientation.

in question. The d value may in turn be calculated from the equations of Table 1.7 if the lattice spacing is known. Table 1.8 gives the angles for GaAs, Ge, and Si when using the Cu Kα line. Some care must be taken at this juncture, since more than one plane can sometimes have the same lattice spacing and hence the same θ value. An example is the (333) and (511) planes. In order to prevent misinterpretation, more than one order can be checked; for example, also look for a (111) reflection. Depending on the actual atomic positions, reflections from some planes may be very weak and difficult or impossible to find. Therefore, before wasting time looking for a particular plane, one should determine whether or not reflections are really expected. For the diamond lattice, they are expected only when

Table 1.7 ***d*** **Values for Cubic Crystals**

Cubic	$d = \dfrac{a}{(h^2 + k^2 + l^2)^{1/2}}$
Tetragonal	$\dfrac{1}{d^2} = \dfrac{h^2 + k^2}{a^2} + \left(\dfrac{l}{c}\right)^2$
Hexagonal	$\dfrac{1}{d^2} = \dfrac{4}{3}\dfrac{h^2 + hk + k^2}{a^2} + \left(\dfrac{l}{c}\right)^2$
Orthorhombic	$\dfrac{1}{d^2} = \left(\dfrac{h}{a}\right)^2 + \left(\dfrac{k}{b}\right)^2 + \left(\dfrac{l}{c}\right)^2$

	Si	Ge	GaAs
d_{100}	5.43A	5.66	5.65
d_{110}	3.83	3.99	3.99
d_{111}	3.13	3.26	3.26
d_{123}	1.56	1.63	1.62

Table 1.8. Bragg angles θ, for X-Ray Diffraction of CuK$_\alpha$ Radiation
(Wavelength λ = 1.54178 Å)

Reflecting Planes (hkl)	Silicon, a = 5.43073 Å (± 0.00002 Å)	Germanium, a = 5.6575 Å (± 0.0001 Å)	GaAs, a = 5.6534 Å (± 0.0002 Å)
111	14°14′	13°39′	13°40′
220	23°40′	22°40′	22°41′
311	28°05′	26°52′	26°53′
400	34°36′	33°02′	33°03′
331	38°13′	36°26′	36°28′
422	44°04′	41°52′	41°55′

$$h^2 + k^2 + l^2 = (4n - 1) \quad \text{for } n = \text{any odd integer}$$
$$h^2 + k^2 + l^2 = 4n \quad \text{for } n = \text{any even integer}$$

Various manufacturers make equipment expressly designed for orientation work. In these, the detector may be set at twice the Bragg angle, and the crystal holder slowly rotated (usually by hand) about one axis only. If the crystal is badly mis-oriented, it is conceivable that the x-ray maximum cannot be found, since no amount of adjustment about one axis can bring a randomly oriented plane into position to reflect into the detector. Some latitude is built into the machines, however, in that the detector usually has a slit or line aperture so that if the beam is deflected slightly to the side, it may still be found. The observed angle will not, however, be exactly correct. Normally, an auxiliary holder is added which allows the crystal to be rotated about two axes so that a true maximum can be found and the alignment can be more accurately performed. Because of these difficulties of aligning the specimen with the beam, the x-ray goniometer is used only if the approximate orientation is already known; otherwise a Laue pattern must first be made.

Laue Method. If a wide band of wavelengths is used instead of monochromatic radiation, reflection spots will occur on the film for the same conditions as before, i.e.,

$$n\lambda = 2d \sin \theta$$

but θ can now be fixed and an appropriate λ from the "white" source will give rise to the various maxima. The interpretation of this method is somewhat involved, and the reader is referred to Refs. 2 and 3 for detailed instructions. Despite a greater complexity, it does allow any orientation to be determined without prior knowledge of growth features or etching characteristics.

Sawing to Orientation. Orientation for sawing can be done either by building a combination crystal jig which can be used in both the orienting machine and the saw, and the crystal transferred back and forth, or by having a sawing jig only. The latter is simpler to construct but requires cutting a test slice, checking the orientation of that slice, and then making the required corrections. If lapping to orientation is required, fixtures with adjustable stops can be used.[30,31]

REFERENCES

1. F. F. Nye, "Physical Properties of Crystals," Oxford University Press, New York, 1960.
2. Charles S. Barrett, "Structure of Metals," McGraw-Hill Book Company, New York, 1952.
3. Elizabeth A. Wood, "Crystal Orientation Manual," Columbia University Press, New York, 1952.
4. Tadami Taoka, Eiichi Furubayashi, and Shin Takeuchi, Gonio-Microscope and Its Metallurgical Application, *Japan J. Appl. Phys.,* **4**:120–128 (1965).
5. Arthur P. Honess, "The Nature, Origin and Interpretation of the Etch Figures on Crystals," John Wiley & Sons, Inc., New York, 1927.
6. S. Mendelson, Stacking Fault Nucleation in Epitaxial Silicon on Variously Oriented Silicon Substrates, *J. Appl. Phys.,* **35**:1570–1581 (1964).
7. M. P. Diment, Orientation Dependence of Stacking Fault Geometry, *Solid State Electron.,* **11**:1177–1178 (1968).
8. R. H. Wynne and Colman Goldberg, Preferential Etch for Use in Optical Determination of Germanium Crystal Orientations, *J. Metals,* **5**:436 (1953).
9. J. W. Faust, Jr., "Etches for Imperfections in Silicon and Germanium," Electrochemical Society Meeting, May 1955.

10. G. A. Wolff, J. M. Wilbur, Jr., and J. C. Clark, Etching and Orientation Measurements of Diamond Type Crystals by Means of Light Figures, *Z. Elektrochem.,* **61**:101–106 (1957).
11. G. H. Schwuttke, Determination of Crystal Orientation by High Intensity Reflectograms, *J. Electrochem. Soc.,* **106**:315–317 (1959).
12. J. W. Faust, Jr., A. Sagar, and H. F. John, Molten Metal Etches for the Orientation of Semiconductors by Optical Techniques, *J. Electrochem. Soc.,* **109**:824–828 (1962).
13. R. Dreiner and R. Garnache, Precision Orientation of Germanium in the [111] Direction Using Alloy Pits, *J. Appl. Phys.,* **33**:888–891 (1962).
14. John G. Gualtieri and Albert J. Kerecman, Analysis of Epitaxially Grown Semiconductor Layers by Means of Light Figures, *Rev. Sci. Instr.,* **34**:108–110 (1963).
15. J. W. Edwards, The Orientation of Cubic Semiconductor Crystals by X-Ray and Optical Measurements, Pt. 2, *Semiconductor Products,* **6**:34 (June 1963).
16. G. R. Cronin, Polarity Determination of GaAs Wafers, *Rev. Sci. Instr.,* **34**:1151–1154 (1963).
17. G. A. Wolff, J. J. Frawley, and J. R. Hietanen, On the Etching of II–VI and III–V Compounds, *J. Electrochem. Soc.,* **111**:22–27 (1964).
18. Isamu Akraski and Hiroyuki Kobayasi, Etching Characteristics and Light Figures of the (111) Surfaces of GaAs, *J. Electrochem. Soc.,* **112**:757–759 (1965).
19. ASTM F 26-66, "ASTM Book of Standards," Part 8, American Society for Testing and Materials, Philadelphia, 1968.
20. D. Coster, K. S. Knol, and J. A. Prins, *Z. Physik,* **63**:345 (1930).
21. R. W. James, "Optical Principles of the Diffraction of X-Rays," G. Bell & Sons, Ltd., London, 1958.
22. E. P. Warekois and P. H. Metzger, X-Ray Method for the Differentiation of (111) Surfaces in $A^{III}B^{V}$ Semiconductor Compounds, *J. Appl. Phys.,* **30**:960–962 (1959).
23. M. S. Abrahams and L. Ekstrom, Etch Pits, Deformation and Dislocations in GaAs, in Harry C. Gatos (ed.), "Properties of Elemental and Compound Semiconductors," Interscience Publishers, Inc., New York, 1960.
24. E. P. Warekois, M. C. Lavine, A. N. Mariano, and H. C. Gatos, Crystallographic Polarity in II–VI Compounds, *J. Appl. Phys.,* **33**:690–696 (1962).
25. S. G. White and W. C. Routh, Polarity of Gallium Arsenide Single Crystals, *J. Appl. Phys.,* **30**:946–947 (1959).
26. H. C. Gatos and M. C. Lavine, Characteristics of the {111} Surfaces of the III–V Intermetallic Compounds, *J. Electrochem. Soc.,* **107**:427–433 (1960).
27. Robert W. Bartlett and Malcom Barlow, Surface Polarity and Etching of β-Silicon Carbide, *J. Electrochem. Soc.,* **117**:1436–1437 (1970).
28. Karl Brack, X-Ray Method for the Determination of the Polarity of SiC Crystals, *J. Appl. Phys.,* **36**:3560–3562 (1965).
29. H. C. Gatos and M. C. Lavine, Etching and Inhibition of the {III} Surface of the III–V Intermetallic Compounds—InSb, *Phys. Chem. Solids,* **14**:169–174 (1960).
30. S. Ipsen Mathiesen, L. Gerward, and O. Pedersen, Preparing Polished Crystal Slices with High Precision Orientation, *Rev. Sci. Instr.,* **45**:278–279 (1974).
31. R. Butz, B. Krahl-Urban, and K. Mench, A Precision Goniometer Polishing Jig, *Rev. Sci. Instr.,* **44**:485–487 (1973).

2

Crystallographic Defects and Their Observation

There are numerous crystallographic defects that may occur during growth and subsequent processing of crystalline semiconductors. Many of them are undesirable at any level, some are helpful in moderation, and some, such as foreign doping atoms, are absolutely necessary. Table 2.1 summarizes these defects (both the good and the bad) and lists some of the methods of detection. The ensuing discussions expand on that table and in many cases give explicit directions for their observation. Occasionally some mention may be made regarding the cause and/or cure of some particular defect, but defect control is not the theme of this chapter. The shape defects, while not reflecting crystallographic imperfections, can still be serious enough to subsequent processing to demand some attention.

2.1 POINT DEFECTS

The first several defects listed in Table 2.1 belong to the category of point defects. Such defects are all local in nature and are characterized by the fact that the imperfect region can be removed and a perfect section substituted without additional lattice distortion.

Vacancies. The simplest point defect is the vacancy (sometimes called a *Schottky defect*) in which a single atom is missing from the lattice. If the crystal has a zinc blende structure (III–V), there can be two kinds of vacancies, one for each component. Further, they may be in various charge states. Vacancies are thermodynamically stable and will therefore be present in all crystals. The equilibrium number N may be calculated from

$$N = N_0 \exp\left(\frac{-E_v}{kT}\right)$$

where N_0 is the number of atomic sites, E_v is the energy required to create a vacancy and is in the order of 2 to 2.5 eV for Ge and 2.5 to 3 eV for Si, k is Boltzmann's constant, and T is the temperature.[1] Such calculations are not very helpful in determining the actual number present, since by rapid cooling after crystal growth, a greater than equilibrium number can be quenched in,[2] and by interactions with various impurities, the equilibrium number can be depressed.[3]

For high-melting-point materials which can be used for field-emission tips in

Table 2.1. Summary of Material Defects

Defect	Method of detection
LATTICE DEFECTS	
LOCAL (Minimum of Long-Range Disorder)	
Vacancies	Density measurements, inference from electrical measurements
Interstitial	Inferred from electrical properties
Antistructure	Inferred from electrical properties
Foreign atom-vacancy complexes	EPR, indirect inference, optical spectra
Foreign atoms (singly)	Electrical properties, spectrographic or wet analysis
Dislocations	Etching, x-ray, electron microscopy
Stacking faults	Etching, x-ray, electron microscopy
Twins	Etching, x-ray, electron microscopy
Lineage	Etching, x-ray, electron microscopy
Single grain boundary	Etching, x-ray, electron microscopy
AGGREGATE	
Polycrystalline regions	Etching, x-ray, electron microscopy
Voids	Microscopic examination
Cracks	Selective etching
Inclusions (separate phase)	Birefringence, x-ray, electron microscopy
Inclusions (compositional variation)	X-ray, electron microscopy, microprobe, etching, Tyndall scattering
MISCELLANEOUS	
Residual lattice strain	X-ray, birefringence, etch rate, electrical properties, mechanical displacement
Mechanical surface damage	Etch rate, dislocation density, decoration, electrical measurements, x-ray
SHAPE DEFECTS (No Atomic Misplacement)	
Faceting	Visual
Habit change	Visual
Variation in surface contour	Profilometer, Light scattering
SURFACE DAMAGE	
Cracks	See Table 2.10
Strain	Bowing, x-ray
Dislocation networks	See Dislocations above
Stacking faults	Etching, x-ray, electron microscope
RADIATION DEFECTS	
Point defects	See vacancies, etc., above
Amorphous layers	Electron diffraction, atomic backscattering

field-ion microscopes, vacancies can be directly observed,[4] but there is no way to estimate their density. If large numbers are present, very precise density measurements can indicate vacancy fluctuations.[5] For example, PbS may have up to 10^{19} vacancies per cubic centimeter. Silicon and germanium, however, even near their melting points, have only about 10^{18} per cubic centimeter and at room temperature have far less. Nevertheless, careful density measurements made on dislocation-free silicon do indicate a considerable variation in vacancy density.[6] The diffusion rate of many impurities depends on the number of vacancies present and may thus be used to infer relative vacancy concentrations.[7] If it can be demonstrated that vacancies are electrically active and separable from other active defects, the measurement of the Hall coefficient may allow the number to be estimated. This approach must be used with caution, however, because there are a number of impurities having activation energies close to those presumed to be due to vacancies. Indeed some of the older data may very well have been obtained from deep-level impurities and not from vacancies. Probably the most definitive way to study vacancies is by use of electron paramagnetic resonance (EPR).[8]

As numbers of vacancies coalesce into clusters, they will eventually collapse and form dislocation loops. These loops are more readily detected than either the single vacancies or the clusters and can, for example, be seen either by etching, in which case shallow depressions is the usual configuration,[9] by transmission electron microscopy in which the complete loop is visible,[10] or by copper-precipitation decoration. Loops are more likely to occur in dislocation-free material, since if there were dislocations, the vacancies could interact with them rather than form clusters.

Interstitials. An interstitial is an extra atom occupying space between the normal lattice sites. For a zinc blende crystal, four distinguishable interstitials are possible, one for each component and each interstitial completely surrounded by group III or V atoms. If an interstitial and a vacancy occur in close proximity, the pair is called a *Frenkel defect.* A combination of optical and EPR spectra, electrical data, and calculations has in the past been used to identify interstitials. However, in some cases atomic backscattering can be used to study them directly.[11,12]

Complexes. From single-atom defects a wide range of more complex multiple defects can be built. There are, for example, two vacancies side by side (divacancy), a vacancy with a trapped electron (referred to as an F center in alkali-halide crystals), a vacancy beside a substitutional impurity (a vacancy–phosphorus atom pair in silicon is an E center), and many more. The fact that most of them were observed in silicon does not imply that it forms more kinds of defects but rather that it has been more extensively studied than other materials. The detection methods are similar to those of vacancies and interstitials. In addition, the rate of precipitation of a fast diffuser can sometimes be used to estimate the number of defects which promote nucleation.[13,14] Such a procedure has been used for Ge (lithium precipitation), but it is not clear whether it is vacancies and/or vacancy-impurity complexes being followed.

Foreign Atoms. Foreign atoms may occur either as interstitials or in the place of normal atoms. In the latter case they are referred to as *substitutional impurities.* In compound semiconductors, one component atom may occur on a site intended for the other, e.g., in GaAs, As may be in a Ga position or vice versa. These are called *antistructure defects.* Widely dispersed foreign atoms cause little lattice disruption and are commonly detected indirectly through their effect on the electrical properties of the semiconductor. These methods are discussed in Chap. 3. In

Table 2.2. Nonelectrical Methods for Impurity-Concentration Determination

Method	Advantages	Disadvantages
Emission spectroscopy	More readily available equipment	Limited sensitivity
Solid mass spectroscopy	Great sensitivity	Very susceptible to surface contamination
Ion mass spectroscope	Combines mass-spectrometer sensitivity with spatial resolution of x-ray microprobe	Equipment complex, calibration difficult
Neutron activation	Great sensitivity	Special equipment, not very useful for III–V compounds
Charged-particle activation	Light-element sensitivity	More difficult than neutron activation
Radio tracer	Great sensitivity	Special equipment, impurity to be studied must be available as a radioactive isotope. Not applicable to routine evaluation
X-ray microprobe	Nondestructive, can be used to examine very small volumes	Limited accuracy
Optical absorption	Equipment widely available	Limited sensitivity, limited applicability, requires special sample preparation

addition, a number of analytical techniques are available which, while not ordinarily as sensitive as resistivity-based measurements, are generally specific to a given atomic species. These are summarized in Table 2.2 and discussed in Chap. 9.

Internal friction, or mechanical damping, is often dependent on defects in the lattice and thus may be used to study them. However, interpretation is difficult and application is limited.[15,16]

Optical-absorption spectra are applicable in some instances. For wavelengths less than those corresponding to the band edge, the attenuation is very high and little data relating to impurities can be obtained. However, for very high impurity concentration, there is some smearing of the tail, as indicated by the shaded portion of Fig. 2.1. For wavelengths longer than A, the general background level of absorption as shown by the dotted line depends on the free carrier density and may be used to estimate shallow donors or acceptor concentrations. Superimposed on that

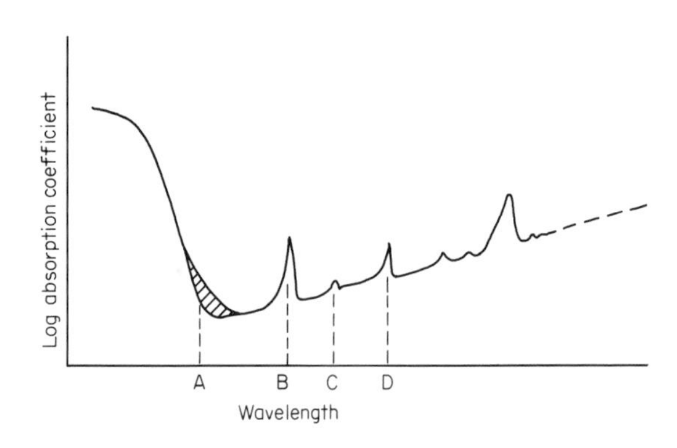

Fig. 2.1. Optical-absorption coefficient vs. wavelength. The rapid drop at A occurs when the band edge is reached, while the peaks at B, C, D, etc., may be due to either impurity or host lattice absorption. The level of the dotted line changes with the number of carriers present, and in general increases as the square of wavelength.

background are absorption peaks, e.g., *B*, *C*, and *D*, which may be due to either the lattice itself or some impurity. For the more common semiconductors, the band edge occurs in the 1- to 2-μm range and the region of possible usable spectra extends to at least 20 or 30 μm. Applications have been primarily to impurities which are not electrically active but which may occur in concentration of several parts per million. Examples of these are oxygen in Si and Ge, carbon in Si, and radiation-induced defects in Si and Ge. In the infrared region, the absorption frequencies are primarily a function of the atomic species and their bonding. Thus, isotopes, e.g., ^{12}C and ^{14}C, are readily separable, and if a carbon-oxygen complex forms, the frequency will shift and it too can be distinguished. There is, of course, the problem of initial identification of a given line. Remember that like other optical spectra, there may be more than one line per impurity, and that with multiple impurities there may be interference between them or between them and the host lattice. For example, one of the Si-Si lines lies very close to one for O-Si and reduces sensitivity. Sometimes by cooling the sample, the offending peak can be shifted or reduced in amplitude relative to the one of interest.

Before such spectra can be used for quantitative analysis, calibration must be done by using samples with known levels of impurities. These calibrations should be approached with caution, since the independent measurement required may respond to the impurity in a different form. For example, a chemical analysis for carbon might detect interstitial and substitutional carbon and the carbon in precipitated silicon carbide, whereas the particular optical-absorption line being studied may be due only to substitutional carbon. Experimental procedures will vary with the material, but ASTM procedure F 121 for determining oxygen in silicon can be used as a guide.

Careful measurement of lattice changes can be used to deduce the amount of impurity present. Such measurements do not, however, distinguish between impurities. Ion scattering can be used, although sensitivity is not very good and impurities must have appreciably higher atomic mass than the host.[17]

2.2 DISLOCATIONS

There are several varieties of dislocations,[18-22] one of which is shown schematically in very simple form in Fig. 2.2*b*. It occurs at the termination of a sheet of extra atoms and is an edge dislocation. If the crystal is subjected to shear as in Fig. 2.2*a*, the crystal will be deformed by bonds along the shear-line shifting, and dislocations will be formed. Dislocations already present will move by glide in a direction perpendicular to the dislocation line and along a slip plane. There will be crystallographically preferred glide and slip directions, so that dislocations so formed will not be random but will be in orderly arrays as indicated in Fig. 2.3.

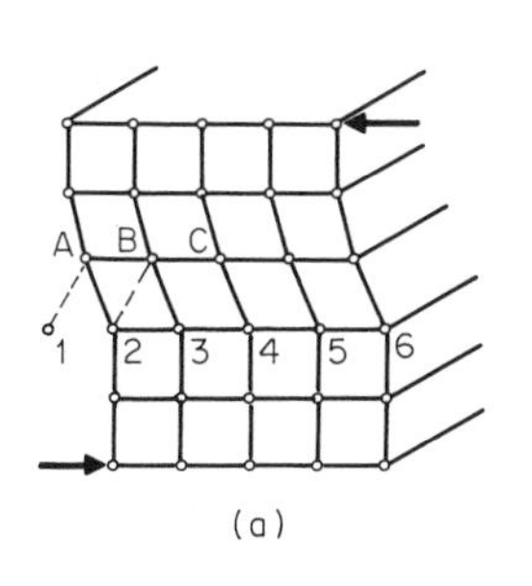

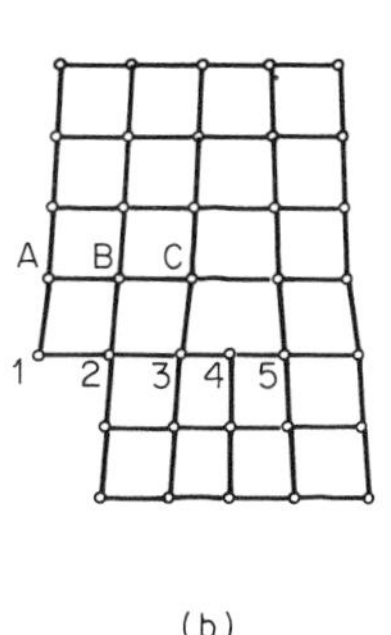

Fig. 2.2. Edge dislocation. The dislocation can be formed by shearing which causes (*a*) bonds *A*-2, *B*-3 to change to (*b*) bonds *A*-1, *B*-2.

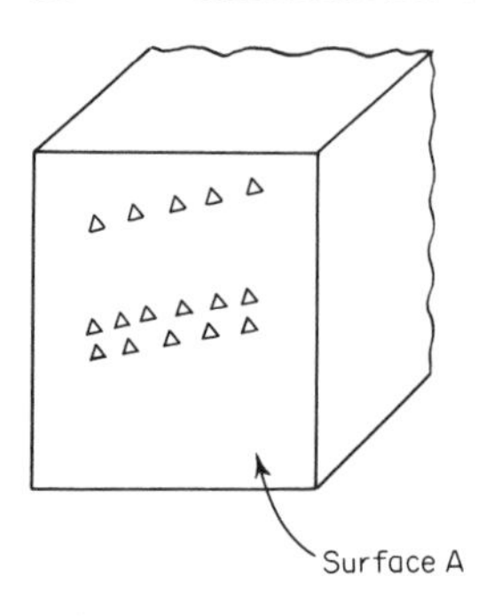

Fig. 2.3. The intersection of edge dislocations with the surface of a crystal. In this case they all lie along slip planes.

Figure 2.4 is a photograph of a slice in which the intersection of the dislocations with a (111) slice surface has been defined by small etch pits (this procedure is described below). In this case three sets of intersecting slip planes are indicated. As can be deduced from their geometry, they are all (111)s, and indeed the {111} family is the most active one in the diamond and zinc blende structures. Such pronounced slip is common and can occur if a slice, or a long crystal ingot for that matter, is nonuniformly heated or cooled so that differential thermal expansion can cause the necessary shear force.

The dislocation could also move by adding or removing a row of atoms at the edge of the extra sheet, as shown in Fig. 2.5. Motion of this kind is perpendicular to the slip plane; it is called *climb* and can occur only by transporting additional atoms or vacancies by diffusion to the dislocation. Typical of this process is the gold-induced climb in silicon. Gold will diffuse interstitially very rapidly and can then become substitutional at the dislocation. Figure 2.6 is a photograph of a portion of a (111) silicon surface which was etched once before the dislocation moved and once afterward. Motion was parallel to the (111) glide planes and therefore due to glide and not climb. The velocity of dislocation travel in semiconductors has been studied by using this same technique, i.e., by carefully polishing a sample, giving it a dislocation etch to show the dislocations, stressing it, e.g., by three-point loading, and again etching.[23]

High-temperature annealing of a crystal which has plastically deformed leads to polygonization. The edge dislocations move to planes perpendicular to the glide plane so that they are arrayed in tilt boundaries which separate near perfect regions of crystal, as illustrated in Fig. 2.7. Polygonization is often observed in silicon after prolonged heat treatments, e.g., after an epitaxial deposition. Based on the previous

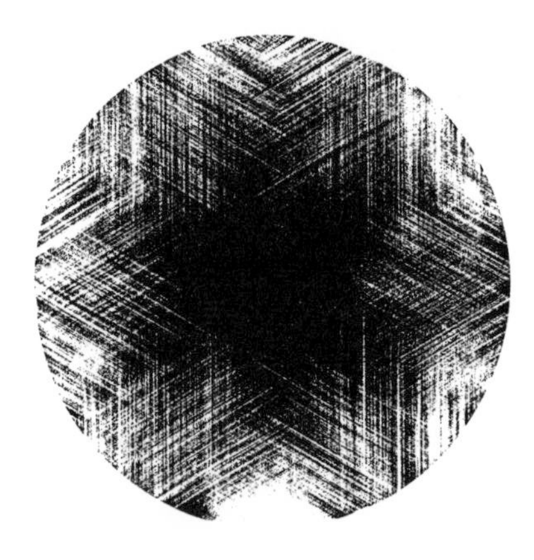

Fig. 2.4. A (111)-oriented silicon slice in which the severe slip has been delineated by etching to show the emergence of dislocations. The lines lie along the intersection of other (111) slip planes with the viewing surface.

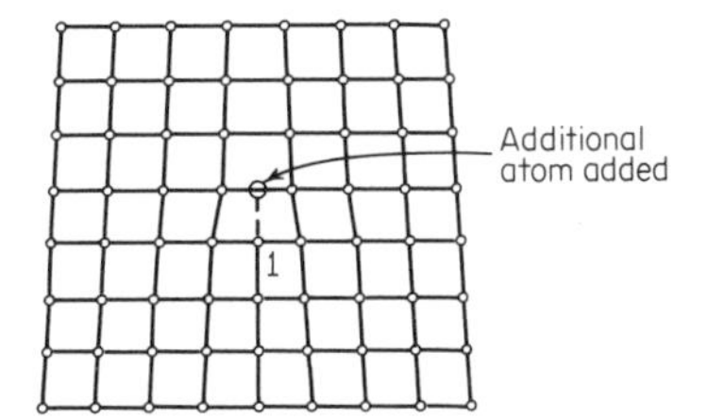

Fig. 2.5. Movement of a dislocation by the diffusion of atoms to the end of the "extra plane." The dislocation could have traveled in the opposite directions by atom 1 diffusing away from its site.

discussions of glide, one might suppose that the orderly alignment of dislocations in rows as was shown in Fig. 2.3 would always indicate the trace of a slip plane. Such lines might, however, be tilt boundaries instead. If the material being studied is a previously well-evaluated one, the etch pits used to demonstrate the dislocation can be used to determine the orientation of the dislocation arrays. (See etch-pit geometry in Chap. 1.) Figure 2.8 shows a close-up view of rows of etch pits along (111)-plane slip lines and another along polygonization boundaries which run perpendicular to the slip surfaces. Note that if the outline of the etch pit is used as a guide, the fact that the two rows run in different crystallographic directions is readily apparent.

Should a segment of an edge dislocation be pinned at each end and the center move by glide, it can keep expanding, eventually give a loop of dislocations, and still retain the segment. More movement will cause the first loop to grow larger and the segment to be again distorted and continue to generate loops. The additional dislocations continually generated from the original segment in this way have been observed in silicon and germanium.[25,26]

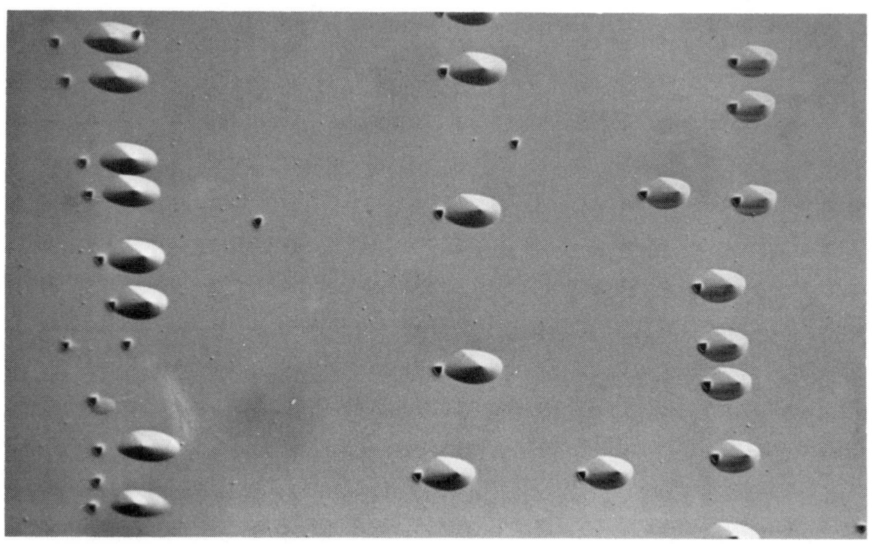

Fig. 2.6. Movement of dislocations in silicon caused by thermal-induced stress. The large pits were formed by high-temperature vapor etching before movement, the small ones by a later room-temperature etch. The sides of the etch pits are traces of (111) planes intersecting the (111) surface of the crystal and are parallel to the slip direction. Hence motion of the dislocation was along a slip direction and therefore not induced by climb. (*Courtesy of Dr. Lawrence D. Dyer, Texas Instruments Incorporated.*)

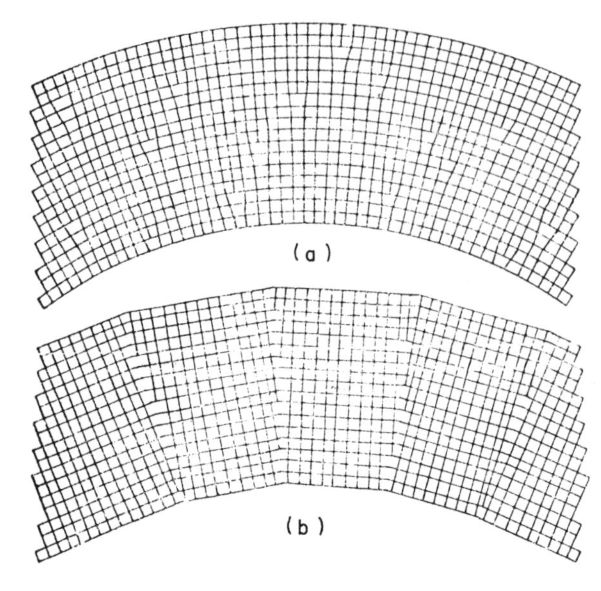

Fig. 2.7. Effect of polygonization. Initially the crystal was deformed as in (*a*) and had many dislocations lying in the slip plane. After polygonization in (*b*), the dislocations have coalesced into low-angle grain boundaries separating slightly misoriented regions of perfect crystal. (*From F. R. N. Nabarro, "Theory of Crystal Dislocations," Oxford University Press, London, 1967.*)

There are a variety of other sources which will also generate loops, spirals, and other geometries. For example, loops are produced from a pinned-edge dislocation by climb. If only one end is fixed, spirals will grow, and if the dislocation has both edge and screw character, climb will produce helices (which are found in germanium and silicon[24] and sometimes are mistaken for screw dislocation). The screw dislocation, however, is very local in nature, whereas a helix may have diameter and pitch of micrometer dimensions. The collapse of a large vacancy cluster will give a dislocation loop; they are found in silicon and germanium.[10,27-30]

The edge dislocation just described is not the only type of dislocation to be found. There are, for example, screw dislocations, more complex edge dislocations involving two extra planes, and several varieties of partial dislocations which occur at the boundaries of stacking errors.[31]

Occurrence. Dislocations can be introduced at nearly every stage of processing. In initial crystal growth, dislocations in the seed will propagate into the new growth as each succeeding atomic layer is added. Should sudden growth-rate changes occur, extra dislocations will generally occur. Misfit dislocations will be generated at the boundary between new and old growth if the new growth is of different lattice spacing from the old. Such circumstances occur during heterocrystal growth, e.g., germanium on gallium arsenide, or even for the same materials if the doping is radically different, as in the case of a lightly doped silicon layer on a heavily doped substrate. Initial growth onto unclean surfaces will lead at best to a higher dislocation density but more often will also produce stacking faults and, in some cases, gross polycrystallinity.

Dislocations introduced after crystal growth are all associated with the subjection of the crystal to excessive mechanical stress. Such stress can originate from unequal heating or cooling, from the diffusion of impurities into the lattice, or from forces applied externally during shaping operations such as sawing, lapping, and polishing. Should precipitation of a second phase occur during either the initial crystal cooldown or subsequent annealing cycles, dislocations are likely to be formed because of differential thermal contraction. In silicon, for example, SiO_2 precipitation can cause dislocation networks, loops, and stacking faults.[32] Also, prismatic loops have

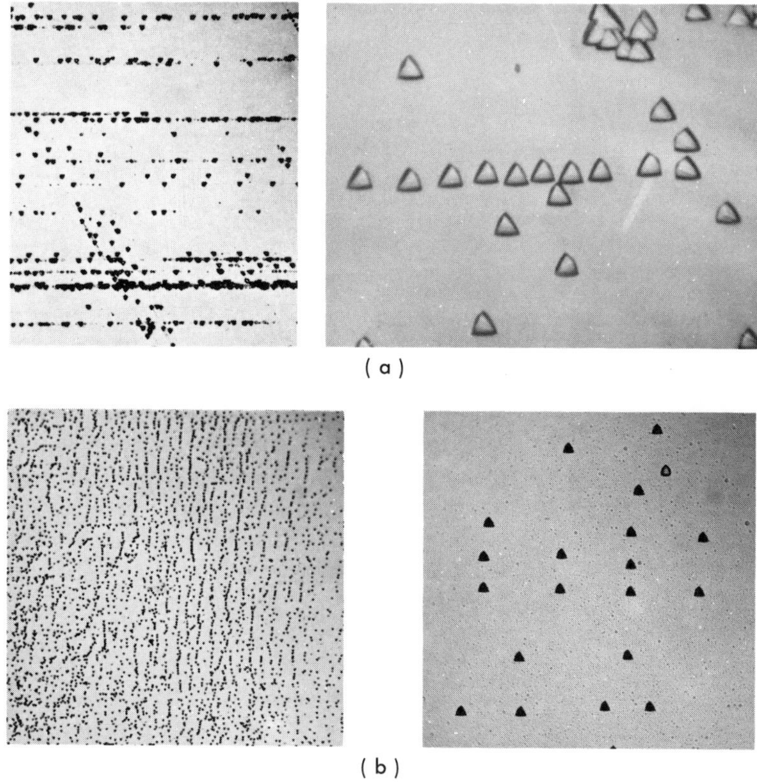

Fig. 2.8. Etch-pit orientation (*a*) along slip lines and (*b*) along polygonization walls. (*Photographs courtesy of Dr. Kenji Morizane, Texas Instruments Incorporated.*)

been reported in erbium-doped (ZnCd)S[33] which presumably arise from inclusions of an erbium compound.

Dislocation Detection by Etch-Pit Formation. The region near a dislocation line usually etches more rapidly than the rest of the crystal and thus develops etch pits, as indicated schematically in Fig. 2.9.[34-37] Depending on the etch, the pit may be conical and rather featureless, or it may reflect the crystallographic structure. These pits afford a simple method of determining the number of dislocations which intersect the surface (dislocations per square centimeter). It is the one most commonly used, and indeed "etch pit" and "dislocation density" are often used synonymously.

The success of a particular etchant can be very much dependent on its exact composition and on a great many other conditions such as surface cleanliness, initial etchant temperature, and the ratio of volume of etchant to semiconductor surface area. Surface damage will also produce etch pits, but they will usually become flat-bottomed as etching proceeds and the damage is removed. Dislocation pits, however, continue to etch rapidly for the length of the dislocation and thus usually maintain pointed bottoms. Because of this shape, a true dislocation pit will appear dark under a bright-field-microscopic examination, since the sloping sides will not reflect incident light back into the objective. The flat-bottomed surface-damage pit will reflect light back and will appear bright. Examples of both types of pits can be seen in Fig. 2.10. The flat-bottomed pits trace out the path of scratches

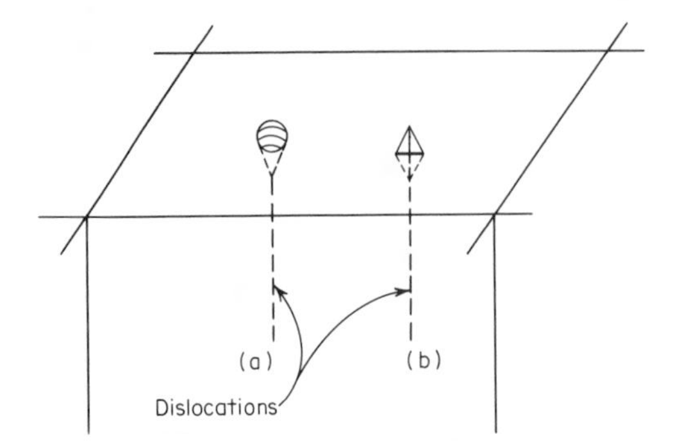

Fig. 2.9. Etch pits formed at the intersection of dislocations with a surface. The shape of the pit is a function of the etchant and crystal orientation.

generated during slice polishing. Sometimes a stair-stepped or terraced pit is observed. This has on occasion been described as indicative of screw dislocations[38] but in actuality is probably due to other effects.[24] To minimize the effect of the surface, i.e., to clean it and to remove surface damage which might complicate interpretation, it is best first to polish the surface and then to subject it to the dislocation etch.[39] Either chemical or mechanical polishing may be used. Ordinarily a surface would not be mechanically polished specifically in preparation for dislocation etching, but if it is already available, it may be used. Chemical polishing is much simpler and in general is recommended. See Chap. 7 for polish formulations and procedures. One exception to not using mechanical polishing occurs if it is desired to look at the density as a function of depth. In that case, angle lapping may be used, followed by a good mechanical polish. This procedure has been applied to the study of extra dislocations generated at epitaxial-substrate interfaces,[40] and to depth-of-damage studies. The latter is discussed in Sec. 2.9.

Silicon has been extensively studied, and reasonably foolproof etchants and techniques have been developed, but even with them there are resistivity ranges which do not satisfactorily respond, and really dependable counting can be done only on (111) faces. For III–V compounds, a dislocation terminating in a group III atom may etch differently from one ending in a group V atom. Further, either type of dislocation can intersect either (hkl) or $(\bar{h}\bar{k}\bar{l})$ planes, and depending on the value of hkl may not develop etch pits at all. For example, InSb dislocation pits develop only for In dislocations intersecting (110) and (111) surfaces [and not $(\bar{1}\bar{1}\bar{1})$ surfaces]. In addition, a variety of non-dislocation-associated etch figures can develop on the various surfaces; so interpretation in terms of dislocations is more difficult than for Si and Ge.

Fig. 2.10. Etch pits from dislocations and from scratch damage. The black ones are due to dislocations.

Under most circumstances, the (111) planes of diamond and zinc blende crystals are the slowest to etch, so that the pits are bounded by them. Thus, on a (111) surface the pits will be triangular. (For more details on orientation see Chap. 1.) However, some etchants may etch most slowly in (100) directions. In that case, a (111)-plane etch pit will still be triangular in cross section, but the sidewalls will make different angles with the face and the outline will be oriented differently as in Fig. 2.11*a*. If either pit occurred on a round (111) slice, they would be indistinguishable, but on a crystal which showed natural faces there could be differences. For example, in Fig. 2.11*b* a tetrahedron is shown with both sketched in. In one case the pit mirrors the face. In the other it appears rotated 60°. It is also occasionally possible to have peaks rather than pits. Such peaks can arise for at least two separate reasons. One is that small amounts of masking material may remain on the surface and prevent etching in that region. If the mask is small, it will soon undercut and give a very small peak which will look under a microscope much like a pit. Usually they can be distinguished by using pseudo-Becke-picture framing (see Chap. 8). Another major reason for raised regions is the fact that impurities insoluble in the dislocation etch may precipitate at dislocations and impede etching. In this case, of course, the dislocation is still being delineated, although for a different reason.

While etching procedures are available for the more common materials, and it is therefore usually unnecessary to consider whether the pits which develop actually occur at dislocation terminations, that problem must be faced with either new materials or new etchants. Such correlation may be made in a variety of ways. The one used originally[35] was to examine the grain boundary between two slightly misoriented regions of a germanium bicrystal. The orientation difference can be accurately determined by x-ray spectrometry, and the number of dislocations expected because of that difference can be calculated. The surface can then be etched, and if the etchant is really delineating edge dislocations, a row of pits spaced accordingly should be observed. If suitable bicrystals are not available, dislocations

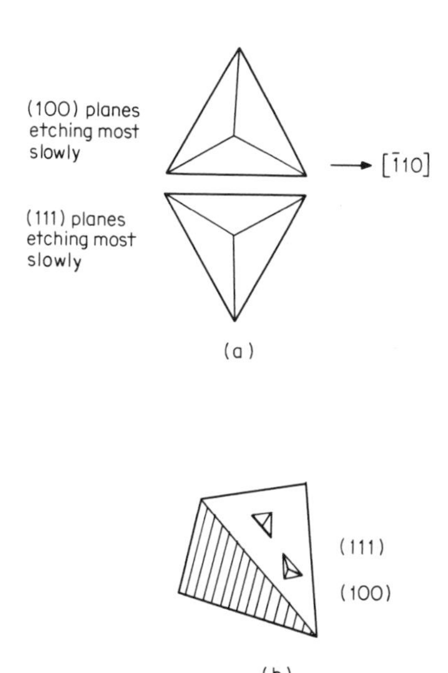

Fig. 2.11. Etch-pit orientation on a (111) surface.

can also be introduced in known amounts by bending. However, without extensive annealing the dislocation distribution is uneven, and not really as predicted. Therefore, before bending is used as a method for introducing a known and calculable amount of dislocations, the sample must be thoroughly annealed.[41] The dislocations generated by bending can also be used to study the effects of etchants on dislocations terminating specifically in either constituent of a compound semiconductor.[42-44]

If a crystal is cleaved, and each face etched, there should be a correspondence of pits if only dislocations are being displayed,[24] since the dislocation pierces both faces. As a variation, a surface can be polished, dislocation-etched, photographed, have more material removed, and be dislocation-etched again. If the etchant is showing dislocations, there should be good correlation, although if the dislocations are not normal to the surface, the pits from the two levels will be laterally displaced. This procedure can be repeated numerous times on the same face and a whole sequence of photographs taken which will trace the path of the dislocations through the crystal. If the photographs are converted to slides, the dislocations will show as dots on a clear background. Then if equal thicknesses are removed each time and if the slides are stacked in sequence, a striking three-dimensional model results.[45]

Unfortunately, while this correspondence method can show that spurious pits are not produced, it will not show whether all dislocations are being etched. X-ray topography, which was not in use when the early etching work was done, can be used to demonstrate a one-to-one correspondence and is probably the best method, though also the one requiring the most specialized equipment and operators. A scanning electron microscope (SEM), using the signal generated from shallow diffused silicon p-n junctions, has enough sensitivity to observe diffusion-induced dislocations, and thus allows a correlation between surface etch pits and the emergence of dislocations.[46,47] (It is of interest to observe, however, that in Ref. 47, etching was used as the standard with which to evaluate the SEM.) Table 2.3 summarizes etchants that have been reported for Si. Of these, Dash and Sirtl are good general-purpose etchants, but for special cases such as (100) faces or heavily doped crystals the others may prove more effective. If the reader is not satisfied with the results of a particular etch, he should by all means try others, always remembering that if he deviates too far in composition, it may not be dislocations which are being delineated. In general, the etchants, unless otherwise noted, will give pits bounded by either (111) planes or some nearby ones having a [110]-zone axis. Overetching in general tends to produce more rounding and distortion. By carefully noting the exact shape of the pit, the direction of the dislocation can sometimes be determined.

Misfit dislocations formed during heavy-concentration planar emitter diffusions into (111)-oriented slices will lie primarily parallel to the surface rather than intersect it as shown by Fig. 2.3. However, etched grooves caused by them can sometimes be seen after several minutes of etching in ultrasonically agitated Dash etch.[56] Dislocations lying deep beneath the surface, either misfits at an epitaxial-substrate junction or the result of contamination, can be seen by angle beveling and looking at their intersection with the bevel surface.

Table 2.4 summarizes the etchants used for Ge. Better results are usually obtained if a polished surface is used, but CP-4, one of the more common etch polishes, will also produce dislocation etch pits. When it is used directly, a 600-grit lapped surface is acceptable. If a chemical prepolish is used with some of the other etchants, the

Table 2.3. Silicon Dislocation* Etchants

Etch	Face	Comments	Reference
Sirtl	(111), (110)	1–7 min	48
Modified Sirtl	(111)	Works better on low-resistivity material than Sirtl. Made by mixing 110 ml of Sirtl with 25 ml HF, 30 ml HNO_3, 100 ml H_2O	
Dash etch	(111), (110)	4 h, works moderately well on (100) faces. Reduce to few min for thin layer	25
Copper etch No. 1	(111)	Most of the metal-ion etchants were developed in an attempt to produce sharper, better-defined pits	49, 50
Copper etch No. 2	(111)		51
Sailor's etch	(111)	2 h ultrasonic, will also show Shockley partials but not stair-rod dislocations	52
ASTM	(111)	4 h (described in full detail below)	ASTM F 47
Mercury	(111), (100)	2 min	53
Dow ("Secco")	(111), (110), (100)	5 min with ultrasonic agitation. Works reasonably well on (100). Gives circular pits	54

*See Ref. 55 for a general discussion of the different kinds of dislocations to be found in Si.

amount of oxide left on the surface will affect the size and number of pits. The removal of the oxide by an HF dip just prior to the dislocation etch allows the formation of numerous very small pits apparently not associated with dislocations.[62] Etching too long will cause some pits to grow larger and will obliterate others, thus giving a false number. Hence normal times may need to be reduced when heavily dislocated material is examined. ASTM standard procedures are available for dislocation etching (ASTM F 47) which are particularly useful in defining the sort of sampling plans that should be used to ensure that representative pit densities are reported.

Dislocation Detection by Decoration. A wide variety of elements will segregate along dislocations. In some cases, they are easier to detect than the dislocation itself. Such procedures are not without pitfalls, however. For example, copper, which is widely used for decorating silicon dislocations,[25] will precipitate along regions of high oxygen content in the form of needles and platelets,[63] and presumably there

Table 2.4. Dislocation Etchants for Germanium

Etchant	Comments	Reference
CP-4	Gives conical pits on (100), (110), and (111) surfaces	34, 36
Superoxol	Develops considerable structure in pits. Diluted 1:1 with H_2O, will develop spiral terraced pits and very small pits in addition to those normally associated with edge dislocations	57, 58
Cyanide	100°C for 3–4 min (111) surface	59
WAg		60
Dash		1
Russian	Room temperature for 8–12 min shows edge dislocations and spiral pits on (100), (110), and (111) surfaces	61

are other impurities that will behave similarly. Should its concentration become too high, it is quite possible to precipitate enough copper to cause additional dislocations.[64] The general procedure when deliberately decorating is to choose a fast-diffusing material, saturate the sample at some relatively high temperature, and then rapidly cool it so that the diffusant solubility decreases rapidly. It will then differentially precipitate at dislocations and can be seen optically. The exact procedure will vary with the material, but for Si[25,63,64] involves heating it along with a source of copper for up to an hour in order for the copper to diffuse through the sample. Care must be taken to ensure that the sample is free of oxide; otherwise entry of the copper will be blocked. The source can be elemental copper in a closed tube with the silicon, but more often it is a few drops of copper nitrate solution allowed to dry on the surface. Note that the diffusion furnace will in most cases become contaminated with copper and therefore cannot be used for normal semiconductor processing. The more common semiconductors are all opaque to visible light; so some form of infrared image converter is required. For silicon, the widely available 1-μm image converters can be used, but germanium requires wavelengths in the vicinity of 1.8 μm. Imaging is much more difficult in that region but vidicons and line-scanning equipment are available which will work with reasonable sensitivity.

Dislocation Detection by X-Ray and Electron Microscope. There are several x-ray methods that have been used for studying dislocations and other crystal defects. Double-crystal spectrometry was one of the earliest of these.[65-69] It depends on the fact that departures from a perfect crystal should make the line width broader, and for randomly located dislocations, the width is proportional to the square root of the dislocation density. For dislocation densities in the 10^3 to 10^6 per square centimeter range, the double-crystal spectrometer is usable, although the sensitivity is not adequate, and other approaches are more rewarding.

Topography[70-87] has developed as the most useful and powerful x-ray method for studying semiconductor defects. There are several forms of topography, but all map local deviations from crystal perfection onto photographic film or an x-ray imaging tube and can show single dislocations* and lattice strain due to impurities, precipitates, etc. Its application to dislocation counting is in the low-concentration region as opposed to the rather high concentrations required if double-crystal spectrometers are to be useful. If topography is used for highly dislocated samples, two deleterious results occur. One is that spatial resolution on the film is not sufficient to separate them, and the other is that if they are too close together, the region between them is also strained and contrast is reduced. The transmission Lang topographic method was the first one applied to semiconductors and was used to show single dislocations in silicon.

To make visualization of the dislocation portrayed by topography somewhat easier, stereo pairs can be taken and viewed conventionally,[73] or the two photographs can be superimposed and printed in separate colors, e.g., red and blue. Then viewing is done with one eye covered with a red filter and the other with a blue filter.[88] For Lang photographs, orientations of (hkl) and $(\bar{h}\bar{k}\bar{l})$ for the reflection plane are normally used, but it is possible to make them by taking one photograph, rotating the crystal a few degrees around the normal to the diffraction planes, and taking

*Actually, copper decoration has comparable or better resolution than x-ray topography, but it has the disadvantage of being destructive and of being applicable to a much narrower range of materials than x-ray topography.

another. Such a procedure has the advantage that different angles can be chosen to enhance the effect[89] and that it is applicable to Borrmann's topography.[90]

The continuous x-ray spectrum can be used for crystallographic studies. In this case, rather than varying the angle to give a diffraction maximum for a given wavelength and crystal spacing, the polychromatic radiation is allowed to fall on the crystal, and the position(s) of the diffracted maximum are determined photographically. If a similar arrangement is used, but with monochromatic radiation (Debye-Scherrer), a few spots will occur for a single crystal, but if the sample has many randomly oriented crystallites, a continuum of spots arranged in circular patterns will result. As the crystallites become less randomly oriented, the circles reduce to shorter and shorter arcs, so that they can be used as some measure of perfection, although it is much more gross than any of the other methods discussed. It is of some use, however, in determining whether vapor-deposited layers have any ordering. It has been suggested that if a source is chosen that has both a sharp high-intensity line and good continuous background, a single-crystal sample will give well-defined Laue spots, while a randomly oriented composite will give a Debye-Scherrer powder pattern. Varying degrees of distortion will then give patterns intermediate between the two extremes. While such a procedure is not very sensitive, it was used to characterize silicon crystals in their early stages of historical development.[91] A similar procedure has been used in metal studies in which rings of widely separated spots are observed using monochromatic radiation in well-annealed samples. After cold working, the individual grains which originally gave spots are distorted and the spots turn into areas which really are a smear of smaller spots and which can sometimes be interpreted in terms of the amount of plastic flow which took place.[92] Even without the continuous-wavelength radiation, crystal imperfections will cause the spot shape to change and in some cases to elongate along certain directions. These "diffuse" reflections are indicative of structural defects such as stacking faults and twinning and have been used in the study of diamonds[93] and radiation-damaged Si.[94]

Transmission electron diffraction can be used in a fashion similar to x-rays to observe dislocations and other defects, and interpretation is similar. It does have the disadvantage of requiring a very thin sample, which usually implies a destructive measurement.

References 95 to 103 are a collection of articles covering x-ray and electron-diffraction studies as they directly relate to silicon. There are also scattered references to Ge, GaAs, and other materials, but the preponderance of journal entries involve Si.

Dislocation Detection for Other Materials. Dislocations in diamond may be detected by x-ray topography[104] or etching in fused KNO_3 or a hot O_2 atmosphere.[105]

GaAs dislocations have been studied by both etching[106-108] and x-ray topograph.[80,109-113] Schell etch will produce dislocation pits on Ga (111) faces. Grocker etch will show pits on both (111) surfaces. The arsenic pits will be triangular and the gallium pits conical. On (111) surfaces, if the etch pits are triangular, they will be roughly bounded by (111) planes. A-B etch develops pits on the arsenic surface and W-R etch only on Ga surfaces.

$Hg_xCd_{(1-x)}Te$[114] can be studied with an etch of HnO_3, H_2O, HCl, and Br_2.

Etching, decoration, and x-ray topography have been used for GaP.[109,115] For the etch, use a concentrated methanol solution of iodine with a trace of bromine. Decoration can be by either copper or zinc precipitation.

Modified CP-4A (2-1-1) will show InSb dislocations terminating on In (111) faces, while $0.2N$ ferric ion in $6N$ HCl works on $(\overline{1}\overline{1}\overline{1})$ faces.[116,117]

Seventy-five percent NaOH, 25 percent Na_2O_2 at 700°C or 6 percent Cl, 26 percent O_2, 68 percent argon at 850°C gives dislocation pits on (111) faces (Si) of β SiC.[118]

Etch pits have been observed in ZnS crystals after etching in diluted H_2O_2 at 80°C, but a one-to-one correspondence to dislocations has not been demonstrated.[119]

For PbSe and PbSeTe dislocation etchant, use a KOH, H_2O, glycerol, and H_2O_2 mixture.[120,121]

Etching in hot HCl for 2 to 6 h defines dislocations (probably both edge and screw)[122] in GaP.

2.3 STACKING FAULTS

Stacking faults,[123-144] as the name implies, are due to errors in the stacking of layers and can occur only when the succeeding layers are different. Figure 2.12*a* is the standard schematic of a stacking fault. It shows one originating when a layer is omitted (intrinsic) and when an extra layer has been inserted (extrinsic). The intrinsic form could presumably form by the collapse of a vacancy cluster as shown in Fig. 2.12*b* and either the extrinsic or intrinsic by a number of crystal-growth errors. For example, in some regions the atoms might nucleate wrong or there might be a thin foreign platelet which when overgrown produces a stacking fault. Further, some kinds of line defects, coupled with climb, can act as sources of extra (or missing) planes. One example is the Bardeen-Herring source, which is thought to contribute to some of the faults observed in silicon. If the fault formed as in Fig. 2.12*b* or from a source like Bardeen-Herring, the outline will be hexagonal or circular (Fig.

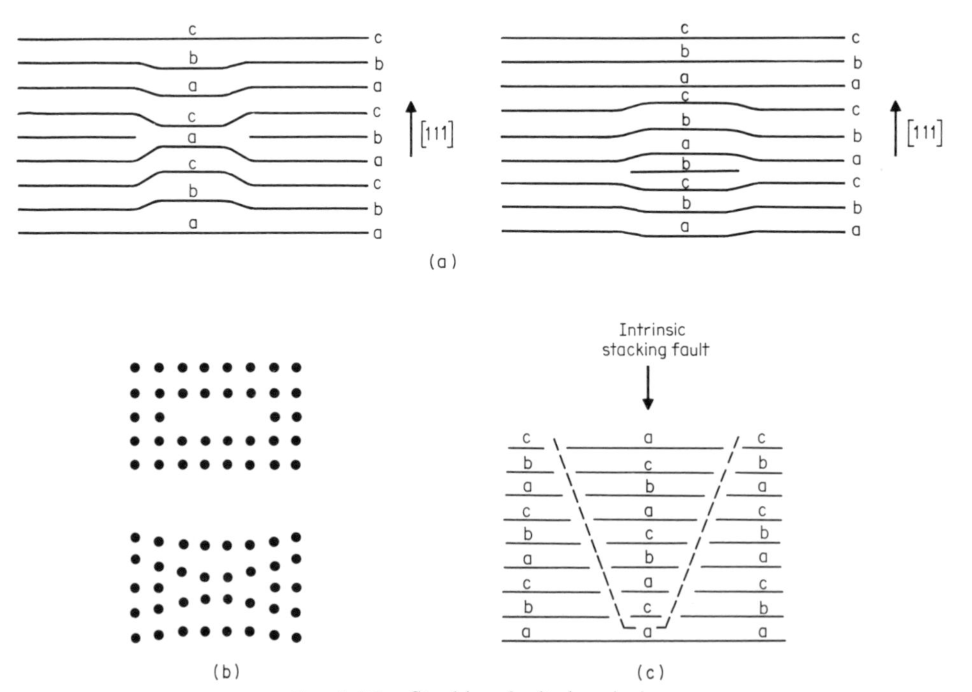

Fig. 2.12. Stacking-fault description.

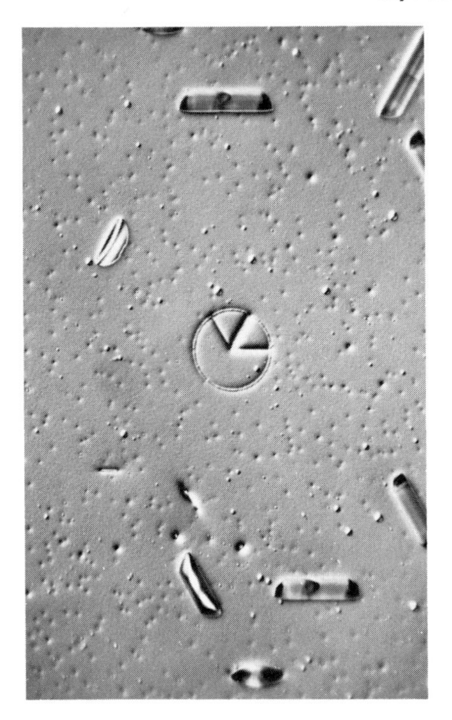

Fig. 2.13. Circular stacking fault in Si. The surface is (111). It was Sirtl-etched and photographed with Nomarski interference. The crystal had been annealed at 1200°C for 4 h. (*Photograph courtesy of Dr. Lawrence D. Dyer, Texas Instruments Incorporated.*)

2.13).[138] For faults introduced during growth, the region around them usually does not deform as shown in Fig. 2.12*a* but rather propagates during any additional crystal growth as a differently stacked region separated by additional faults on various (111) planes as illustrated in Fig. 2.12*c*. Their traces intersecting the surface will give the stacking-fault outlines shown in Fig. 2.14. Should two of the same type be nucleated close together, the boundary will disappear where they touch and will produce the pattern shown in Fig. 2.15*a*. Should the two be of opposite type, an overlap of the

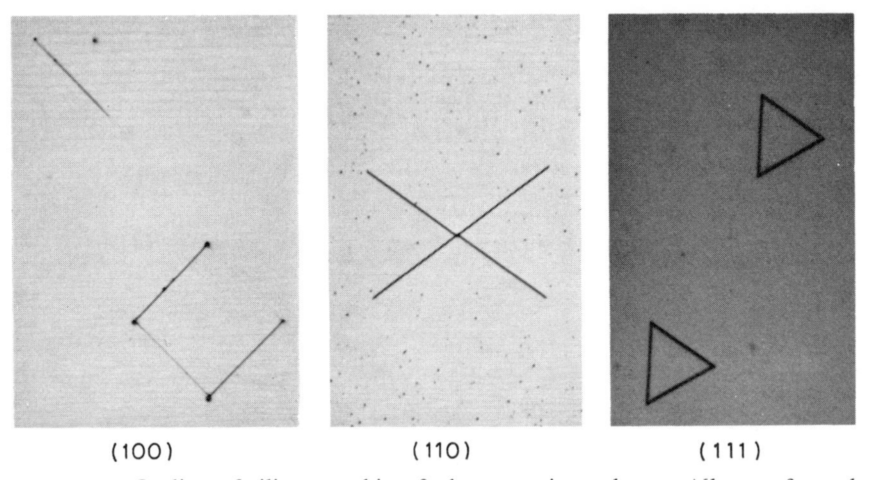

Fig. 2.14. Outline of silicon stacking faults on various planes. All were formed during initiation of epitaxial overgrowth and were delineated by Sirtl etch. (*Photographs courtesy of Kenneth E. Bean, Texas Instruments Incorporated.*)

pattern will appear as in Fig. 2.15*b*. Stacking faults can also be linear; e.g., they can be very thin and comprised of two closely spaced faults which, when viewed on edge, appear as a line.

Occurrence. Stacking faults in semiconductors are commonly observed in epitaxial growths[40,124-133,144] and are occasionally seen in bulk melt-grown material which has been annealed[32,136-138] and in surfaces which were mechanically damaged and

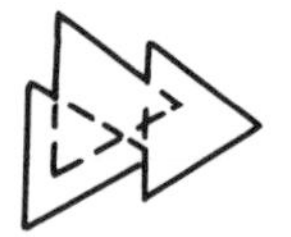

(a)

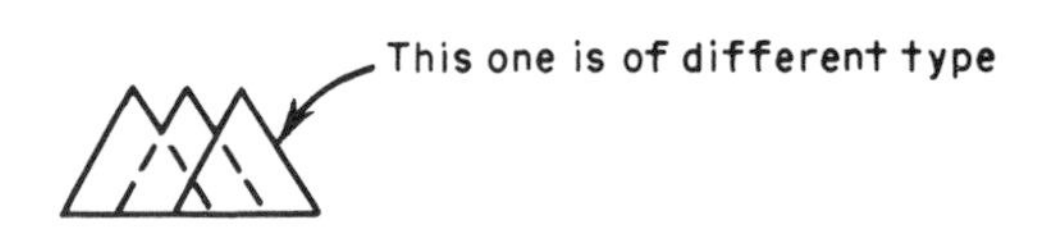

(b)

(c)

Fig. 2.15. Composite patterns produced by multiple stacking faults growing together. When epitaxial layers are thick and/or the density is high, very complex outlines composed of many faults can occur. (*a*) The same type combining. (*b*) Two of one type and one of the other. (*c*) Photograph showing these combinations.

subsequently annealed. In the latter case they are predominantly line faults rather than the closed-figure forms widely seen in epitaxial material. In some cases such faults occur after oxidation but without previous mechanical damage. The origin of these faults is possibly the strain arising from either oxygen or fluorine[123] complexes formed during heat treatment. Faults observed in epitaxial layers can nucleate at slip lines, scratches, and regions of impurity segregation, and from particulate matter on the initial growth surface. Most of them originate at the initial growth surface, and because of their well-defined geometric growth behavior, they will all be the same size when viewed from the top surface provided that the layer thickness is uniform. If smaller-sized faults are also visible, additional nucleation occurred during the growth cycle, probably from particulate matter in the feed stream.

Observation. Stacking faults may be observed by x-ray topography, by surface etching, or in many cases, by direct viewing of a natural growth surface using some form of interference contrast.[*,129,144] The latter is possible because of small step-height differences and has the advantage of being completely nondestructive. There is, however, the possibility that some of the faults will anneal out during subsequent heat cycling. In that case the outline will still be present and might be counted.[145] Etching would not define them and would not give an erroneous number. Next to interference contrast, etching is the handiest and most commonly used method of examining stacking faults. It produces grooves where the fault planes intersect the surface, and thus gives the outlines which were shown in Fig. 2.14. Table 2.5 summarizes the most appropriate etchants. Usually they just define the outline, but Sailor's etch appears to be sensitive to the type of partial dislocation at the corner of the fault and hence can assist in interpretation. The stair-rod dislocations $\frac{1}{6}\langle 110\rangle$ at the corners of the conventional stacking faults are not delineated, but the $\frac{1}{6}\langle 112\rangle$ dislocations which terminate the linear faults are enhanced.[127] Occasionally a triangular etch figure will be observed which will have an etch pit at each corner. It is possible that the figure is not really defining a triangular stacking fault but

*Such as the Nomarski interference attachment available for most microscopes.

Table 2.5. Etchants for Stacking Faults

Application	Etch	Notes	Reference
Si (111)	Dash	15–20 min at room temp	127
	Sirtl	15–30 s	147
	Iodine		127
	Sailor's	Up to 4 h at room temp enhances delineation of $a/6\ \langle 112\rangle$ partial dislocations	127
(100)	Dash, Sirtl		127
(110)	Dash		127
Ge (111)	WAg		40
GaAs (111) both faces (100) (110)		2 ml H_2O, 1 ml HF, 8 mg $AgNO_3$, 1 g CrO_3, 10 min at 65°C	108
β-SiC (111)	Fused salt		146

rather the intersection of linear faults and the etch pit delineates the dislocations associated with them. Square faults in (100) material apparently cannot have stair rods at all four corners, but the dislocations that are there usually etch along with the fault outline. Thus the (100) outlines usually show dislocation pits at the corners.[134]

X-ray topography and transmission electron-beam diffractions are both applicable to stacking faults.

2.4 TWINS

When two contacting regions are of different orientation but still are oriented so that at the interface each portion shares the same crystallographic plane, the two regions are twins and the interface between them is the twin plane. Nearest-neighbor positions are maintained for the atoms on each side of the interface but next-nearest-neighbor positions are violated. Crystal symmetry and energy considerations dictate which planes will serve as twin planes. For diamond and zinc blende structures (111) twin planes are experimentally observed, although from symmetry alone, (112)s would also be possible. Should a crystal twin, and then twin again, the additional boundaries are sometimes planar and are referred to as higher-order twins.[148,149] The (221) in particular has been observed in Si. These boundaries, of course, have more lattice distortion associated with them than do the original twins.

Occurrence. Twins usually develop in semiconductors during growth as a result of temperature fluctuations or chance contamination. Metals quite often show twins owing to mechanical deformation, but the brittle semiconductors will only occasionally twin in that manner, and then only after deformation followed by heat treatment.

Detection. When twin planes intersect the surface, the lines can often be directly observed because of slightly different rates of growth of the two orientations. Sandblasting will usually produce an easily discernible difference, because as the material fractures, the cleavage planes make different angles with the surface when the crystallographic orientation changes. This in turn produces a sharp change in the reflectivity as the twin plane is crossed. A similar effect is produced by a selective etch which will provide etch pits, since the pits will have different orientations and will then reflect differently. Examples are shown in Fig. 2.16. Etches suitable for crystal orientation are thus useful for the optical differentiation of twinned regions.

(a) (b)

Fig. 2.16. Germanium block with a twin boundary. The (*a*) side was given a selective etch and is very clear. Side (*b*) was ground with a 120-grit wet paper and requires a more grazing light to show the boundary. In the latter case, wetting the surface will make the effect more noticeable.

Table 2.6. Etchants for Twin Definition

Si	CP4A
Ge	CP4, Superoxol, white-etch, aqueous solution of 10% KOH, and 10% potassium ferricyanide. Etching will ordinarily give either a step or a groove, depending on whether there is only one, or a multiplicity of twins. If optical microscopy does not have the resolving power to separate grooves, replication-transmission microscopy can be used
InSb	1 HF, 3 HNO_3, 6 H_2O
GaSb	1 HF, 3 HNO_3, 6 H_2O

Etches recommended for dislocation definition will produce grooves or steps at twin boundaries which can then be used for boundary delineation. For very closely spaced twins such as occur in dendrites, cleavage will usually produce steps at the twin planes which can then be observed optically or by replication and electron-beam microscopy.[150] Etchants suggested for some of the more common materials are listed in Table 2.6. If required, the orientation on each side of the line can be determined by the methods discussed in Chap. 1. If such orientations are consistent with the presence of a twin plane (see Fig. 2.17), it can be inferred. When the separation plane (and not just its trace on the surface) is well defined, its orientation may be determined to see whether or not it is consistent with the allowable twin planes for the material in question. From a practical standpoint, any long, straight boundary is probably a twin,* since otherwise the boundary tends to be jagged or curved. It might, however, be a grain or lineage boundary, or even a long scratch. In the latter two cases surface roughening or etching will not give a change of reflectivity upon crossing the boundary. Etching a grain or lineage boundary will produce myriads of etch pits along the boundary which can be interpreted in light of ASTM F 47 (discussed in Secs. 2.5 and 2.6).

2.5 LINEAGE

Lineage is used to describe a mosaic of regions with small angular deviations from one another. It now is seldom observed in silicon and germanium because of the close control of growth conditions.

Occurrence. The origin of lineage is not well understood, but in growth from

*It may not be a first-order plane, however.

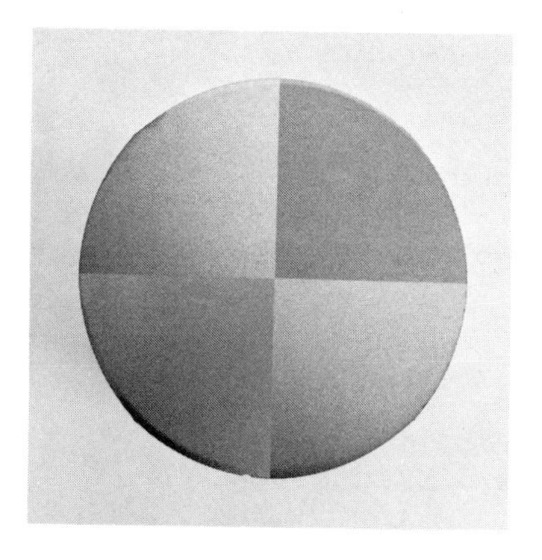

First order	Second order
{111} → {115}	{114} → {114}
{110} → {110}, {114}	{221} → {221}
{100} → {221}	

Fig. 2.17. Possible sets of diamond-lattice orientations having twin relationship. Neither list is all-inclusive, but they do include the more commonly observed one. The photograph shows a slice from a silicon boule which started growing in the [100] direction and simultaneously twinned at four positions to (221) planes. These grew together to produce the second-order twin boundaries shown criss-crossing the slice. (*Courtesy of K. E. Bean, Texas Instruments Incorporated.*)

the melt it is assumed that if the growth front is dendritic the dendrites may be rather easily bent because of their high temperature. Then, when the space between them fills in, a mosaic occurs. Crystals grown from defective seeds will continue to show lineage, and thick epitaxial growths onto an elastically deformed substrate can ultimately cause misorientation of the layer.

Detection. When the deviations are minimal, a double-crystal spectrometer or other x-ray technique may be required, but for most purposes, standard dislocation etching will suffice. The intersection of two slightly misoriented blocks will give a line of dislocations which increases in linear density as the misorientation increases. Therefore, the dislocation etchants listed in Sec. 2.2 can be used to search for lineage. ASTM F 47 for silicon has defined "lineage" as being present when the density of dislocations as indicated by etch pits exceeds 25 per millimeter along a line of 0.5 mm minimum length. When the density increases to the point where the etch pits are no longer individually distinguishable, it is common to refer to it as a "grain boundary." Slip may give similar dislocation densities but is due to an entirely different set of misfortunes. When the geometric pattern of the dislocations is in accord with the expected slip pattern, slip should of course be suspected. In the case of heteroepitaxy, if the lattice spacings of the overlayer and the host are not identical, as the various regions which were nucleated at different localities on the surface grow together, there will be a high density of dislocations even when each region is identically oriented with its neighbors.

2.6 GRAINS AND GRAIN BOUNDARIES

Grain boundaries occur where regions with gross differences in orientation join. These various orientations can arise when randomly nucleated regions grow together or when regions again come in contact after several intervening twinning steps. The existence of grains or their boundaries can be inferred from the presence of the other, so that detection may be based on finding either of them.

Detection. Because of the large differences in orientation there will be high dislocation densities, as was just described for the junction of mosaic blocks. Thus, if the surface is subjected to a dislocation etch, grooves at the boundary will result. The orientation of the etch pits over the surface should also be examined, and if they show different orientation at different places, grains or twins are indicated. On a gross scale, large areas of etch pits will reflect light differently depending on their orientation, and will facilitate locating individual grains. Most solutions used for chemically polishing a given semiconductor will also reveal grain boundaries,

Table 2.7. Etchants for Delineating Grain Boundaries

Material	Etchant*	Time, min	Temp
Si	Sirtl	0.5	Room
	1-3-6	6	Room
	1-3-10	60	Room
Ge	1-1-1	1.5	Room
	CP4	2	Room
	WAg	5	Room
	0.1 ferricyanide	20	80°

*See Chap. 7 for etch formulations and safety precautions.

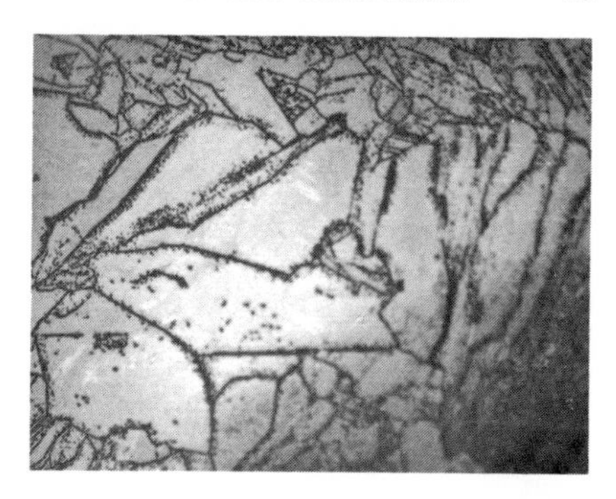

Fig. 2.18. Grain boundaries of a silicon boule delineated by selective deposition of SiO as the boule was held above the melt for a few minutes before cooling. Magnification approximately 2×.

but delineation can be enhanced by a proper etchant applied after the chemical polishing. Specific directions are given in Table 2.7 for the more common semiconductors.

Thin layers suspected of having grain boundaries must be treated with care; otherwise the whole layer may be removed during the delineation process. If the thickness is known, it can be reconciled with the etch rate to give a maximum allowable etch time. When this is not possible, one should start with one-tenth to one-fourth of the recommended times and see if definition is adequate.

Polycrystalline areas may also be delineated by lapping the surface in an aqueous slurry of 1,800-grit abrasive, or by sandblasting the surface with an abrasive whose particle size is less than 5 μm. Either preparation will present a surface in which the variously oriented areas appear as different shades of gray. It may be necessary to vary the way the light is reflected from its roughened surface in order to see the polycrystalline areas. Grain boundaries are also sometimes decorated by selective depositions. As an example, Fig. 2.18 shows the bottom of a silicon polycrystal with an accumulation of SiO along the boundaries. When the density of grains becomes large, the x-ray techniques for observing high densities of dislocations are applicable, as well as the changes in optical constants.[215]

2.7 INCLUSIONS

Inclusions are small volumes of a separate phase included in a matrix. They may be either a separate phase but identical composition, e.g., cubic SiC in hexagonal SiC, or separate compositions, such as silicon phosphide in silicon.

Occurrence. Inclusions may be introduced during crystal growth, diffusion, annealing, irradiation, or bombardment with high-energy particles. The determination of how a particular inclusion originated must be made based on its content and when first observed. Small regions of α-SiC in β-SiC crystals can logically be assumed to originate in crystal growth, but SiP platelets in Si observed after a phosphorus diffusion probably grew during diffusion.[151]

It is also possible to have a second phase present which originated during growth from the melt and propagates as long columns parallel to the growth axis. Such inclusions are found in some metallic systems, and have been reported in AlSb.[152]

Detection. Inclusions can usually be made visible by etching, which will leave them standing out in relief, after which their composition can be identified by x-ray microprobe analysis and/or electron diffraction. Cathodoluminescence combined with an electron-beam scan can also be used to map out inclusions which are very close to the surface. Thin samples may be examined by electron-transmission microscope. Note that thinning operations necessary for electron microscopy often leave residual surface deposits which may be mistaken for precipitates. Tyndall scattering (ultramicroscopy) can be used to visually observe included particles smaller than the resolution limits of ordinary microscopes. If the material to be examined is transparent, commercially available oil-immersion ultracondensers and standard microscopes can be used.[153] For the more common nontransparent semiconductors, infrared instrumentation must be used and the resolution obtained with high-quality visible optics will not be realized. However, the scattering can still be a measure of the number of particles in a large volume.[155] Tyndall scattering is ordinarily viewed at right angles to the incident light beam, but it is also possible to use other optical arrangements to keep the direct beam separate from any scattered light.

When one phase is birefringent and the other not, examination in polarized light will allow separation. For materials such as ZnS which may grow in a layered structure with the hexagonal and cubic forms interleaved, the relative amount of hexagonal type can be estimated by the degree of birefringence observed in a light beam traversing the crystal perpendicular to the layers.[156] Cross sectioning and etching can be used to delineate these and other gross inclusions. The sample may be powdered when high enough concentrations of the second phase are present and examination may be done by standard x-ray powder techniques. In some materials, e.g., diamond, impurities precipitate in the form of oriented platelets and may be so numerous that separate diffraction peaks corresponding to the structure of the oriented precipitate can be observed during x-ray topography.[157] X-ray topography can also be used to search for precipitates by detecting the microstrain associated with deformation around the precipitate. In general, such strain will produce contrast which is independent of the plane of observation, whereas if there is no precipitate, the contrast will be strongly dependent on the reflection plane used.[158] Opaque inclusions can be observed by infrared microscopy. Should the material be transparent in the visible range (e.g., CdS[159]), observation is much easier, since ordinary optical microscopy can be used. Viewing should be from several aspects so that the true shape and orientation of the precipitates are determined.

Particles of materials which alloy with the semiconductor at low temperatures may move about in thermal gradients in the same manner that traveling solvent-crystal growth occurs and leave distinctive tracks.[160,161] Since the Seebeck coefficient will be different for materials of different composition, a heated probe similar to those described in Chap. 5 can be used to map out large included grains if fine resolution is not required.[162]

Inclusions should be suspected if a diffusion concentration profile based on analytical measurements (rather than electrical) shows an abrupt increase as the surface is approached. Table 2.8 summarizes the application of these methods to various materials.

2.8 LATTICE STRESS AND STRAIN

Stress and strain can arise from work damage on the surface; from internal forces due to dislocations, excess vacancies, and impurities of radii different from the host material; from growth around included foreign material; from thermal gradients; and from dissimilar materials bonded together (e.g., SiO_2 on Si). Crystals grown from the melt sometimes have large stresses because the outer layers cooled before the interior and partially relieved the resulting stresses by plastic flow. Then after

Table 2.8. Guide to Inclusion Detection

Matrix	Inclusion	Mode of observation	Reference
Si	Au	Electron diffraction	169, 170
	As	X-ray topography	164
	B, O, Cu	X-ray topography	158, 167
	SiP	X-ray topography	151, 164
	Oxygen	Light scattering	154, 155
	Unidentified	Transmission electron microscopy	166, 167
Ge	GeAs	X-ray powder diffraction	163
GaAs	Te	Cathodoluminescence	165
	Zn	Infrared microscopy	168, 171
CdS		Optical microscopy	159
Diamond	Ni	X-ray topography	157
ZnS	Hex ZnS	Birefringence	156
PbSnTe	Metallic	Surface etching	172

the whole crystal has equilibrated, the interior is in tension and the outer layer in compression.[59] Vapor-deposited layers, whether amorphous, polycrystalline or single-crystal, often develop severe internal stresses during deposition.[173] Sawing, grinding, and lapping cause surface stresses which in turn cause appreciable bowing in thin slices unless the damage is approximately uniform on both sides.[174] Structures involving semiconductor-dielectric-metal sandwiches such as in device fabrication are particularly susceptible to differential-expansion-induced stress because of the great disparity of expansion coefficients often encountered. Of special interest is the case of partial covering of the semiconductor by a layer of different properties. At the layer edge severe strain can occur, as shown in Fig. 2.19. This is typical of behavior at the oxide windows of silicon planar devices.[176] When concentrations of impurities are diffused into the surface, strain occurs and will sometimes propagate damage well away from the diffused region.[177,178]

Direct quantitative values for strain can be obtained from lattice-spacing changes or from birefringence measurements. Qualitative values can be surmised from etch-rate behavior. Stress is calculable from a variety of stress-deflection measurements, and rough estimates of its value if caused by thermal mismatch can sometimes be made by observing the sample temperature required to change the sign of

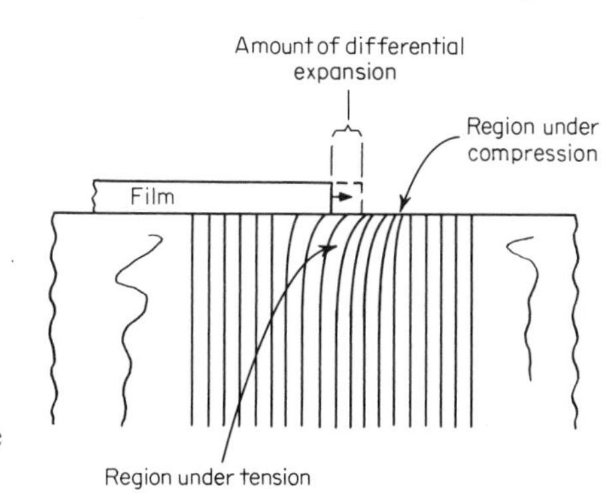

Fig. 2.19. Effect of an expanding film on the substrate lattice spacing.

deflection. Once either stress or strain is determined, the other can be obtained by using appropriate stress-strain relationships. For single-crystal materials the problem is somewhat complicated because those relations will depend on the crystallographic direction in which the stress is applied. Should only approximate values be desired, some intermediate value of the appropriate constants can be used and calculations greatly simplified. Most cases of interest to the semiconductor technologist will involve isotropic glassy layers on single-crystal substrates, but polycrystalline layers (e.g., aluminum metallization) may also be of interest. For the case of polycrystalline layers, strain associated with variously oriented anisotropic grains will be different for each grain and for special cases can be calculated,[175] but ordinarily such calculations are not required.

X-Ray Observation.[176-183] Since the Bragg angle θ is given by $n\lambda = 2d \sin \theta$, where d is the lattice spacing, λ the x-ray wavelength, and n the order, changes in the lattice spacing can be determined from changes in θ. In order to get maximum sensitivity, the highest order possible should be used and the equipment must be well aligned. Two modes of operation are possible. In one, the observed lattice spacing in the material being studied is compared with spacings in presumably unstrained samples. In the other, spacings of different planes of some particular (hkl) family are compared in the same sample. In either method geometrical errors will be introduced which might be interpreted as residual stress. Reference 181 discusses methods for estimating and minimizing these errors. Should the sample have been subjected to plastic flow, some microregions will be under compression and others under tension. If the flow was unidirectional, error can be introduced in that the overall stress (macrostress) can be zero, but the x-ray contribution from the small local microstressed regions will produce a line shift.[180] Heavily faulted regions will also cause some line shift unless measurements are made by the second method. The position of contrast in x-ray topographs can be used to determine whether the strain is compressive or tensile[182] and has been widely used to study diffusion-induced strain in silicon.[176,177,178,183]

The first method is in principle quite simple and is often used for thin polycrystalline layers. A diffractometer can be used with the surface of the layer aligned to the instrument. A quick scan can be made to check the preferred orientation of the sample and see which planes have enough intensity to be used, after which careful measurements can be made about the chosen θ position. The measurement of θ for an unstrained sample should if possible be made on the same equipment to be used for strain measurements, but regardless, care must be taken to ensure that the reference sample really is unstrained. Data taken from powder samples should not be used, since appreciable strain is introduced during the powdering unless it was prepared without any mechanical grinding by direct precipitation or deposition. A very pure reference sample should also be used because a high concentration of impurities can cause considerable lattice strain. For this reason all old data for semiconductors are suspect and should be used with caution. Figure 2.20 shows typical data for an aluminum film on a Si-SiO_2 sandwich and illustrates the variations in θ to be expected.

After measuring θ and finding d, the strain S_z is given by

$$S_z = \frac{d_{\text{unstrained}} - d_{\text{meas}}}{d_{\text{unstrained}}} \tag{2.1}$$

This S is the strain perpendicular to the surface, and before it is translated into

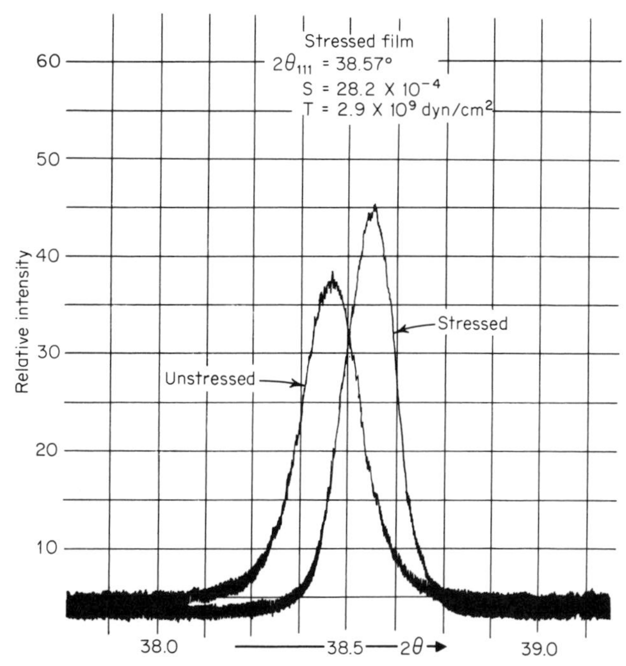

Fig. 2.20. X-ray diffraction traces of stressed and unstressed aluminum films deposited over a thermally grown SiO_2 layer. Intensity ranges for the two samples were adjusted to permit easy comparison of peak position. (*Adapted from P. B. Ghate, "Failure Mechanism Studies on Multilevel Metallization Systems for LSI," RADC F30602-70-C-0214, 1971.*)

applied stress, some boundary conditions must be known. For example, if the sample is long and thin, and the only stress T_x is in the long direction,

$$T_x = \frac{E_f S_z}{\nu_f} \tag{2.2}$$

where E_f and ν_f are Young's modulus and Poisson's ratio* for the film. Should the stress be uniform in the surface plane, Eq. (2.2) becomes[184]

$$T_x = \frac{E_f S_z}{2\nu_f} \tag{2.3}$$

For film stresses in circular samples (e.g., metallization on a silicon slice) arising from differential thermal expansion, and assuming no shearing forces, Eq. (2.3) is appropriate.

Observation by Mechanical-Deformation Measurements. If the material to be measured is glassy, as, for example, an oxide grown or deposited on a silicon slice, x-ray methods are restricted to studying the substrate strain only, but from that, stress in the film can also be estimated.[185] A method more appropriate to thin-film technology and to the problems relating to stresses developed during device processing involves the deposition of the film in question onto a relatively thick substrate

*ν can be assumed equal to 0.3 and will lead to errors of only a few percent for most materials.

with known properties. Then, for small deflections, no plastic flow in the substrate, no slippage between layer and substrate, and both materials isotropic in their elastic properties, the deflection of the composite (Fig. 2.21) is given by

$$d = \frac{3w_f r^2 T_f(1 - \nu_f)}{E w_s^2} \tag{2.4}$$

when the sample is a circular disk and $d \ll w_s$. w_f is the film thickness and is much less than the substrate thickness w_s, T_f is the stress in the film, E_s and ν_s are, respectively, Young's modulus and Poisson's ratio for the substrate, and r is the distance from the center to the point of measurement.[187] Equation (2.4) can be rewritten as

$$d = Kr^2 \tag{2.5}$$

and shows that the surface becomes a paraboloid of revolution. The deflection can be measured by direct mechanical profiling using either point-by-point microscope examination, a profilometer, or optical methods. Of the latter, the Newton-ring interferometer is probably the simplest to use and will also show whether or not the stress is uniform. If the substrate and film are both truly isotropic and the stresses uniform over the surface, the rings should be circular and centered with respect to the slice. Should thev not be, and such circumstances often occur, the substrate probably had residual stress T_i in it. In any event, the simple expression of Eq. (2.4) cannot be used, and such samples should ordinarily be rejected. Sensitivities of 2.5×10^8 dyn/cm^2 for Si substrates are possible.[188] Should a cubic single-crystal slice be used whose orientation is other than (111), Young's modulus will show periodic azimuthal variations which if large enough will prevent radial symmetry of the deflection. When the surface is highly reflective, its parabolic shape will produce a mirror whose focal length can be calculated in terms of Eq. (2.4).[189] Thus, by measuring f of the mirror, T_f can be calculated.

Instead of circular wafers (slices) it is sometimes more convenient to use long, narrow strips. In that case the stress is given by[186]

$$T_f = \frac{E w_s^2}{6\, w_f R(1 - \nu)} \tag{2.6}$$

where R is the radius of curvature of the strip (the deflection is now cylindrical instead of parabolic).

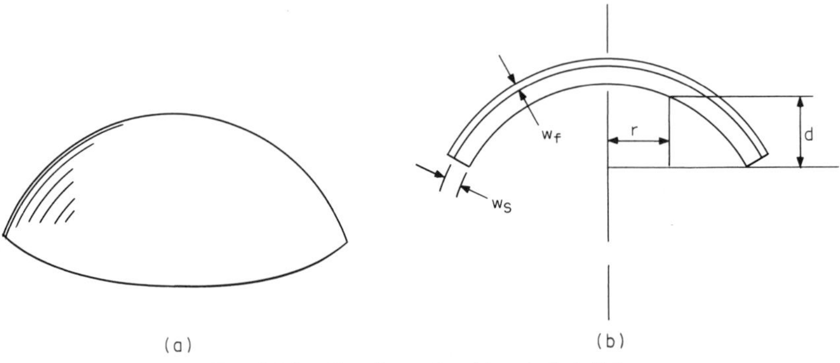

Fig. 2.21. Bowing of a bimaterial disk.

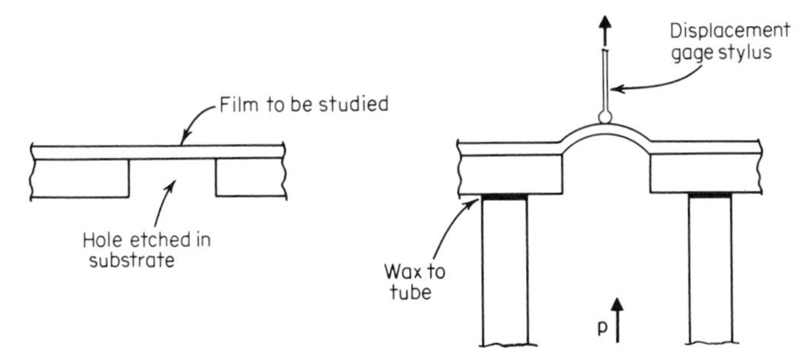

Fig. 2.22. Method of measuring deflection of thin unsupported films.

The stresses to be expected in deposited films are generally due either to differences in expansion coefficients between the film and its substrate or to intrinsic stress developed during deposition. Since most films are deposited at elevated temperatures and measurements are made near room temperature, contributions from both usually occur.

For a film or thin diaphragm with no backing and having a uniform pressure applied to deflect it, the tensile stress can be calculated from the deflection.[208] If T_i is zero, the pressure-deflection curve is parabolic; otherwise there will be a straight-line portion near zero with a slope proportioned to T_i. An experimental arrangement for these measurements is indicated in Fig. 2.22. If the pressure is increased until rupture, breaking strength may be calculated.[209] Depending on the diameter, the rupture may be explosive; so adequate shielding should be used.

Etch Rate. The rate of attack of etchants is usually dependent on lattice strain, but such rates are difficult to calibrate. It is, however, often used to detect the presence or absence of strain. (The lattice strain associated with dislocations makes certain etchants effective as dislocation delineants.) Similarly, the strain caused by cracks and other mechanical damage allows the depth of damage to be ascertained by etching until the rate slows down and becomes constant. If there is residual stress in a brittle semiconductor at room temperature, heating the material to a high enough temperature to allow stress relief will generate dislocations which can be delineated by an appropriate etchant.

2.9 MECHANICAL SURFACE DAMAGE

Surface damage, like beauty, is dependent on the eyes of the beholder and will vary according to the methods of detection. The exact nature of the damage is still the subject of some controversy, but it is substantially as shown in Fig. 2.23. Surface damage might then be measured in terms of surface roughness, misorientation of the surface due to the cracks, the number of dislocations, or the amount of residual elastic strain. Further, since electrical effects are introduced by the mechanical disruption, they too can be used to measure damage. Table 2.9 summarizes the variety of measurements that have been used, and while there is reasonable agreement

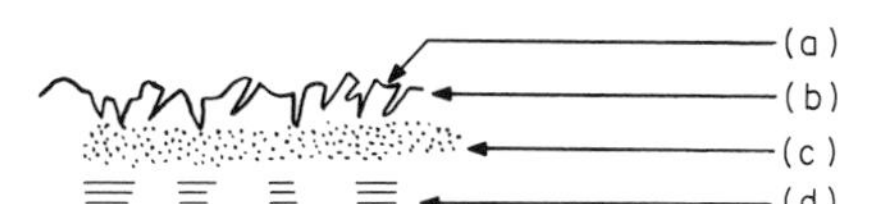

Fig. 2.23. Nature of mechanical damage in semiconductor materials. (*a*) The rough surface as depicted by a profilometer. (*b*) Region of cracks. (*c*) Dislocation networks. (*d*) Elastic strain.

Table 2.9. Methods of Determining Depth of Damage

Method	Reference
Constancy of etch rate	205
Angle section combined with selective etch	190
Incremental polish combined with selective etch	191
Electron microscope	192
Open-circuit photomagnetoelectric voltage	193
X-ray topography	109
X-ray double crystal spectrometer linewidth	194, 195
Diode reverse leakage current	193
Solar cell open-circuit voltage	
I-V curves of electrolyte-semiconductor barrier	196
Infrared reflection coefficient	197
Electron paramagnetic resonance	198
Diffusion-length measurement (Morton-Haynes geometry)	199
Filament lifetime by photoconductive decay	194
Impurity precipitates	64
Residual elastic stress	200
Ion backscattering	200

among the methods, in many instances there may be substantial differences. Those directly related to device performance [such as photoelectromagnetic (PEM) and diode reverse leakage] have usually been considered to be better indexes, but because of the likelihood that the dislocations of region *c* of Fig. 2.23 will interact deleteriously with impurities introduced during subsequent processing steps, the dislocations themselves are good indicators. Surface damage occurs as the result of mechanical abrasion during such operations as sawing, lapping, polishing, and cavitroning. It can also be generated by careless handling with mechanical implements such as tweezers.

Constancy of Etch Rate. The etch rate of an abraded semiconductor surface is higher than that of a "damage-free" one and thus can be used (and indeed was one of the earliest reported methods[205]) to measure depth of damage. The thickness removed before the etch rate becomes constant is taken as the depth of damage. One disadvantage of this method arises from the fact that the damage is not uniform, and thus the initial etching is also uneven, since it will quickly etch out deep grooves where the damage was deepest. However, the amount of material removed is usually calculated by assuming uniform material over the whole surface, taking the weight loss after each etch step, and converting it to an equivalent thickness. Such a procedure thus gives only an average depth and underestimates the maximum damage depth. Some precautions to be observed are: the back and sides of the slice must be protected during etching or else the study should be made simultaneously on both sides; a constant etch temperature should be maintained throughout all etching; and if a fresh batch of etch is required during the study, the etch rate between the two should be correlated.

Sectioning Methods. There are several sectioning methods for studying surface damage. One is to bevel the surface using the same procedures as are common in thickness measurements (see Chap. 6). The beveled surface is then carefully polished to remove any damage resulting from the beveling operation. After that, the surface can be subjected to the appropriate etches to delineate damage.[195,201,202,203] A better way if relatively large areas of uniformly damaged material are available is to break the sample into several pieces, mechanically polish or etch each piece for varying times in a polishing etch in order to remove different amounts of material, and then subject them to a dislocation etch. An example of this procedure is shown in Fig. 2.24. The amount removed can be determined by weighing, direct thickness measurements, or masking a small section of the surface during etching and subsequently measuring the step height. To minimize the effects of the material removal required for sectioning, the damaged side of the slice can be mounted face down and the reverse side mechanically polished. Removal and examination then proceed from the back and progress into the damaged region.[204]

Another method of obtaining a cross-sectional view of slice damage is to cleave the slice along a plane perpendicular to the surface, and examine the cleaved region by x-ray topography. In order to minimize distortion, the diffracting plane should be chosen so that the diffracted beam is as near normal to the cleaved surface as

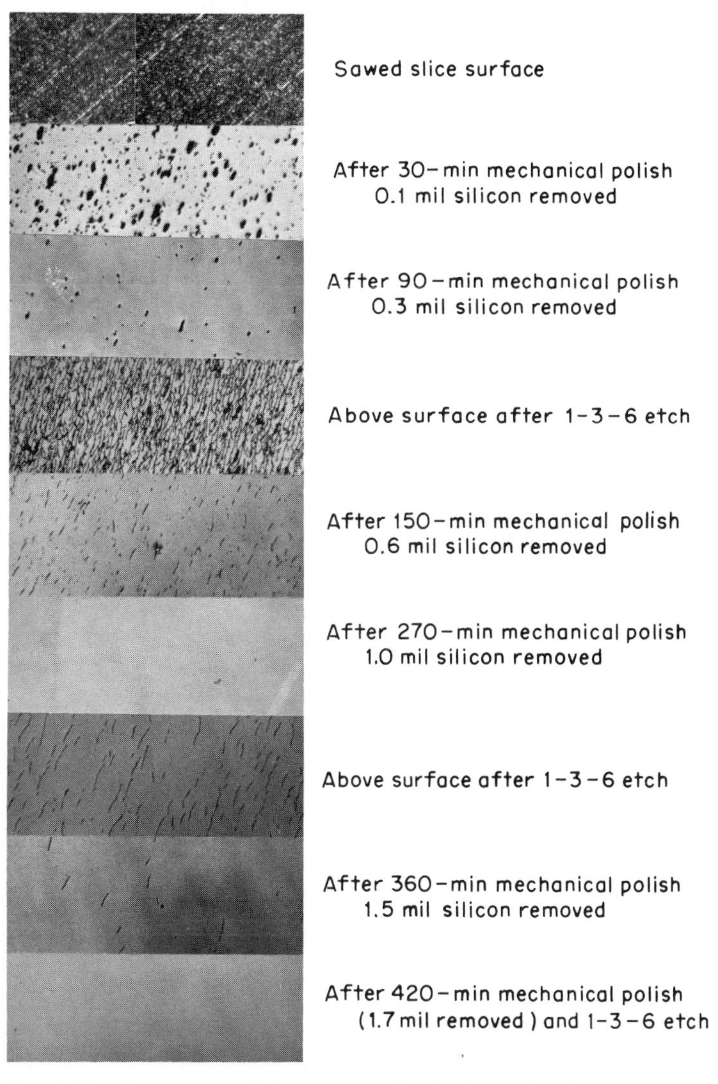

Fig. 2.24. Step sequence in damage-depth determination. Magnification 32× in all cases. (*Photographs courtesy of Jimmie B. Sherer, Texas Instruments Incorporated.*)

possible. Further, the slice should be oriented so that the incident beam, the diffracted beam, and the normal to the cleaved surface all lie in the same plane.[109]

Effect on Current Carriers. Mechanical damage induces extra free carriers which increase conductivity and affect reflectivity. It also provides recombination centers which reduce lifetime and increase surface recombination velocity. Thus if any of these properties are monitored as a function of material removal, damage depth can be estimated (see Table 2.9 for reference).

Decoration. Copper decoration can be used to indicate residual damage remaining after annealing.

Strain Measurements. Strain associated with mechanical damage may be detected directly by x-ray, birefringence, or warping.*

Even a very carefully mechanically polished surface may leave residual strain which is difficult to detect directly but upon subsequent heating will produce a variety of readily observable secondary defects. For example, silicon surfaces, mechanically polished and subsequently oxidized, when Sirtl-etched will show a large incidence of line-stacking faults. Scratching or indenting will, upon heating, produce dislocation loops which fan out from the damaged site.[24] A silicon slice subjected to an 800°C chlorine etch after a four-point-probe resistivity measurement will often have holes etched through the slice where the probes contacted the slice even though prior examination of similar areas showed no discernible damage. Germanium, and to some extent silicon, will show "crow's feet" after damage by indenting, followed by heat treating and etching.

EPR. Electron-paramagnetic-resonance lines have been observed which are related to surface damage, but they have not been used extensively to study damage depth.[198]

2.10 SHAPE DEFECTS

Crystal-shape defects are primarily of two kinds. Either they are unwanted growths, usually polycrystallites or multiple twins which project above the surface of otherwise planar layers of vapor-grown crystals, or else they are a nonplanar but crystallographically perfect surface which did not faithfully reproduce the original surface contour. The spurious growths can be traced to chance contamination either left on the slice or brought in during the crystal-growing operation. The nonreplication or shift of steps, holes, hills, etc., due to natural growth processes can be minimized by choice of growth conditions.[206,207] They may be detected by light scattering (either visually from an intense source such as a microscope light, or by instruments[214] designed especially for the purpose) or, in the case of shifts of ledges ("pattern shift") of buried layers (epitaxy), by sectioning and staining. The nonreplication can be followed during growth by alternately producing n- and p-layers which can later be exposed by sectioning and staining.

2.11 RADIATION DAMAGE[210]

Irradiation of semiconductors by x- and gamma rays, and particles such as neutrons, electrons, protons, and heavier ions produce a variety of damaging effects to both the semiconductor itself and devices made from it. The reader may be faced with the problem either of irradiating and then searching for damage or of examining

*See previous section.

material or devices and deducing what environment they have seen.

The damage will be either bulk or surface, but the device performance will reflect not only those effects but also the effects of the generation of excess carriers during irradiating by x- or gamma rays. The bulk damage is atomic displacement caused by the colliding incident particle and may be as minimal as a few vacancy-interstitial pairs (and the interstitials usually anneal out rather rapidly) or as drastic as local melting or amorphous-layer formation. These defects in turn introduce deep electrical levels into the semiconductor, which causes the resistivity to increase and the lifetime to decrease. Surface effects generally involve charge buildup in overlaying insulating layers (thermal oxide of Si planar devices) and can cause heavy surface inversion or accumulation. The bulk and surface charges are generally lasting, though much of it can be annealed out, while the excess carrier generation is but a transient effect and lasts after cessation of irradiation only as long as it takes the carriers to recombine by normal processing.

The units in common usage are given in Table 2.10. However, since the various sources differ appreciably in both particle and energy spectrum, one of the major problems is in correlating various investigators' results.

Observation. Once a radiation defect has been introduced, it is not different from one produced by other means. Hence the methods described throughout this chapter are appropriate, as well as those described in Chaps. 3 and 4. The more heavily damaged regions, such as local melting, can be studied by etching techniques.[211] Since most defects are of the point variety, EPR and infrared-absorption spectra are widely used.[212]

To study the effects on finished devices, special test structures are often used. For example, if a diode is constructed which has one side of the junction much thinner than the diffusion length of the minority carriers, e.g., a solar-cell-type structure, short-circuit current will be proportional to the diffusion length in the other

Table 2.10. Units Used in Describing Radiation

Type	Descriptive units
Neutron	Flux: in particles/(cm^2-s). Fluence: (time integral of flux) in particles/cm^2 Integrated flux: same as fluence abbreviated as n/cm^2 or as nvt (from neutron density $\times$ velocity $\times$ time) Energy:* in MeV (million electron volts)
Electrons	Flux: in particles/(cm^2-s); or charge/(cm^2-s) Fluence: in particles/cm^2, N/cm^2; or charge/cm^2 Energy:* in MeV
Protons	Same as electrons
Heavy ions	Same as electrons, plus atomic weight
X-ray Gamma	Amount of ionization produced in air 1 roentgen (R) = 2.09×10^9/electrons/1.293×10^{-3} g of air (1 esu/cm^3 of air) Amount of energy deposited in unit volume of a given material, cal/cm^3 Amount of energy deposited in 1 g of a given material, cal/g 1 rd = 100 cal/g Rate(γ): in rd/s, R/s, or cal/(cm^3-s)

*If the particles are not monoenergetic, a graph of their number vs. energy is also required.

side. Therefore, lifetime can be continuously monitored during irradiation by x- or gamma rays. Properly designed, diodes can be very sensitive to radiation damage. For instance, with small-volume avalanche diodes, the electrical effect of a single neutron collision has been reported.[213]

REFERENCES

1. R. G. Rhodes, "Imperfections and Active Centres in Semiconductors," The Macmillan Company, New York, 1964.
2. J. Melngailis and S. O'Hara, Diffusion of Vacancies during Quenching of Ge and Si, *J. Appl. Phys.*, **33**:2596–2601 (1962).
3. R. A. Swalin, "Thermodynamics of Solids," John Wiley & Sons, Inc., New York, 1972.
4. Erwin W. Müller, Direct Observation of Crystal Imperfections by Field Ion Microscopy, in J. B. Newkirk and J. H. Wernick (eds.), "Direct Observation of Imperfections in Crystals," Interscience Publishers, Inc., New York, 1962.
5. W. W. Scalon, Stoichiometry in Compound Semiconductors, in Harry C. Gatos (ed.), "Properties of Elemental and Compound Semiconductors," Interscience Publishers, Inc., New York, 1960.
6. National Bureau of Standards, personal communication.
7. James W. Gilpin and James C. Boatman, Vacancy Distribution in Silicon Ribbon Material and Resultant Diffusion Anomalies, in Rolf R. Haberecht and Edward L. Kern (eds.), "Semiconductor Silicon," Electrochemical Society, New York, 1969.
8. James W. Corbett, Electron Radiation Damage in Semiconductors and Metals, *Solid State Phys.,* Suppl. 7, 1966.
9. J. R. Patel, R. F. Tramposch, and A. R. Chaudhuri, Growth and Properties of Heavily Doped Germanium Single Crystals, in Ralph O. Grugel (ed.), "Metallurgy of Elemental and Compound Semiconductors," Interscience Publishers, Inc., New York, 1961.
10. G. E. Brock and B. K. Bischoff, Arsenic Precipitation in Germanium, in Geoffrey E. Brock (ed.), "Metallurgy of Advanced Electronic Materials," Gordon and Breach, Science Publishers, Inc., New York, 1963.
11. E. Rimini, J. Haskell, and J. W. Mayer, Beam Effects in the Analysis of As-doped Silicon by Channeling Measurements, *Appl. Phys. Lett.,* **20**:237–239 (1972).
12. J. A. Davies, J. Denhartag, L. Eriksson, and J. W. Mayer, Ion Implantation of Silicon, *Can. J. Phys.*, **45**:4053–4061 (1967).
13. F. Morin and H. Reiss, Precipitation of Lithium in Germanium, *Phys. Chem. Solids,* **3**:196–209 (1957).
14. R. Weltzin and R. A. Swalin, Application of Precipitation Techniques to the Study of Defects in Germanium, *J. Phys. Soc. Japan,* **18**(Suppl. III):136–141 (1963).
15. P. D. Southgate, Mechanical Damping of Germanium and Silicon Containing Impurity Oxygen, in M. Désirant and J. L. Michiets (eds.), "Solid State Electronics and Telecommunications," vol. 1, Academic Press, Inc., New York, 1960.
16. S. I. Tan, B. S. Berry, and B. L. Crowder, Elastic and Anelastic Behavior of Ion-implanted Silicon, *Appl. Phys. Lett.,* **20**:88–90 (1972).
17. S. Chau, L. A. Davidson, and J. F. Gibbons, Use of High Energy Ion Beams for the Analysis of Doped Surface Layers, in Charles P. Marsden (ed.), "Silicon Device Processing," NBS Special Publication 337, 1970.
18. A. H. Cottrell, "Dislocations and Plastic Flow in Crystals," Oxford University Press, Oxford, 1953.
19. W. T. Read, Jr., "Dislocations in Crystals," McGraw-Hill Book Company, New York, 1953.
20. John Price Hirth and Jens Lothe, "Theory of Dislocations," McGraw-Hill Book Company, New York, 1968.

21. F. R. N. Nabarro, "Theory of Crystal Dislocations," Oxford University Press, London, 1967.
22. V. L. Indenbom, Dislocations in Crystals, *Soviet Phys. Cryst.,* **43:**112–132 (1958).
23. O. W. Johnson, High Stress Low-Temperature Dislocation Kinetics of Ge, *J. Appl. Phys.,* **36:**3247–3250 (1965).
24. W. C. Dash, Dislocations in Silicon and Germanium Crystals, in Harry C. Gatos (ed.), "Properties of Elemental and Compound Semiconductors," Interscience Publishers, Inc., New York, 1960.
25. W. C. Dash, Copper Precipitation on Dislocations in Silicon, *J. Appl. Phys.,* **27:**1193–1195 (1956).
26. Eugene S. Meieran, Observation of Frank-Reed Sources in Silicon, *J. Appl. Phys.,* **36:**1497–1498 (1965). See also *J. Appl. Phys.,* **27:**1193; **35:**1956.
27. Mitsuru Yoshimatsu, Tetsuro Suzuki, Teruo Kobayashi, and Kazatake Kohra, Loop Shaped Images Observed in X-ray Diffraction Micrographs of Silicon Single Crystals, *J. Phys. Soc. Japan,* **17:**583–584 (1962).
28. A. J. R. De Kock, Vacancy Clusters in Dislocation Free Silicon, *Appl. Phys. Lett.,* **16:**100–102 (1970).
29. T. S. Plaskett, Evidence of Vacancy Clusters in Dislocation-free Float-Zone Silicon, *Trans. AIME,* **233:**809–812 (1965).
30. Junji Matsui and Tsutomu Kawamura, Spotty Defects in Oxidized Floating-zoned Dislocation-free Silicon Crystals, *Japan. J. Appl. Phys.,* **11:**197–205 (1972).
31. J. Hornstra, Dislocations in the Diamond Lattice, *Phys. Chem. Solids,* **5:**129–141 (1958).
32. William J. Patrick, The Precipitation of Oxygen in Silicon, and Its Effect on Surface Perfection, in Charles P. Marsden (ed.), "Silicon Device Processing," NBS Special Publication 337, 1970.
33. M. S. Abrahams and A. Dreeben, Formation of Dislocations around Precipitates in Single Crystals of (Zn, Cd)S:Er, *J. Appl. Phys.,* **36:**1688–1692 (1965).
34. F. L. Vogel, W. G. Pfann, H. E. Corey, and E. E. Thomas, Observation of Dislocation in Lineage Boundaries in Germanium, *Phys. Rev.,* **90:**489–490 (1953).
35. F. L. Vogel, Jr., Dislocations in Low-Angle Boundaries in Germanium, *Acta Met.,* **3:**245–248 (1955).
36. W. G. Pfann and F. L. Vogel, Jr., Observations on the Dislocation Structure of Germanium Crystals, *Acta Met.,* **5:**377–384 (1957).
37. V. R. Regel, A. A. Urusovskaya, and V. N. Kolomiichek, Revealing the Emergence of Dislocations on the Surface of a Crystal by the Etch Method, *Soviet Phys. Cryst.,* **4:**895–917 (1960).
38. T. L. Johnston, C. H. Li, and C. I. Knudson, Spiral Etch Pits in Silicon, *J. Appl. Phys.,* **28:**746 (1957).
39. J. W. Faust, Jr., The Influence of Surface Preparation on Revealing Dislocations in Germanium, *Electrochem. Tech.,* **1:**377–378 (1963).
40. T. B. Light, Imperfections in Germanium and Silicon Epitaxial Films, in John B. Schroeder (ed.), "Metallurgy of Semiconductor Materials," Interscience Publishers, Inc., New York, 1962.
41. F. L. Vogel, Jr., Dislocations in Plastically Bent Germanium Crystals, *J. Metals,* **8:**946–949 (1956).
42. J. D. Venables and R. M. Broudy, Dislocations and Selective Etch Pits in InSb. *J. Appl. Phys.,* **29:**1025–1028 (1958).
43. M. C. Lavine, H. C. Gatos, and M. C. Finn, Characteristics of the {111} Surfaces of the III–V Intermetallic Compounds, *J. Electrochem. Soc.,* **108:**974–980 (1961).
44. Morio Inoue, Iwao Teramoto, and Shigetoshi Takayanagi, Cd and Te Dislocations in Cd Te, *J. Appl. Phys.,* **34:**404–405 (1963).
45. Dr. Lyndon Taylor, personnal communication.
46. W. Czaja and G. H. Wheatley, Simultaneous Observation of Diffusion-induced Dis-

location Slip Patterns in Si with Electron Beam Scanning and Optical Means, *J. Appl. Phys.*, **35**:2782–2783 (1964).
47. W. Czaja and J. R. Patel, Observation of Individual Dislocations and Oxygen Precipitates in Silicon with a Scanning Electron Beam Method, *J. Appl. Phys.*, **35**:1476–1482 (1965).
48. E. Sirtl and A. Adler, Chromic-Hydrofluoric Acid as a Specific System for the Development of Etch Pits on Silicon, *Z. Metallk.*, **52**:529 (1961).
49. D. Navon, R. Bray, and H. Y. Fan, Lifetime of Injected Carriers in Germanium, *Proc. IRE,* **40**:1342–1347 (1952).
50. R. V. Jensen and S. M. Christian, "Etch Pits and Dislocation Studies in Silicon Crystal," RCA Industry Service Laboratory, *Bull.* L13-1023, Mar. 5, 1956.
51. W. J. Feuerstein, Etch Pit Studies on Silicon, *Trans. AIME,* **212**:210–212 (1958).
52. Allegheny Electric Chemical Co., *Tech. Bull.* 6, June 1958.
53. F. L. Vogel, Jr., and L. Clarice Lovell, Dislocation Etch Pits in Silicon Crystals, *J. Appl. Phys.*, **27**:1413–1415 (1956).
54. F. Secco d'Aragona, Dislocation Etch for (100) Planes in Silicon, *J. Electrochem. Soc.*, **119**:948–951 (1972).
55. S. O'Hara and G. H. Schwuttke, Dislocation Reactions in Silicon Web-Dendrite Crystals, *J. Appl. Phys.*, **36**:2475–2479 (1965).
56. M. L. Joshi and F. J. Wilhelm, Observation of Diffusion-induced Dislocation Lines in Silicon through Optical Microscopy, *J. Appl. Phys.*, **36**:2593–2594 (1965).
57. S. G. Ellis, Dislocations in Germanium, *J. Appl. Phys.*, **26**:1140–1146 (1955).
58. H. A. Schell, Etch Figures on Germanium Single Crystals, *Z. Metallk.*, **47**(9):614–620 (1956).
59. E. Billig, Some Defects in Crystals Grown from the Melt, I. Defects Caused by Thermal Stresses, *Proc. Roy. Soc. London,* **A235**:37–55 (1956).
60. R. H. Wynn and C. Goldberg, Preferential Etch for Use in Optical Determination of Germanium Crystal Orientation, *J. Metals,* **5**:436 (1955).
61. V. N. Vasilevskaya and E. G. Miselyuk, The Problem of Visualization of Dislocations in Germanium by Etching, *Soviet Phys. Solid State,* **3**:313–318 (1961).
62. J. W. Faust, The Influence of Surface Preparation on Revealing Dislocations in Germanium, *Electrochem. Tech.*, **1**:377–378 (1963).
63. G. H. Schwuttke, Study of Copper Precipitation Behavior in Silicon Single Crystals, *J. Electrochem. Soc.*, **108**:163–167 (1961).
64. D. J. D. Thomas, Surface Damage and Copper Precipitation in Silicon, *Phys. Stat. Solids,* **3**:2261–2273 (1963).
65. Arthur H. Compton and Samuel K. Allison, "X-Rays in Theory and Experiment," D. Van Nostrand Company, Inc., Princeton, N.J., 1948.
66. A. D. Kurtz, S. A. Kulin, and B. L. Averbach, Effect of Dislocations on the Minority Carrier Lifetime in Semiconductors, *Phys. Rev.*, **101**:1285–1291 (1956).
67. J. R. Patel, R. S. Wagner, and S. Moss, X-Ray Investigation of the Perfection of Silicon, *Acta Met.*, **10**:759–764 (1962).
68. J. R. Carruthers, R. B. Hoffman, and J. D. Ashner, X-Ray Investigation of the Perfection of Silicon, *J. Appl. Phys.*, **34**:3389–3393 (1963).
69. Boris W. Batterman, X-Ray Integrated Intensity of Germanium Effect of Dislocations and Chemical Impurities, *J. Appl. Phys.*, **30**:508–513 (1959).
70. A. R. Lang, A Method for the Examination of Crystal Sections Using Penetrating Characteristic X-Radiation, *Acta Met.*, **5**:358–364 (1957).
71. A. R. Lang, Direct Observation of Individual Dislocations by X-Ray Diffraction, *J. Appl. Phys.*, **29**:597–598 (1958).
72. A. R. Lang, The Projection Topography: A New Method in X-Ray Diffraction Microradiography, *Acta Cryst.*, **12**:249–250 (1959).
73. A. R. Lang, Studies of Individual Dislocations in Crystals by X-Ray Diffraction Microradiography, *J. Appl.* Phys., **30**:1748–1755 (1959).

74. G. H. Schwuttke, X-Ray Diffraction Microscopy Study of Imperfections in Silicon Single Crystals, *J. Electrochem. Soc.,* **109:**27–32 (1962).
75. A. R. Lang, Application of Limited Projection Topographs and Direct Beam Topographs in Diffraction Topography, *Brit. J. Appl. Phys.,* **14:**904–907 (1963).
76. G. H. Schwuttke, New X-Ray Diffraction Microscopy Technique for the Study of Imperfections in Semiconductor Crystals, *J. Appl. Phys.,* **36:**2712–2721 (1965).
77. Kazutake Kohra, Mitsuru Yoshimetsu, and Ikuzo Shimizu, X-Ray Observation of Lattice Defects Using a Crystal Monochrometer, in J. B. Newkirk and J. H. Wernick (eds.), "Direct Observations of Imperfections in Crystals," Interscience Publishers, Inc., New York, 1962.
78. A. E. Jenkinson and A. R. Lang, X-Ray Diffraction Topographic Studies of Dislocations in Floating-Zone Grown Silicon, in J. B. Newkirk and J. H. Wernick (eds.), "Direct Observations of Imperfections in Crystals," Interscience Publishers, Inc., New York, 1962.
79. A. R. Lang, Crystal Growth and Crystal Perfection: X-Ray Topographic Studies, *Discussions Faraday Soc.,* No. 38, pp. 292–297, 1964.
80. Eugene S. Meieran, Reflection X-Ray Topography of GaAs Deposited on Ge, *J. Electrochem. Soc.,* **114:**292–295 (1967).
81. N. Kato and A. R. Lang, A Study of Pendellösung Fringes in X-Ray Diffraction, *Acta Cryst.,* **12:**787–794 (1959).
82. W. W. Webb, X-Ray Diffraction Topography, in J. B. Newkirk and J. H. Wernick (eds.), "Direct Observations of Imperfections in Crystals," Interscience Publishers, Inc., New York, 1962.
83. G. Borrmann and K. Lehman, X-Ray Wave Fields and Lattice Imperfections in Silicon, in J. B. Newkirk and J. H. Wernick (eds.), "Direct Observations of Imperfections in Crystals," Interscience Publishers, Inc., New York, 1962.
84. Volkmar Gerold, X-Ray Methods for Detection of Lattice Imperfections in Crystals, in William M. Inveller, (ed.), "Advances in X-Ray Analysis," vol. 3, Plenum Press, New York, 1960.
85. G. H. Schwuttke, Direct Observation of Imperfections in Semiconductor Crystals by Anomalous Transmission of X-Rays, *J. Appl. Phys.,* **33:**2760–2767 (1962).
86. Volkmar Gerold, Direct Observation of Dislocations in Germanium by an X-Ray Method, in J. B. Newkirk and J. H. Wernick (eds.), "Direct Observation of Imperfections in Crystals," Interscience Publishers, Inc., New York, 1962.
87. Mitsuru Yoshimatsu, Atsushi Shibata, and Kazutka Kohra, A Modification of the Scanning X-Ray Topographic Camera (Lang's Method), in Gavin R. Mallett, Marie Fay, and William M. Mueller (eds.), "Advances in X-Ray Analysis," vol. 9, Plenum Press, New York, 1966.
88. A. E. Jenkinson, Projection Topographs of Dislocations, *Philips Tech. Rev.,* **23:**82–88 (1961–1962).
89. Kyoichi Haruta, New Method of Obtaining Stereoscopic Pairs of X-Ray Diffraction Topographs, *J. Appl. Phys.,* **36:**1789–1790 (1965).
90. F. W. Young, Jr., T. O. Baldwin, A. E. Merlini, and F. A. Sherrill, A Camera for Borrmann Stereo X-Ray Topographs, in Gavin R. Mallett, Marie Fay, and William M. Mueller (eds.), "Advances in X-Ray Analysis," vol. 9, Plenum Press, New York, 1966.
91. H. K. Herglotz, Simple Method of Determining Crystal Perfection, *J. Electrochem. Soc.,* **106:**600–605 (1959).
92. See, for example, P. B. Hirsch and J. N. Kellar, A Study of Cold-worked Aluminum by an X-Ray Micro-Beam Technique: I. Measurement of Particle Volume and Misorientations, *Acta Cryst.,* **5:**162–167 (1953).
93. W. A. Wooster, "Diffuse X-Ray Reflections from Crystals," Oxford University Press, New York, 1962.
94. Richard A. Coy and James E. Thomas, Jr., Observation of X-Ray Diffuse Scattering

from Neutron-irradiated Silicon, *J. Appl. Phys.*, **42**:1236–1239 (1971).

95. Ilan A. Blech, Eugene S. Merran, and Harry Sello, X-Ray Surface Topography of Diffusion-generated Dislocations in Silicon, *Appl. Phys. Lett.*, **7**:176–178 (1965).
96. K. Akiyama and J. Yamarguchi, Dislocations of Silicon Single Crystals Grown by the Floating Zone Method, in Marvin S. Brooks and John K. Kennedy (eds.), "Ultra-purification of Semiconductor Materials," The Macmillan Company, New York, 1962.
97. J. K. Howard and D. P. Miller, Vacancy Related Defects in Antimony-doped Silicon, in William M. Mueller, Gavin Mallett, and Marie Fay (eds.), "Advances in X-Ray Analysis," vol. 7, Plenum Press, New York, 1964.
98. A. Authier and A. R. Lang, Three-Dimensional X-Ray Topographic Studies of Internal Dislocation Sources in Silicon, *J. Appl. Phys.*, **35**:1956–1959 (1964).
99. M. L. Joshi and J. K. Howard, Defects Induced in Silicon through Device Processing, in Charles P. Marsden (ed.), "Silicon Device Processing," NBS Special Publication 337, 1970.
100. P. Wang, F. X. Pink, and D. C. Gupta, Structural Faults in Epitaxial and Buried Layers in Silicon Device Fabrication, in Charles P. Marsden (ed.), "Silicon Device Processing," NBS Special Publication 337, 1970; T. Kato, H. Koyama, T. Matsukawa, and R. Shimizu, SEM Contrast Mechanism of Stacking Faults in an Epitaxial Silicon Layer, *J. Appl. Phys.*, **45**:3732–3737 (1974); Masao Tamura and Yoshimitsu Sugita, Misfit Dislocations in (100), (112), and (113) Homoepitaxial Silicon Crystals, *J. Appl. Phys.*, **44**:3442–3444 (1973).
101. Mitsuru Yoshimatsu, Studies on Lattice Defects of a Pair of Rods and Platelets in Silicon Single Crystals Observed by X-Ray Diffraction Micrography, *J. Phys. Soc. Japan,* **18**(Suppl. II):335–340 (1963).
102. Kazutake Kohra and Shigeru Nakano, X-Ray Studies on Lattice Defects in Silicon by Diffraction Microscopy and Double Crystal Spectrometer, *J. Phys. Soc. Japan,* **18**(Suppl. II):341–346 (1963).
103. E. S. Meieran and K. E. Lemons, A Study of Defects Due to Surface Processing in Silicon by Means of X-Ray Extinction Contrast Topography, in William M. Mueller, Gavin Mallett, and Marie Fay (eds.), "Advances in X-Ray Analysis," vol. 8, Plenum Press, New York, 1965.
104. R. Berman (ed.), "Physical Properties of Diamonds," Clarendon Press, Oxford, 1965.
105. M. Omar and M. Kenawi, The Etching of Diamonds by Low Pressure Oxygen, *Phil. Mag.*, **2**:859–863 (1957).
106. M. S. Abrahams, Dislocation Etch Pits in GaAs, *J. Appl. Phys.*, **35**:3626–3627 (1964).
107. M. S. Abrahams and L. Ekstrom, Etch Pits Deformation and Dislocations in GaAs, in Harry C. Gatos (ed.), "Properties of Elemental and Compound Semiconductors," Interscience Publishers, New York, 1960).
108. M. S. Abrahams and C. J. Buiocchi, Etching of Dislocations on the Low-Index Faces of GaAs, *J. Appl. Phys.*, **36**:2855–2863 (1965).
109. G. A. Rozgonyi and S. E. Haszko, Reflection X-Ray Topography of GaAs and GaP Cleavage Faces, *J. Electrochem. Soc.*, **117**:1562–1565 (1970).
110. J. K. Howard and R. H. Cox, The Crystalline Perfection of Melt-grown GaAs Substrates and Ga(As,P) Epitaxial Deposits, in Gavin R. Mallett, Marie Fay, and William M. Mueller (eds.), "Advances in X-Ray Analysis," vol. 9, Plenum Press, New York, 1966.
111. G. H. Schwuttke and H. Rupprecht, X-Ray Analysis of Diffusion-induced Defects in Gallium Arsenide. *J. Appl. Phys.*, **37**:167–173 (1966).
112. E. D. Jungbluth, X-Ray Diffraction Topographs of Imperfections in Gallium Arsenide by Anomalous Transmission of X-Rays, *J. Electrochem. Soc.*, **112**:580–583 (1965).
113. G. O. Krause and E. C. Teague, Observation of Misfit Dislocations in GaAs-Ge Heterojunctions, *Appl. Phys. Lett.*, **10**:251–253 (1967).
114. Sidney G. Parker and Jack E. Pinnell, Revelation of Dislocations in (Hg, Cd)Te by an Etch Technique, *J. Electrochem. Soc.*, **118**:1868–1869 (1971).

115. M. Gershenzon and R. M. Mikulyak, Structural Defects in GaP Crystals and Their Electrical and Optical Effects, *J. Appl. Phys.,* **35:**2132–2141 (1964).
116. J. W. Allen, On the Mechanical Properties of Indium Antimonide, *Phil. Mag.,* **2:**1475–1481 (1957).
117. H. C. Gatos and M. C. Lavine, "Chemical Behavior of Semiconductors: Etching Characteristics," Lincoln Laboratory Technical Report 293, January 1962 (AD 401398). This compendium lists etchants for many different materials.
118. Robert W. Bartlett and Malcom Barlow, Surface Polarity and Etching of Beta-Silicon Carbide, *J. Electrochem. Soc.,* **117:**1436–1437 (1970).
119. Schlomo I. Ben-Abraham, Geometry of Etch-Pits in ZnS Crystals, *J. Appl. Phys.,* **36:**2096–2098 (1965).
120. Marriner K. Norr, A Dislocation Etch for Lead Selenide Crystals, *J. Electrochem. Soc.,* **109:**1113–1114 (1962).
121. Marriner K. Norr, John V. Gilfrich, and Bland Houston, A Chemical Polish That Reveals Compositional Variations in $PbSe_{(1-X)}$ Te_X, *J. Electrochem. Soc.,* **114:**632–633 (1967).
122. C. M. Wolfe, C. J. Nuese, and N. Holonyak, Jr., Growth and Dislocation Structure of Single-Crystal $Ga(As_{1-X}P_X)$, *J. Appl. Phys.,* **36:**3790–3801 (1965).
123. C. M. Drum and W. Van Gelder, Stacking Faults in (100) Epitaxial Silicon Caused by HF and Thermal Oxidation and Effects on p-n Junctions, *J. Appl. Phys.,* **43:**4465–4468 (1972).
124. J. M. Charig, B. A. Joyce, D. J. Stirland, and R. W. Bicknell, Growth Mechanism and Defect Structures in Epitaxial Silicon, *Phil. Mag.,* **7:**1847–1860 (1962).
125. G. R. Booker and R. Stickler, Crystallographic Imperfections in Epitaxially Grown Silicon, *J. Appl. Phys.,* **33:**3281–3290 (1962).
126. H. J. Queisser, R. H. Finch, and J. Washburn, Stacking Faults in Epitaxial Silicon, *J. Appl. Phys.,* **33:**1536–1537 (1962).
127. T. L. Chu and J. R. Gavaler, Stacking Faults in Vapor Grown Silicon, *J. Electrochem. Soc.,* **110:**388–393 (1963).
128. T. L. Chu and J. R. Gavaler, Imperfections in Vapor Grown Silicon, in Geoffrey E. Brock (ed.), "Metallurgy of Advanced Electronic Materials," Gordon and Breach, Science Publishers, Inc., New York, 1963.
129. R. H. Dudley, Nondestructive Method for Revealing Stacking Faults in Epitaxial Silicon, *J. Appl. Phys.,* **35:**1360–1361 (1964).
130. S. Mendelson, Stacking Fault Nucleation in Epitaxial Silicon on Variously Oriented Silicon Substrates, *J. Appl. Phys.,* **35:**1570–1581 (1964).
131. R. H. Finch, H. J. Queisser, G. Thomas, and J. Washburn, Structure and Origin of Stacking Faults in Epitaxial Silicon, *J. Appl. Phys.,* **34:**406–415 (1963).
132. S. Mendelson, Growth and Imperfections in Epitaxially Grown Silicon on Variously Oriented Silicon Substrates, in Maurice H. Francombe and Hiroshi Sato (eds.), "Single Crystal Films," pp. 251–281, The Macmillan Company, New York, 1964.
133. S. Mendelson, Twin Boundaries and Stacking Faults in Silicon, *Acta Met.,* **13:**555–558 (1965).
134. G. Dionne, Nature of Stacking-Fault Defects in Epitaxial Silicon Layers, *J. Appl. Phys.,* **39:**2940–2941 (1968).
135. V. A. Phillips, Lattice Resolution Observations on the Structure of Twin Boundaries, Faults, and Dislocations in Epitaxial Silicon, *Acta Met.,* **20:**1143–1156 (1972).
136. T. Iizuka and M. Kikuchi, X-Ray Observation of Gold-induced Dislocation Loops in Silicon Crystals, in R. R. Hasiguti (ed.), "Lattice Defects in Semiconductors," University of Tokyo Press, Tokyo, and Pennsylvania State University Press, University Park, 1968.
137. T. Iizuka, Gold-induced Dislocation Loops in Silicon Crystals, *Japan. J. Appl. Phys.,* **5**(11):1018 (1966).
138. Lawrence D. Dyer and Fred W. Voltmer, Circular and Hexagonal Stacking Faults in

Bulk Silicon Crystals, *J. Electrochem. Soc.,* **120**:812–817 (1973).
139. H. J. Queisser and P. G. G. van Loon, Growth of Lattice Defects in Silicon during Oxidation, *J. Appl. Phys.,* **35**:3066 (1964).
140. G. R. Booker and R. Stickler, Two-Dimensional Defects in Silicon after Annealing in Oxygen, *Phil. Mag.,* **11**:1303 (1965).
141. M. L. Joshi, Stacking Faults in Steam-oxidized Silicon, *Acta Met.,* **14**:1157 (1966).
142. A. W. Fisher and J. A. Amick, Defect Structure on Silicon Surfaces after Thermal Oxidation, *J. Electrochem. Soc.,* **113**:1054–1060 (1966).
143. Yoshimitsu Sugita, Teruo Kato, and Masao Tamura, Effect of Crystal Orientation on the Stacking Fault Formation in Thermally Oxidized Silicon, *J. Appl. Phys.,* **42**:5847–5849 (1971).
144. J. M. Charig, B. A. Joyce, D. J. Stirland, and R. W. Bicknell, Growth Mechanism and Defect Structures in Epitaxial Silicon, *Phil. Mag.,* **7**:1847–1860 (1962).
145. G. Dionne, Discrepancies in the Number of Stacking Faults Revealed by Various Methods in Epitaxial Silicon, *J. Appl. Phys.,* **38**:3417–3418 (1967).
146. W. K. Liebmann, Orientation of Stacking Faults and Dislocation Etch Pits in β-SiC, *J. Electrochem. Soc.,* **111**:885–886 (1964).
147. ASTM F 80, "ASTM Book of Standards," Part 8, American Society for Testing and Materials, Philadelphia, 1968.
148. J. A. Kohn, Twinning in Diamond-Type Structures: High Order Twinning in Silicon, *Am. Mineralogist,* **41**:778–784 (1956).
149. J. A. Kohn, Twinning in Diamond-Type Structure: A Proposed Boundary-Structure Model, *Am. Mineralogist,* **43**:263–284 (1958).
150. J. W. Faust and H. F. John, A Comparison of Etching and Fracturing Techniques for Studying Twin Structures in Ge, Si, and III–V Intermetallic Compounds, *J. Electrochem. Soc.,* **107**:562–565 (1960).
151. P. F. Schmidt and R. Stickler, Silicon Phosphide Precipitates in Diffused Silicon, *J. Electrochem. Soc.,* **111**:1188–1189 (1964).
152. H. C. Gorton, Dendritic Inclusions in AlSb Grown Junction Diodes, *J. Electrochem. Soc.,* **107**:248–249 (1965).
153. G. D. Miles, Observations of Impurity Precipitates in Magnesium Oxide Single Crystals by Ultra-Microscopy, *J. Appl. Phys.,* **36**:1471–1475 (1965).
154. W. Kaiser, Electrical and Optical Properties of Heat-treated Silicon, *Phys. Rev.,* **105**:1751–1756 (1957).
155. G. H. Schwuttke, O. A. Weinreich, and P. H. Keck, A Sensitive Method for Measuring Optical Scattering in Silicon, *J. Electrochem. Soc.,* **105**:706–709 (1958).
156. F. J. Baum and F. J. Darnell, Birefringence Studies in Electroluminescent Zinc Sulfide, *J. Electrochem. Soc.,* **109**:165–166 (1962).
157. K. Kamiya and A. R. Lang, X-Ray Diffraction and Absorption Topography of Symmetric Diamonds, *J. Appl. Phys.,* **36**:579–587 (1965).
158. G. H. Schwuttke, Boron-induced Microstrains in Dislocation-free Silicon Crystals, *J. Appl. Phys.,* **34**:1662–1664 (1963).
159. Arthur Dreeben, Precipitation of Impurities in Large Single Crystals of CdS, *J. Electrochem. Soc.,* **111**:174–179 (1964).
160. E. Biederman, Traveling Solvent Defects on Silicon Wafers, *J. Electrochem. Soc.,* **114**:207–208 (1967) (note p. 676 of vol. 115).
161. C. H. L. Goodman, Melting Phenomena with Silicon, *Solid State Electron.,* **3**:72–73 (1961).
162. J. H. Westbrook, A. U. Seybolt, and A. J. Peat, A Thermal Probe for Segregation Detection, *J. Electrochem. Soc.,* **111**:888–891, (1964).
163. Thomas E. Seidel, Hall Measurements on Tunnel Diode Material, in Ralph O. Grubel (ed.), "Metallurgy of Elemental and Compound Semiconductors," Interscience Publishers, Inc., New York, 1961.

164. M. L. Joshi, Precipitates of Phosphorus and Arsenic in Silicon, *J. Electrochem. Soc.*, **113**:45–48 (1966).
165. H. C. Casey, Jr., Investigation of Inhomogeneities in GaAs by Electron-Beam Excitation, *J. Electrochem. Soc.*, **114**:153–158 (1967).
166. A. N. Knopp and R. Stickler, Transmission Electron Microscope Investigations on Diffused Silicon Wafers with Relation to Electrical Properties of Controlled Rectifiers, *J. Electrochem. Soc.*, **111**:1372–1376 (1964).
167. G. R. Booker and R. Stickler, Small Particles in Silicon, *J. Electrochem. Soc.*, **111**:1011–1012 (1964).
168. J. F. Black and E. D. Jungbluth, Decorated Dislocations and Sub-Surface Defects Induced in GaAs by the In-Diffusion of Zinc, *J. Electrochem. Soc.*, **114**:188–192 (1967).
169. M. L. Joshi, Effect of Fast Cooling on Diffusion-induced Imperfections in Silicon, *J. Electrochem. Soc.*, **112**:912–915 (1965).
170. E. D. Wolley and R. Stickler, Formation of Precipitates in Gold Diffused Silicon, *J. Electrochem. Soc.*, **114**:1287–1292 (1967).
171. J. F. Black, The Occurrence and Identification of Precipitates in Zinc Diffused GaAs, *J. Electrochem. Soc.*, **114**:1292–1297 (1967).
172. J. F. Butler and T. C. Harmon, Metallic Inclusions and Cellular Substructure in $Pb_{1-x}Sn_xTe$ Single Crystals, *J. Electrochem. Soc.*, **116**:260–263 (1969).
173. Carroll F. Powell, Stresses in Deposits, in Carroll F. Powell, Joseph H. Oxley, and John M. Blocher, Jr. (eds.), "Vapor Deposition," pp. 191–218, John Wiley & Sons, Inc., New York, 1966.
174. William C. Dash, Distorted Layers in Silicon Produced by Grinding and Polishing, *J. Appl. Phys.*, **29**:228–229 (1958).
175. R. W. Vook and F. Witt, Thermally Induced Strains in Evaporated Films, *J. Appl. Phys.*, **36**:2169–2171 (1965).
176. Ilan A. Blech and Eugene S. Meieran, Enhanced X-Ray Diffraction from Substrate Crystals Containing Discontinuous Surface Films, *J. Appl. Phys.*, **38**:2913–2919 (1967).
177. S. M. Fairfield and G. H. Schwuttke, Strain Effects around Planar Diffused Structures, *J. Electrochem. Soc.*, **115**:415–422 (1968); and references contained therein.
178. E. D. Jungbluth and H. C. Chiao, Intense Interjunction Strain in Phosphorus in Diffused Silicon, *J. Electrochem. Soc.*, **115**:429–433 (1968).
179. B. G. Cohen and M. W. Focht, X-Ray Measurement of Elastic Strain and Annealing in Semiconductors, *Solid State Electron.*, **13**:105–112 (1970).
180. B. D. Cullity, Sources of Error in X-Ray Measurements of Residual Stress, *J. Appl. Phys.*, **35**:1915–1917 (1964).
181. A. K. Singh and C. Balasingh, Effect of X-Ray Diffractometer Geometrical Factors on the Centroid Shift of a Diffraction Line for Stress Measurement, *J. Appl. Phys.*, **42**:5254–5260 (1971).
182. Brian R. Laun, Measurement of Elastic Strains in Crystal Surfaces by X-Ray Diffraction Topography, in John B. Newkirk, Gavin R. Mallett, and Heinz A. Pfeiffer (eds.), "Advances in X-Ray Analysis," vol. II, Plenum Press, New York, 1968; G. A. Rozgonyi and T. J. Ciesielka, X-ray Determination of Stress in the Thin Films and Substrates by Automatic Bragg Angle Control, *Rev. Sci. Instr.*, **44**:1053–1057 (1973).
183. Eugene S. Meieran and Ilan Blech, X-Ray Extinction Contrast Topography of Silicon Strained by Thin Surface Films, *J. Appl. Phys.*, **36**:3162–3167 (1965).
184. R. W. Hoffman, Physical Properties of Thin Films, in G. Haas and R. E. Thun (eds.), "Physics of Thin Films," vol. 2, Academic Press, Inc., New York, 1966.
185. G. H. Schwuttke and J. K. Howard, X-Ray Stress Topography of Thin Films on Germanium and Silicon, *J. Appl. Phys.*, **39**:1581–1591 (1968).
186. C. M. Drum and M. J. Rand, A Low-Stress Insulating Film on Silicon by Chemical Vapor Deposition, *J. Appl. Phys.*, **39**:4458–4459 (1968).
187. R. Glang, R. A. Holmwood, and R. L. Rosenfield, Determination of Stress in Films

on Single Crystalline Silicon Substrates, *Rev. Sci. Instr.,* **36**:7–10 (1965).

188. J. A. Aboaft, Stresses in SiO_2 Films Obtained from the Thermal Decomposition of Tetraethylorthosilicate-Effect of Heat-Treatment and Humidity, *J. Electrochem. Soc.,* **116**:1732–1736 (1969).
189. Richard Lathlaen and Donald A. Diehl, Stress in Thin Films of Silicon Vapor-deposited Silicon Dioxide, *J. Electrochem. Soc.,* **116**:620–622 (1969).
190. E. N. Pugh and L. E. Samuels, Etching of Abraded Germanium Surfaces with CP-4 Reagent, *J. Electrochem. Soc.,* **109**:409–412 (1962).
191. R. Stickler and J. W. Faust, Jr., Comparison of Two Different Techniques to Determine the Depth of Damage, *Electrochem. Tech.,* **4**:399–401 (1966).
192. R. Stickler and G. R. Booker, Transmission Electron Microscope Investigation of Removal of Mechanical Polishing Damage on Si and Ge by Chemical Polishing, *J. Electrochem. Soc.,* **111**:485–488 (1964).
193. T. M. Buck and F. S. McKim, Depth of Surface Damage due to Abrasion on Germanium, *J. Electrochem. Soc.,* **103**:593–597 (1956).
194. T. M. Buck, Damaged Surface Layers: Semiconductors, in Harry C. Gatos (ed.), "The Surface Chemistry of Metals and Semiconductors," John Wiley & Sons, Inc., New York, 1960.
195. E. N. Pugh and L. E. Samuels, Damaged Layers in Abraded {111} Surfaces of InSb, *J. Appl. Phys.,* **35**:1966–1969 (1964).
196. A. Uhlir, Electrolytic Shaping of Germanium and Silicon, *Bell System Tech. J.,* **35**:333–347 (1956).
197. Charlie E. Jones and A. Ray Hilton, The Depth of Mechanical Damage in Gallium Arsenide, *J. Electrochem. Soc.,* **112**:908–911 (1965).
198. G. Feher, Electron Spin Resonance Experiments on Donors in Silicon, I. Electronic Structure of Donors by the Electron Nuclear Double Resonance Technique, *Phys. Rev.,* **114**:1219–1244 (1959).
199. D. Baker and H. Yemm, The Depth of Surface Damage Produced by Lapping Germanium Monocrystals, *Brit. J. Appl. Phys.,* **8**:302–303 (1957).
200. T. M. Buck and R. L. Meek, Crystallographic Damage to Silicon by Typical Slicing, Lapping, and Polishing Operations, in Charles P. Marsden (ed.), "Silicon Device Processing," NBS Special Publication 337, 1970.
201. E. N. Pugh and L. E. Samuels, Damaged Layers in Abraded Section Surfaces, *J. Electrochem. Soc.,* **111**:1429–1431 (1964).
202. E. N. Pugh and L. E. Samuels, Dislocation Cracks in Abraded Germanium Surfaces, *J. Electrochem. Soc.,* **109**:1197–1199 (1962).
203. E. N. Pugh and L. E. Samuels, Dislocation Arrays Produced in Germanium by Room-Temperature Deformation, *Phil. Mag.,* **8**:301–310 (1963).
204. Ronald L. Meek and M. C. Huffstutler, ID-Diamond Sawing Damage to Germanium and Silicon, *J. Electrochem. Soc.,* **116**:893–898 (1969).
205. P. R. Camp, A Study of the Etching Rate of Single-Crystal Germanium, *J. Electrochem. Soc.,* **102**:586–593 (1955).
206. W. R. Runyan, Earl G. Alexander, and S. E. Craig, Jr., Behavior of Large-Scale Surface Perturbations during Epitaxial Growth, *J. Electrochem. Soc.,* **114**:1154–1158 (1967).
207. C. M. Drum and C. A. Clark, Geometric Stability of Shallow Surface Depressions during Growth of (111) and (100) Epitaxial Silicon, *J. Electrochem. Soc.,* **115**:664–669 (1968).
208. J. W. Beams, Mechanical Properties of Thin Films of Gold and Silver, in C. A. Neugebauer et al. (eds.), "Structure and Properties of Thin Films," John Wiley & Sons, Inc., New York, 1959.
209. W. R. Runyan, J. M. Willmore, and L. E. Jones, Growth Techniques and Mechanical Properties of Optical Quality Silicon, *IRIS,* **4**:49–58 (1958).
210. Frank Larin, "Radiation Effects in Semiconductor Devices," John Wiley & Sons, Inc., New York, 1968.

211. M. Bertolotti, T. Papa, D. Sette, and G. Vitali, Electron Microscope Observation of High-Energy-Neutron-Irradiated Germanium, *J. Appl. Phys.,* **36:**3506–3512 (1965).
212. James W. Corbett, Electron Radiation Damage in Semiconductors and Metals, *Solid State Phys.,* Suppl. 7, 1966.
213. R. Gereth, R. H. Haitz, and F. M. Smits, Effects of Single Neutron-induced Displacement Clusters in Special Silicon Diodes, *J. Appl. Phys.,* **36:**3884–3894 (1965).
214. W. J. Patrick and E. J. Patzner, The Detection of Surface Defects on Silicon Wafers by Scattered Light Measurements, in Howard R. Huff and Ronald R. Burgess (eds.), "Semiconductor Silicon 1973," pp. 482–490, Electrochemical Society, Princeton, 1973; H. J. Ruiz, C. S. Williams, and F. A. Padovani, Silicon Slices Analyzer Using a He-Ne Laser, *J. Electrochem. Soc.,* **121:**689–692 (1974).
215. Ch. Kühl, H. Schlötterer, and F. Schwidefsky, Optical Investigation of Different Silicon Films, *J. Electrochem. Soc.,* **121:**1496–1500 (1974).

3

Resistivity and Carrier-Concentration Measurements

Resistivity, carrier concentration, and impurity concentration are all interrelated, and for most impurities in most semiconductors, the interrelations are well known. Thus a determination of any of the three will usually suffice. In many cases, special requirements are imposed, e.g., measurement in a very small area, and while in principle most of the methods to be discussed can be used, in practice some are more suitable than others. Table 3.1 lists a number of these requirements and some of the more applicable methods.

3.1 GENERAL CONSIDERATIONS

Almost all the methods to be described can be used for either polycrystalline or single-crystal material, but a considerable amount of additional interpretation may be required for polycrystalline samples. The grain boundaries behave differently from the individual crystallites, and it is not always possible to tell a priori whether the apparent resistivity will be greater or less than that of similar material without grain boundaries (i.e., single-crystal material), particularly if there is a possibility of precipitates along the boundaries. Interpretation difficulties also arise when phases with widely differing resistivity are present in the sample (e.g., β-silicon carbide crystallites in an amorphous silicon carbide matrix) or if there is only a single phase but it is anisotropic. Multiple layers of alternating high-low resistivity or p-n types will make any material appear anisotropic, since measurements with the current parallel to the layers will almost always be different from those with the current flow perpendicular. Such samples can be analyzed only by some of the profiling methods discussed in a later section.

The surface preparation can affect apparent resistivity. If an inversion layer forms and a probe measuring system is being used, the probes may not punch through the layer and only it will be measured. If they do punch through and contact the bulk, the layer and the bulk will form a parallel circuit. While some conflicting observations are reported for mechanically abraded surfaces[1-3], it appears that at least for low-resistivity silicon and germanium, an increase in the measured resistivity occurs at room temperature (probably because of microcracks). If the surface-damage depth (approximately proportional to the diameter of the abrasive used) is an appreciable fraction of the total thickness of the sample, errors of from several

Table 3.1. Resistivity Methods for Special Requirements

n on n^+ layer p on p^+ layer	MOS Spreading resistance 3-point probe
n on p layer p on n layer	All of the above Four-point probe sheet-resistance measurement coupled with thickness
Depth profiling	1. *C–V* measurements 2. *C–V* coupled with etched steps 3. Differential resistance in avalanche 4. Angle lap coupled with *a* Spreading resistance *b* 3-point probe *c* 4-point probe 5. Sequential thin-layer removal and sheet-resistance measurement
Lateral profiling	Electroplating (qualitative only) Electropolishing (qualitative only) Spreading resistance 2-point probe Close-spaced 4-point probe Photovoltaic probe
Very high resistivity	Hall measurements Forward-biased 4-point probe
Very small areas	Spreading resistance Voltage breakdown *C–V* measurement
Noncontacting	Eddy current, microwave, capacitative coupled probes
Small irregular-shaped sheets	van der Pauw

to several hundred percent can occur. This problem is most likely when profiling thin layers but should always be considered, since an abraded surface is often used to make contacting easier. Conversely, exceedingly high resistivity materials may be lowered in resistivity because of additional carriers arising from damage-induced defects.

By their very nature, resistivity measurements are geometry-dependent and quite sensitive to boundary conditions. Because of this sensitivity, many correction factors have been calculated. Some are included in the following pages, and many others are referenced. Most semiconductor materials have rather high temperature coefficients of resistivity; so if precise measurements are desired, or if the ambient varies widely, suitable corrections should be made. Curves for silicon and germanium are included later in the text. Since the coefficient can change appreciably with impurity content, some caution should be exercised in extrapolations involving materials for which few data are available.

When the material to be measured is being electrically isolated by a p-n junction, the p-n junction may not afford complete electrical isolation, and in some cases, the current flow used for measurement can debias the junction and allow additional current to flow across it.[4]

3.2 BASIC METHODS

Direct Method. The oldest way of finding the resistivity ρ(ohm-centimeter) is to use a rectangular sample of known dimensions to measure the resistance R and use the relation $R = \rho L/A$, where L is the sample length and A its cross section. A disadvantage is that ρ will also contain a contact-resistance term, which for semiconductors can be appreciable. The effect can be minimized by plotting measured resistivity vs. applied voltage.[5] This resistivity will ordinarily decrease as the voltage increases and finally become relatively constant. If injection from the contacts has not become excessive at that point, the resistivity is probably no more than a few percent high. Injection difficulties can arise only with long-lifetime materials such as silicon and germanium but should seldom be a problem.

Two-Point Probe. The effect of contact resistance can be eliminated by use of the two-point probe of Fig. 3.1 if the specimen cross section is relatively uniform. Measurement restrictions are that the current must be kept low enough to prevent heating of the sample, the voltmeter must have a high input impedance, and measurements must be made far enough away from the contacts that any minority carriers injected will have already recombined. The requirements for the contacts are not very stringent and vary from plating or solder to spring-loaded crumpled-metal mesh. However, if the contacting is very poor, the equipotential lines will be distorted near the ends. To minimize this effect, ASTM F 43 recommends that the maximum cross-sectional dimension be not more than one-third the length of the sample and assumes the measurement will be made at the midpoint of the bar. If long bars are used and profiling is done by moving the two voltage points along it, readings taken at points closer than one maximum cross-sectional dimension from the ends are suspect. As an example of the effect of poor contacts, Fig. 3.2 shows the error introduced when the current enters and leaves at the specified corners (which should be the worst case) and the probes displaced varying amounts from the centerline.[6,7] If there are abrupt fluctuations in either cross section or resistivity, the equipotential surfaces may not be perpendicular to the axis of the crystal and thus lead to errors. To simplify contacting the sample, and to minimize error due to disturbing the equipotential lines,[6] ears protruding from the sides of the sample can be used (see also Fig. 5.4.). These same kinds of configurations are also useful in test structures for evaluating diffused layer resistivity.[7]

If more than one set of measurements is to be made on a given sample, reproducibility will be improved if a positioning jig is used so that all measurements to be compared are actually made on the same volume of semiconductor. However, continued contacting at the same point will cause mechanical damage to the surface and erroneous readings.

Automatic two-point-probe instruments have been developed. Some even measure

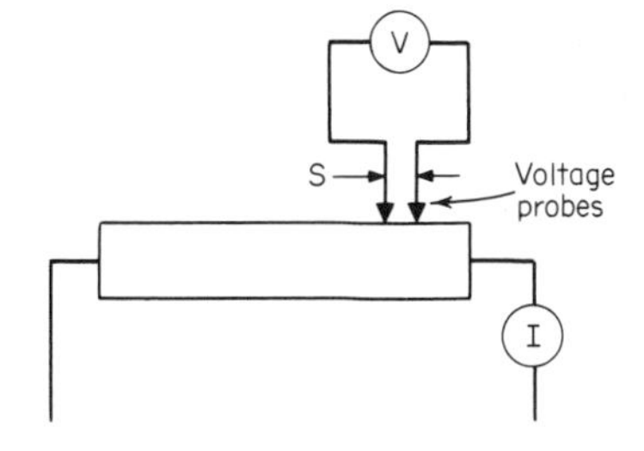

Fig. 3.1. Two-point resistivity geometry. The voltage probes have fixed spacing and are moved in unison along the surface.

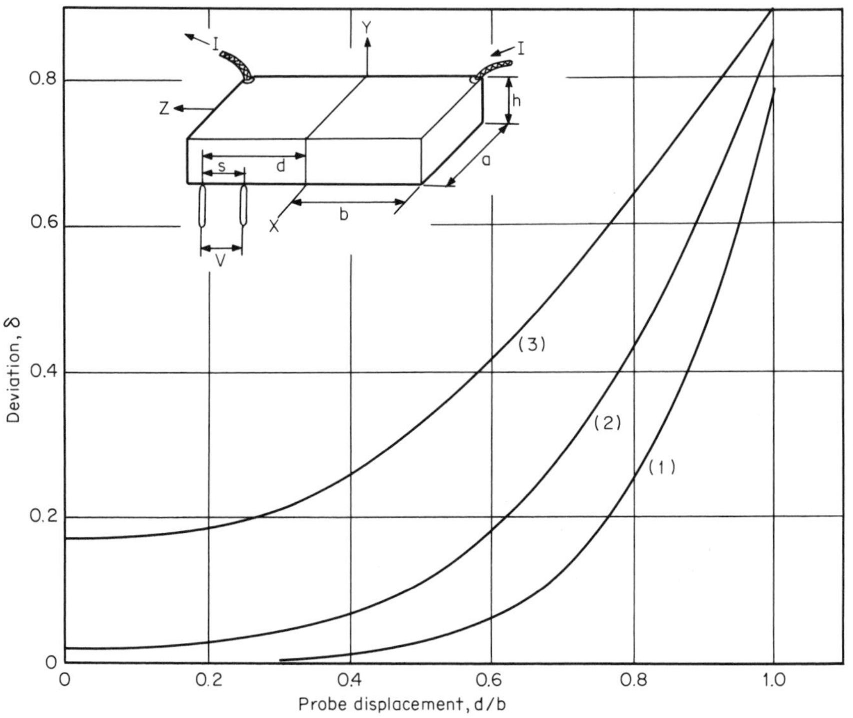

Fig. 3.2. Graph of the deviation δ caused by worst-case contacts vs. the probe displacement d/b. Curve (1) is for $a = 4.0$, $h = 4.0$, $l = 20.0$, and $s = 10.0$. Curve (2) is for $a = 6.0$, $h = 6.0$, $l = 20.0$, and $s = 1.0$. Curve (3) is for $a = 10.0$, $h = 10.0$, $l = 20.0$, and $s = 1.0$. (*Adapted from Swartzendruber.*[6])

the crystal diameter at the point of resistivity measurement, and provide automatic data printout of the position of measurement, crystal diameter at that point, and resistivity.[8] Advantages of such systems are speed of evaluation and a minimization of human errors.

Most Czochralski-grown silicon crystals have a radial-resistivity variation as well as a longitudinal one. Because of the radial component, the two-point method and the four-point method to be described in the next section will not give comparable results when measurements are made the length of the crystal. That is, V/I of Fig. 3.1 is given by

$$\frac{V}{I} = \frac{S}{2\pi \int_0^{R_0} [r\,dr/\rho(r)]} \tag{3.1}$$

where R_0 is the radius of the ingot. However the four-point-probe voltage will depend primarily on $\rho(R_0)$. By making both readings and assuming the functional form of the radial gradient, its magnitude can be estimated.[9] The two-point method can be used for slice radial-resistivity measurements by cutting a test bar as recommended by ASTM F 81 and shown in Fig. 3.3.

A single movable probe can be used and the voltage measured between it and a current lead or other suitable reference. By making several readings, dV/dx can be plotted and the resistivity calculated from

$$\rho = \frac{A}{I}\frac{dV}{dx} \tag{3.2}$$

where x is the distance along the surface.

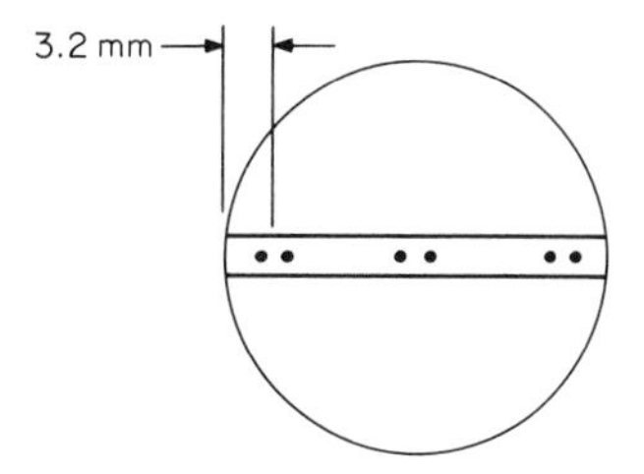

Fig. 3.3 Bar and probe orientation for measuring radial-resistivity variation by the two-point-probe method. The maximum bar width should be less than one-tenth of the slice diameter. (*From ASTM F81.*)

Linear Four-Point Probes. In the semiconductor industry the most generally used technique for the measurement of resistivity is the four-point probe.* The method is normally nondestructive; however, the probe points may damage certain semiconductor materials when excessive probe pressure is applied. The usual geometry is to place the probes in a line and use equal probe spacing. Current is passed through the outer two probes and the potential developed across the inner two probes is measured, though any of the other five combinations of current and voltage probes can in principle be used (see, for example, Table 3.3), and many combinations of unequal spacings have been considered.† For probes resting on a semi-infinite medium (Fig. 3.4) the resistivity is

$$\rho = \frac{2\pi(V/I)}{[1/S_1 + 1/S_3 - 1/(S_1 + S_2) - 1/(S_2 + S_3)]} \tag{3.3}$$

where S is the probe spacing in centimeters. When the probes are equally spaced, $S_1 = S_2 = S_3$, and Eq. (3.3) reduces to

$$\rho = 2\pi S \frac{V}{I} \tag{3.4}$$

The limitation of current for accurate measurements will be discussed later, but in general it is small fractions of amperes. Further, it is often convenient to preset

*The four-point probe is by no means new. Indeed, it was used as early as 1916 to measure the earth's resistivity and is referred to in geophysics texts as "Wenner's method" [F. Wenner, A Method of Measuring Earth Resistivity, *Bulletin of the Bureau of Standards,* **12**:469–478 (1916)].

†See, for example, C. A. Heiland, "Geophysical Exploration," Prentice-Hall, Inc., Englewood Cliffs, N.J., 1940.

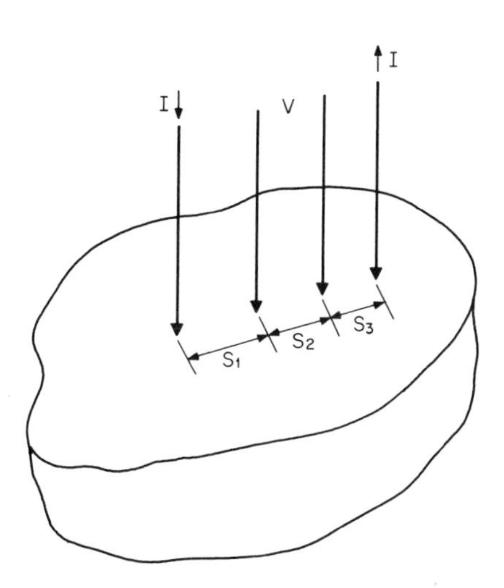

Fig. 3.4. Linear four-point resistivity probe.

the current to $2\pi S$ milliamperes or microamperes so that the resistivity in ohm-centimeters will be numerically equal to the measured voltage in milli- or microvolts, respectively. Alternately, the probe spacing can be 0.159 cm (0.062 in), so that $2\pi S$ equals 1, in which case ρ is given numerically by V/I.

More often than not, a large enough sample to be considered infinite is not available, and in those cases Eq. (3.4) is not directly applicable. However, since the four-point probe offers the most convenient mode of resistivity measurement, a variety of corrections have been developed. They are summarized in Table 3.2 and mostly fall into three categories:

1. Those to be used on cylindrical samples, i.e., unsliced crystals
2. Those which are primarily applicable to slices
3. Those for rectangular parallelepipeds

In general, the solutions are such that simple multiplicative correction factors can be applied to Eq. (3.4) to give satisfactory accuracy. The corrections are available from the references cited in Table 3.2, sometimes as curves and sometimes in tabular form. Figure 3.5 includes some of the more common ones. It is recommended that

Table 3.2. Linear Four-Point-Probe Formulas and Correction Index

Configuration	Comments	Reference
Thick sample, boundaries $>10S$ from probes	No corrections required, $\rho = 2\pi S(V/I)$	10
Thick sample, near edge	For nonconducting boundaries the meter may read as much as 100% high	10
Circular rod	Applicable to pulled crystals	11, 12
Half-cylinder		13
Rectangular bars of infinite length but with cross-sectional dimensions comparable with S	If the sample becomes very thin, see Ref. 27	14
Thin sample, with $W < 0.1S$, and boundaries $>20S$ from probes	$\rho = 4.53\ W(V/I)$, $R_s = 4.53\ (V/I)$	10
Intermediate-thickness slice with lateral boundaries $>20S$ from probe (conducting and nonconducting bottom surface)	This and following cases are applicable to slice measurements	10, 15
Circular sheet with radius $<20S$ and probes centered on slice		17, 20–26
Circular sheet with radius $<20S$ and probes displaced from center but lying along a radius		21–25
Circular sheet with radius $<20S$ and probes displaced from center, lying along a radius but perpendicular to radius		21–25
Rectangular sheet with probes symmetrically placed		27
Rectangular sheet with probes displaced from the center		27
Infinite sheets with holes near probes	Not likely in semiconductor slices, might occur in measuring metallization R_s near feedthrough	28

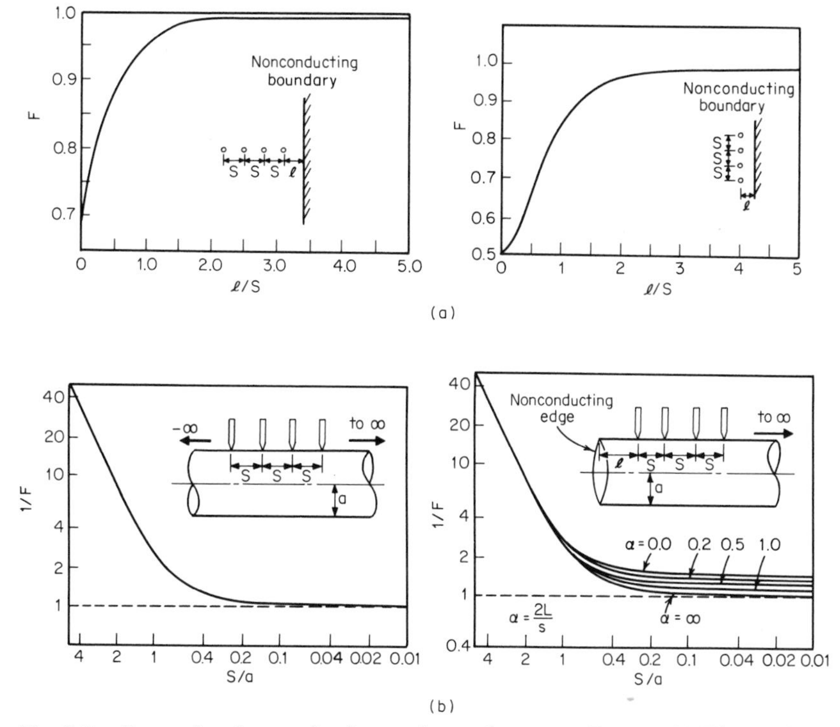

Fig. 3.5. Correction factors for four-point probes. $\rho = F\rho_{\text{meas}}$. (*a*) Flat surface. (*From Valdes.*[10]) (*b*) Circular cylinder. (*From Murashima et al.*[12])

those of particular interest be plotted on a scale large enough to be easily legible and kept with the resistivity test set.*

An isolated thin slice of resistivity ρ_1 and infinite lateral expanse can be considered as a special case of a two-layered structure in which the bottom layer is infinitely thick and has infinite resistivity. If a metallic backing is on the slice, it can be approximated by a zero-resistivity bottom layer. Corrections for these two limits are given in Refs. 10 and 15. When the second layer has an intermediate resistivity ρ_2, the measured resistivity is given by[16]

$$\rho_{\text{meas}} = \rho_1\left(1 + 4\sum_{n=1}^{\infty}\left[\frac{k^n}{\sqrt{1 + (2nt/15)^2}} - \frac{k^n}{\sqrt{4 + (2nt/15)^2}}\right]\right) \qquad (3.5)$$

where the top-layer thickness is t, and

$$k = \frac{\rho_2 - \rho_1}{\rho_2 + \rho_1}$$

For some combinations of three-layered structures, similar expressions are available but are seldom used. For an electrically isolated slice, ρ approaches $0.73\,(W/S)\rho_{\text{meas}}$

*For methods of obtaining more accurate interpolations directly from small curves such as those in Figs. 3.5 through 3.7, see Robert L. Wolke, An Interpolator for Reading Plots in Technical Journals, *Rev. Sci. Instr.*, **44**:1418 (1973), and J. S. Blakemore, Comments on "An Interpolator for Reading Plots in Technical Journals," *Rev. Sci. Instr.*, **45**:466 (1974).

(where W = slice thickness) as W/S becomes less than 1. If the back of the slice is covered with a conducting layer, e.g., a metal layer, or a very low resistivity substrate in the case of epitaxial layers, dependable results are possible only if W/S is greater than about 0.5. Thus, in order to measure layers a few micrometers thick accurately, very close probe spacing is required. In an attempt to circumvent this difficulty, other four-point-probe geometries have been used with limited success. These are described in later sections.

If thin slices have finite extent, two sets of corrections are required. They are usually considered to be independent of each other and are given in terms of a measured V/I. That is,

$$\rho = F_1 F_2 \rho_{\text{meas}} \tag{3.6}$$

where F_1 is the correction for edge effects and F_2 takes into account the slice thickness.[17,18] For thicknesses greater than the probe spacing interaction between thickness and edge effects does not allow a simple set of independent corrections. Sheet resistance R_s in ohms per square is often used in evaluating thin conducting layers. R_s equals V/I when the contacts extend the full length of opposite sides of a square of material and are independent of the size of the square. For four-point probes $R_s = F^* (V/I)$. Table 3.3 gives F^* for all possible combinations of current and voltage probes.[19] Figure 3.6 gives F_1, and Fig. 3.7 shows F_2 for circular samples.

Noncollinear Probe Spacing. As mentioned earlier, the probe array need not be linear and in principle can be of any configuration. The one most commonly used is square (Table 3.4). However, several others which have advantages for special applications have been investigated. These are summarized in Table 3.4.

Square Array. Table 3.5 summarizes the various corrections developed for the square array. They are not as extensive as for the linear case but are still adequate for most circumstances. The basic equation[17] for the square array resting on the surface of a semi-infinite medium is

$$\rho = \frac{2\pi S}{2 - \sqrt{2}} \frac{V}{I} \tag{3.7}$$

van der Pauw Method.[34,35] Rather than depend on miscellaneous corrections for finite sheets, it is possible, by placing four contacts on the periphery of the sample, to determine R_s directly. The geometry is shown in Fig. 3.8.

Table 3.3. Four-Point-Probe Correction for Use When Measuring Thin Slices

1 ↓ 2 ↓ 3 ↓ 4 ↓

Current probes	Voltage probes	Correction factor F^* for thin layer
1-4	2-3	$(\pi/\ln 2) \simeq 4.532$
1-2	3-4	$2\pi/(\ln 4 - \ln 3) \simeq 21.84$
1-3	2-4	$2\pi/(\ln 3 - \ln 2) \simeq 15.50$
2-4	1-3	$2\pi/(\ln 3 - \ln 2) \simeq 15.50$
3-4	1-2	$2\pi/(\ln 4 - \ln 3) \simeq 21.84$
2-3	1-4	$(\pi/\ln 2) \simeq 4.532$

$R_s = F^* (V/I)$, $\rho = F^* W(V/I)$.
Adapted from Rymaszewski.[19]

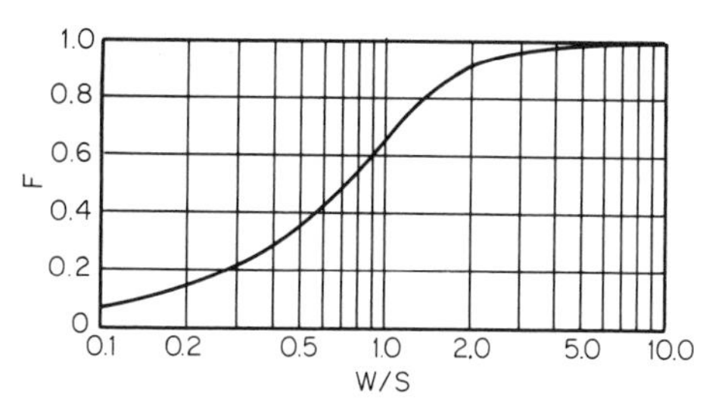

Fig. 3.6. Correction factor for a thin slice with nonconducting surfaces. $\rho = F\rho_{\text{meas}}$. [*Adapted from G. Knight, Measurement of Semiconductor Parameters, in Lloyd P. Hunter (ed.),* "Handbook of Semiconductor Parameters," *McGraw-Hill Book Company, New York, 1956. Used by permission.*]

$$R_s = \frac{\pi}{2\ln 2}(R' + R'')f\left(\frac{R'}{R''}\right) \tag{3.8}$$

where $f(R'/R'')$ is van der Pauw's function and is shown in Fig. 3.9. R' is the potential difference between the contacts C and D per unit current through the contacts A and B, and R'' is the potential difference between the contacts A and D per unit current through the contacts B and C.

If the contacts are placed so that they are symmetrical about a line through any pair of nonadjacent contacts, $R'/R'' = 1$ and van der Pauw's function also becomes 1. In addition to requiring contacts on the periphery, the method also must have very small contacts, a uniform thickness sample, and no isolated holes in the interior of the sample. When symmetrical contacts are used, any deviation in the ratio of R'/R'' is a measure of resistivity inhomogeneity and is often used for that purpose. However, if the variation of resistivity is not too great, the value read will be very close to the average obtained by integrating over the whole area. For large variation, the van der Pauw average will be lower than the integrated value.[36]

Delta Four-Point Probe.[37] This and the following configuration have been devel-

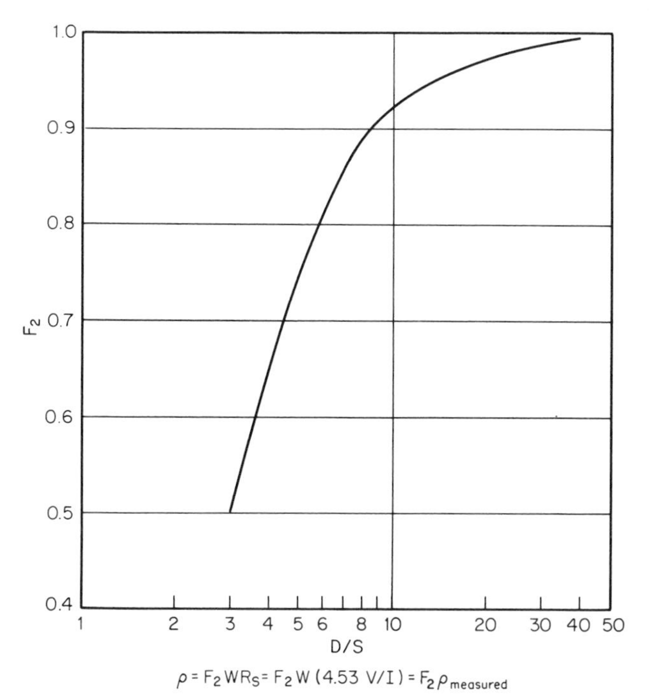

Fig. 3.7. Correction factor F_2 for probes centered in a circular slice of finite diameter D. (*Adapted from Smits.*[20])

Table 3.4. Noncollinear Four-Point-Probe Formulas and Guide

Description	Geometry	Resistivity given by	Advantages
Square array	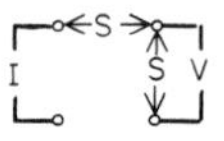	$\rho = \frac{2\pi S}{2 - \sqrt{2}} \frac{V}{I}$ $= 10.7\, S \frac{V}{I}$	Given probe spacing will fit in smaller area
Square array		$R_s = \frac{2\pi}{\ln 2} \frac{V}{I}$ $= 9.06 \frac{V}{I}$	Given probe spacing will fit in smaller area
Rectangular	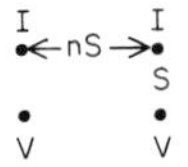	$\rho = \frac{2\pi S}{2 - (2/\sqrt{1 + n^2})} \frac{V}{I}$	
Random placement at periphery of uniform-thickness plate	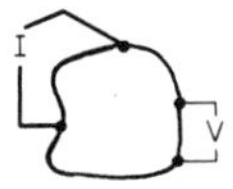	See van der Pauw	Irregularly shaped samples can be measured without precalculated correction factors
Delta		See Delta	Can be used to measure thin high-resistivity layers on low-resistivity layers
Over-under	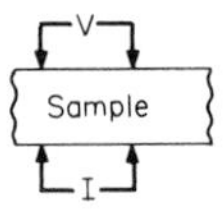	See Over-under	Can be used to measure thin high-resistivity layers on low-resistivity layers

Table 3.5. Square-Array Correction Index

Configuration	Reference
Thick sample, near edge	29
Intermediate-thickness sheet	17
Thin semi-infinite sheet with probes near edge	30
Probes in center of thin circular slice of finite radius	26, 24, 30
Probes displaced from center of thin circular slice along radius. Voltage probes parallel to radius	24
Probes displaced from center of thin circular slice along a radius. Voltage probes perpendicular to radius	24
Probes symmetrical about radius perpendicular to straight side of semicircular slice	30
Probes symmetrically placed on quarter slice	30
Square array symmetrically placed on square slice	31, 32
Probes on edge of hole in infinite slice	33

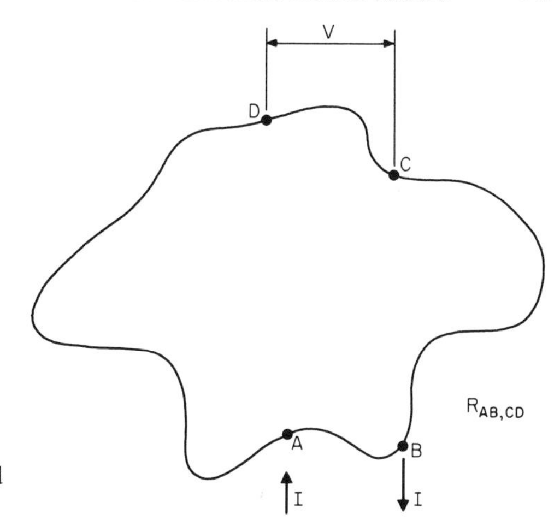

Fig. 3.8. A sample of arbitrary shape used with the van der Pauw method.

oped in an effort to allow direct measurement of thin high-resistivity layers which are in direct contact with low-resistivity substrates. The probe geometry is shown in Table 3.4. The resistivity is a complex function of probe spacing, layer thickness, substrate resistivity, and substrate thickness, and interpretation requires an elaborate set of correction factors. The useful resistivity range depends on the layer thickness and probe spacing. For 10-μm layer thickness and a spacing of 24 μm, a layer resistivity twenty times the substrate resistivity can be measured.

Over-Under Probe.[38] The placement of the probes is shown in Table 3.4. Current flows through probes 1 and 3 and voltage is measured between probes 2 and 4. As in the previous configuration, the solutions are complex. As an example of the useful range, for layer resistivities of the order of 0.1 Ω-cm, the probe spacing must be less than 50 μm. Higher resistivities can tolerate wider probe spacing.

3.3 TWO- AND FOUR-POINT PROBE INSTRUMENTATION

Basic Electrical Circuitry. The electrical circuitry for a four-point probe can be quite simple, and requires only a probe, ammeter, voltmeter, and source of current. However, the current and voltage circuits do not have a common ground, and thus one must float relative to the other. In addition, to minimize effects of pickup, rectification at the probes, and Seebeck voltages, provisions are usually made for reversing current flow. However, if the sample, probes, leads, etc., are all properly shielded, there will be no pickup and the readings will be the same unless the current is high enough to cause heating. This circuit suffers primarily because the current source is not constant so that continual adjustment is required. A somewhat better one uses the high output impedance of a transistor as a current source. If more sophistication is desired, constant-current sources can be used,[39] and for production, equipment with preset currents is a necessity.

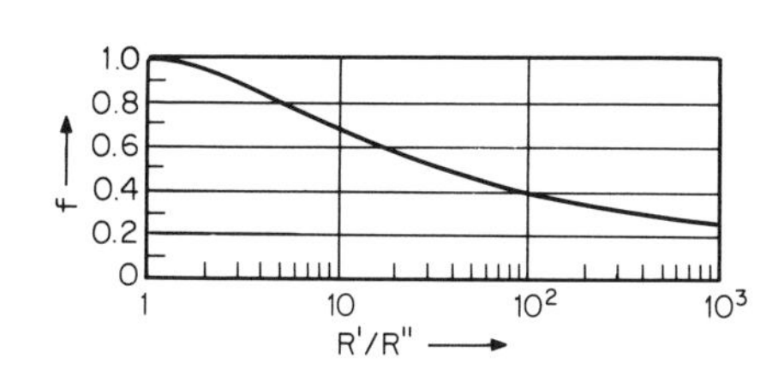

Fig. 3.9. $f(R'/R'')$ vs. R'/R''. To be used with Eq. (3.8). (*From van der Pauw.*[34])

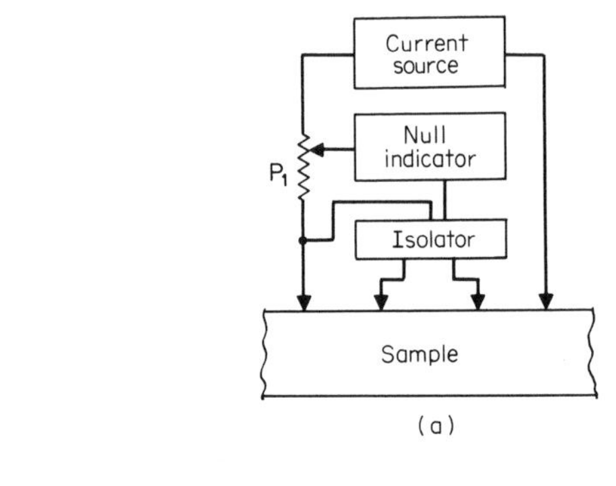

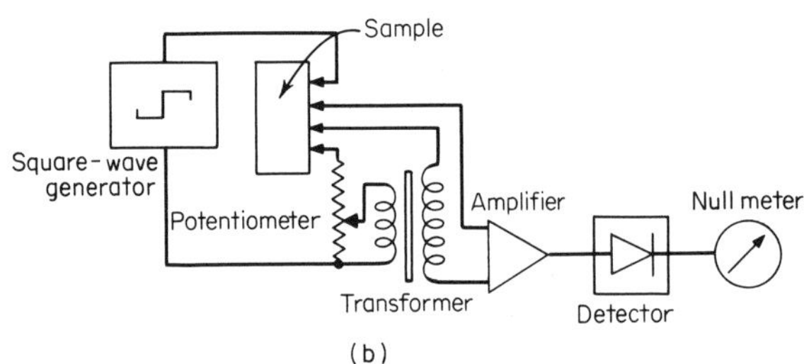

Fig. 3.10. Ratio-reading ac meter using transformer isolation. (*Adapted from Ref. 43.*)

Various ac meters have been built.[15,41,42] They have the advantage of eliminating thermoelectric effects and of allowing tuned voltmeters to reduce system noise. Rectification at the contacts will cause waveform distortion and should be avoided, since it may cause appreciable error. It is possible to apply a forward dc bias to all probes, lower contact resistance, and thus lessen errors due to the voltmeter's loading the probes.[41,42] There are certain restrictions which must be imposed in order that accuracy is not impaired. The bias current cannot be large enough to flood the area between the probes with excess carriers, nor can the peak alternating current be larger than the dc bias; otherwise rectification can occur during some portions of the cycle.

Since the resistivity depends on V/I, rather than measuring them independently, some equipment has been designed to determine the ratio directly. The general scheme is as shown in Fig. 3.10*a*, in which the inner-probe voltage is balanced against the voltage generated by the current flowing through the potentiometer P_1. Practi-

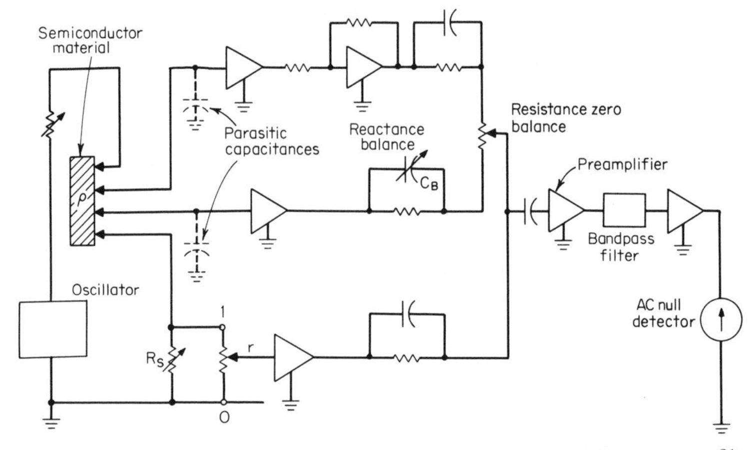

Fig. 3.11. Ratio-reading ac resistivity meter. (*Adapted from Logan.*[21])

cally, isolation is most easily effected by using alternating current and a transformer. Such a circuit is shown[43] in Fig. 3.10*b*.

An alternate approach in Fig. 3.11 uses operational amplifiers for isolation but still is ac, so that capacitor coupling can be used where appropriate to prevent difficulties with drift and dc levels.[21] In circuitry such as this, it is not necessary to have a constant-current source or even to know its approximate value. From a practical standpoint, however, it is best to monitor the current to make sure that it is in the range recommended by the manufacturer. Otherwise, gross errors can occur, as, for example, if the probes should not make good contact, very little current would flow, and system noise would be balanced against the potentiometer voltage.

A different approach for comparing the probe voltage with a voltage proportional to the current is to use a capacitor switched between points *A–B* and *C–D* of Fig. 3.12. When the voltage across R_1 is equal to that of V_{probe}, the capacitor will neither charge nor discharge as it is switched.[44] In the event that it is desired to use this method and observe the effects of current in each direction (i.e., if there is rectification), the circuit can be suitably modified.[45]

Instead of *I–V* measurements, the four-terminal network of either the two- or four-point probe connected to various bridges may be used.[46,47] Also, for dc measurements, a low-impedance potentiometer may be used between ground and one of the voltage probes to provide a virtual ground.[46]

Special Circuitry. In addition to the basic circuitry just described, circuit modifications can be used to perform some of the corrections discussed earlier, and/or to make data collection more rapid.

For W/S less than 0.5, the slice-thickness correction factor is linear in W/S. Thus, a simple potentiometer attenuator somewhere in the voltmeter circuit can be used to make the calculation directly.[41] Nonlinear corrections, e.g., for a finite-diameter slice, can be approximated by a series of linear steps, and again may be done by potentiometer. Alternatively, operational amplifiers can be used and the various correction networks inserted in the feedback loop.[48]

Many instruments use meters, but digital readout minimizes operator error and is preferred for routine operations. Further, complete timing and sequencing controls, coupled with automatic temperature compensation and punched-card (or magnetic-tape) output, can be combined with any of the equipment described to provide fully automatic operation and data reduction.[49-51] When profiling surfaces, for example, along the length of a crystal, the output of the meter can be fed into printout equipment so that distance and resistivity values can be automatically tabulated.

If a single stylus (or an electron beam) is moved in discrete steps or is used continuously,

$$\rho = \frac{A}{I}\frac{dV}{dx} \tag{3.9}$$

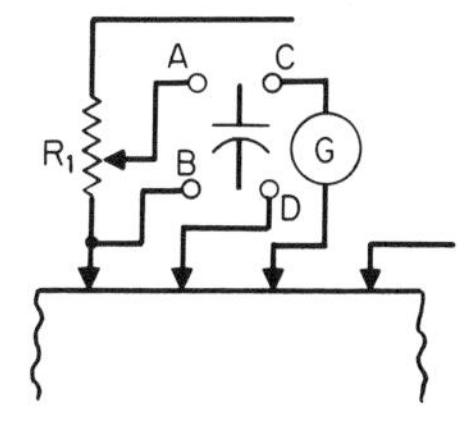

Fig. 3.12. Chopper method for comparing the inner-probe voltage with the *IR* drop across resistor R_1.

Thus, by differentiating the stylus voltage V, $\rho(x)$ is directly determined and can be plotted.[52,53] For point-by-point movement of the stylus across the surface of very high resistivity materials, pulsed current may be used to avoid heating, and bridge balancing of the probe can be used to minimize the effects of detector impedance on the measured voltage.[54]

Probes. Probe design centers around accurately maintaining spacing, providing proper loading, and minimizing contact resistance. In general, a wide variety of metals with high Young's moduli are quite satisfactory. Silicon carbide[56] has been suggested because of its great hardness; however, its contact resistance can be very high. Tungsten needles can be used and are easily brought to a fine point by electrolytic shaping.

Nonjeweled phonograph needles make good probe points, since they are hard and well pointed. "Burning in" is sometimes used to reduce contact resistance, and consists of heating the probe tip and semiconductor surface enough with a short current pulse (e.g., a capacitor discharge) to cause local alloying. For example, copper-plated osmium-tipped probes[57] and tin-plated phosphorus bronze needles[58] have been used with GaAs, Duralumin for p-type silicon, and phosphorus bronze for n-type silicon.[59] A Tesla-coil discharge also can be used with a wide variety of points to improve the contacting. For high-temperature operation, the probe must maintain strength and not react with the semiconductor being measured. Tungsten carbide tips, for example, have been used on some materials at temperatures of 950°C.[60]

In order to minimize damage to the surface, liquid-metal probes can be built. Mercury columns[61] and globules held on the end of metallic pins[62] have been used, but the material to be measured must not amalgamate with the mercury. Liquid gallium is also applicable, but measuring temperatures must be a little above normal room temperature, since the gallium melting point is 29.8°C. Concern over the possibility of probe-semiconductor chemical interaction should not be restricted to the high-temperature range. Chance contaminates left on the surface can cause etching, and high humidity combined with high voltages may cause an electrolytic transfer of probe material to the semiconductor surface.[63]

Numerous guides and loading mechanisms have been devised,[64-70] but in general they are either spring-loaded or at the end of pivoted arms with dead weights. The amount of probe loading depends on the material being measured and the tip diameter, but typically germanium will require 25 to 100 g and silicon 100 to 200 g for tip radii of 0.2 mil. Ordinarily, guide bearings should be as near the probe ends as possible. Most designs allow individual motion of each probe, but some have pairs rigidly mounted. For high-temperature operation, the guides and support (head) can be made of ceramic.

Since errors in probe spacing can cause significant errors in the resistivity readings, considerable care must be taken in both the initial spacing and the spacing maintenance. Precision boring combined with quality bearings can produce the accuracy required, but a simple and rather unique alternate probe-head arrangement involves the use of two sections of threaded dielectric rod as guides, with the probes being laid in the grooves and held in place by pressure pads. The spacing is then as good as the accuracy of the thread.[71] The actual spacing can be determined by indenting metal foil backed with paper and measuring the distance between imprints. This procedure gives crisp imprints which are easy to observe but does not take into account the possibility of skidding which may occur when a loose probe contacts

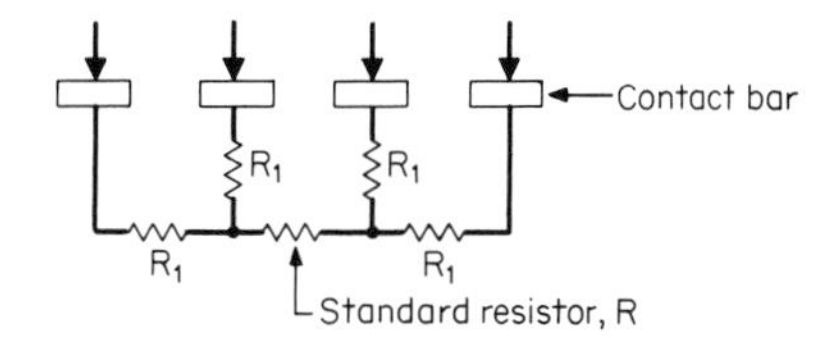

Fig. 3.13. Calibrator for four-point probes. R_1 should be many times R. Its value will depend on specific characteristics of electronics and smiconductor. For silicon R_1 should be approximately $500R$.

the hard semiconductor surface. The small impressions made in the semiconductor itself can be measured, though they are usually somewhat indistinct. If an inverted-stage microscope is available, the probe may be pressed down on a microscope slide and viewed directly.

In order to facilitate measurement-sampling plans, the equipment is sometimes designed to step about on a slice surface to prearranged positions and to repeat them quite accurately. If linear profiling is desired, table motion in only one direction will suffice, and a variety of mechanical arrangements are possible.[63,72] As an alternative to raising the probes each time the sample is moved, rolling-ball probes (e.g., ball-point pens) have been used, and the resistivity read continuously.[73]

Calibration. It is important to be able to calibrate the equipment, and two procedures are available. The first is to keep a sample of the material(s) to be measured, and periodically read its value. When doing this, certain precautions should be observed: (1) Either always measure at the same temperature, or check the temperature and correct for it. (2) Either have a sample free of inhomogeneities or always measure at the same spot on the sample, preferably by using permanent jigging. (3) Resurface the sample as necessary, since after many probe applications in the same region, considerable chipping can occur. The second procedure uses a resistor of known value connected either between probes, or instead of probes, to check the voltmeter and ammeter calibration circuitry. A somewhat better way uses the resistor network of Fig. 3.13. Now, any excessive loading of the voltage probes or failure of the constant-current source will show up as a faulty reading. In neither case, however, will it detect any shift in probe spacing.

3.4 ERRORS IN TWO- AND FOUR-POINT MEASUREMENTS

Sample Size. One of the more obvious errors in determining resistivity arises from failure to consider all corrections to account for limited geometry. It should be remembered, however, that in determining slice resistivity, if the slice thickness is less than a probe spacing (the usual case), the calculated value for ρ varies directly with the assumed value for slice thickness. Thus errors in thickness measurement translate directly into resistivity error. If no corrections are to be made, the probe spacing can be initially chosen to minimize the errors for any expected range of variation.[74]

Substrate Leakage. If the sample being measured is isolated from a substrate by a p-n junction, e.g., an n-on-p epitaxial layer, substrate leakage current can introduce errors.[4] The current may arise either from defective junctions or from debiasing. The latter is more likely to occur as the sheet value of the layer increases. For silicon, reasonable measurements can be made if the sheet resistance is less than 1,000 Ω/square and the measuring current is low enough. If there is doubt about the current range, R_s should be measured as a function of probe current and operation restricted to the region of R_s independent of current.

Probe Spacing. Probe spacing enters directly into four-point bulk-resistivity calculations, so that if the spacings are equal, any error in determining that spacing translates into the same error in resistivity. If the spacing between each probe of a linear array is slightly different from the nominal S,

$$\frac{d\rho}{\rho} = \frac{1}{4S}(3\Delta x_1 - 5\Delta x_2 + 5\Delta x_3 - 3\Delta x_4) \tag{3.10}$$

where probes 2 and 3 are the voltage probes and Δx_i is the linear displacement of the ith probe from its nominal position.[75] If the Δx_i are measured for each resistivity determination, Eq. (3.10) can be used to make corrections as required. As an example, if one of the voltage probes of a 10-mil spaced set is displaced 1 mil toward the other one, the measured resistivity value will be approximately 12 percent low.

When the probe wander is random, independent, and has a standard deviation of δx, the standard deviation of the resistivity is

$$\delta\rho = \frac{2.06\ \delta x}{S} \tag{3.11}$$

For two-point measurements, the standard deviation is

$$\delta\rho = \frac{1.41\ \delta x}{S} \tag{3.12}$$

so that for comparable probe design, a two-point probe is more accurate.

For very thin slices, if the calculations are based on V/I measurements and not on a previously bulk-calibrated machine, S does not affect the readings as long as all spacings are equal. When the spacing is not equal, but known, an additional correction factor F_s can be introduced such that[76]

$$\rho = F_s \rho_{\text{meas}}$$

where

$$F_s = \frac{2 \ln 2}{\ln\left[\dfrac{(S_1 + S_2)(S_2 + S_3)}{S_1 S_3}\right]} \tag{3.13}$$

For small variations,

$$F_s \cong 1 + 1.082\left(1 - \frac{S_2}{\bar{S}}\right) \tag{3.14}$$

where $\bar{S}$ is the mean value of separation. When other multiplicative corrections must be applied, F_s can be added to the list. If the probe displacements are random and independent, each with a standard deviation δx, the relative standard deviation of F_s, if all probe spacings are nominally equal, is given by

$$\delta F_s = (\sqrt{5}\ \delta S)(\bar{S} \ln 4) \tag{3.15}$$

For square arrays, first-order probe-displacement error can be eliminated by averaging the value obtained from two separate measurements using different pairs of current probes but with one current probe common between the two measurements.[77]

Light. Light shining on the surface may introduce spurious photovoltages which will cause instrumentation problems.

Temperature Effects. Since semiconductors have a relatively large temperature coefficient of resistivity, a few percent error can be introduced either by failure to compensate for varying ambient or by unknowingly heating the sample during the measurement itself. The latter is most likely to occur in low-resistivity samples where large currents are required in order to obtain readily measurable voltages. The National Bureau of Standards recommended procedure suggests placing the sample on a large copper block* which has a thermometer placed in it. Figure 3.14 gives C_T for Si and Ge.[78] Note that for 10 Ω-cm and greater, a 5° temperature difference will produce a 4 percent difference in the resistivity reading.

Thermoelectric Effects. Temperature gradients in the sample, whether caused by the ambient or by excessive probe current, will generate a thermoelectric voltage. The use of alternating or low current will minimize the effect.

Probe Injection. For long-lifetime material, the contacts may inject enough carriers to cause conductivity modulation and seriously affect resistivity readings.

AC Pickup. DC sets are likely to have errors introduced through contact rectification of miscellaneous induced stray currents. Operation in a shielded room is sometimes required or, better still, the probe assembly is enclosed in a carefully shielded box. Readings should be taken with current flow in each direction and the corresponding resistivity values averaged. If the two values differ by more than a few percent, more shielding should be used.

Instrument Current. From a pure instrumentation standpoint, higher currents make the voltage measurement easier and less susceptible to noise. However, the high currents cause local heating and, sometimes, conductivity modulation. Because these effects are difficult to evaluate analytically, it is helpful to plot measured

*A thin mica spacer can be used to electrically insulate the slice from the copper and still maintain good thermal contact.

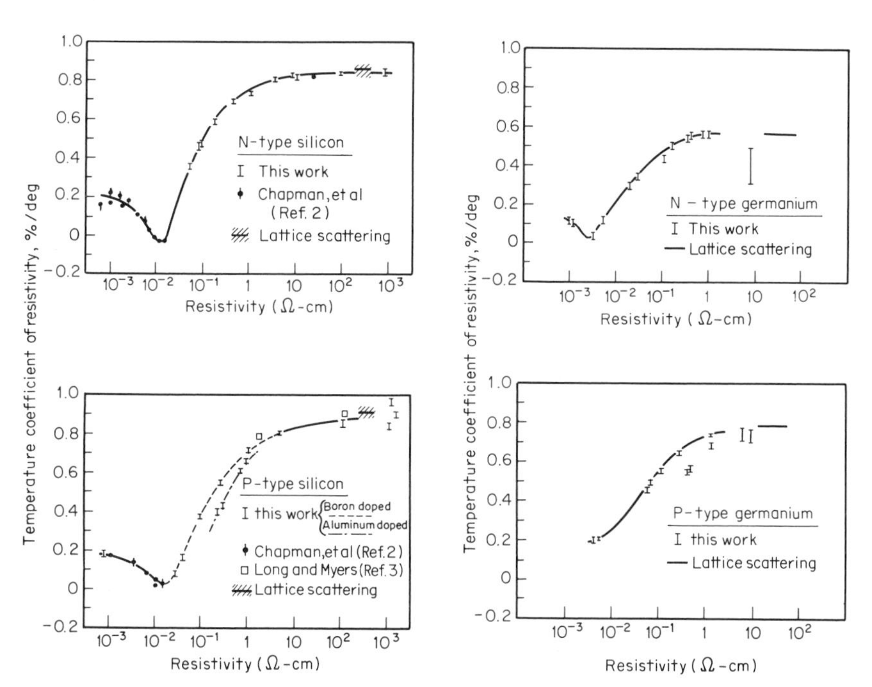

Fig. 3.14. Temperature coefficient of resistivity of silicon and germanium in the extrinsic region. $\rho(T) = \rho_{23°}[1 - C_T(T - 23)]$, where T is the temperature of measurement. (*From Bullis et al.*[78])

resistivity vs. current for each range to be covered. Usually there will be a broad region where the resistivity is independent of current. From the boundary of this region, safe operating currents can be drawn.[79,80]

Applied Voltage. If the electric field becomes too high, a mobility decrease occurs which will make the resistivity reading too high.

3.5 SPREADING RESISTANCE

In order to minimize the corrections that must be used with probe arrays and limited volumes of material, spreading resistance of a point-contact probe is sometimes used.[58,81-87] For a flat circular contact of radius a uniformly affixed to the surface of a semi-infinite medium of resistivity ρ, the spreading resistance R_{sp} is[88]

$$R_{sp} = \frac{\rho}{4a} \tag{3.16}$$

If the contact is a hemisphere of radius r embedded in the solid, as could happen if a needle were pressed down into the material,

$$R_{sp} = \frac{\rho}{2\pi r} \tag{3.17}$$

For values of other geometries see Refs. 88 to 91. For a brittle semiconductor material, the probe would normally be expected to flatten, and the geometry is probably a flat circular contacting area with radius[90,91]

$$a = 1.1\left[\frac{Fr}{2}\left(\frac{1}{E_1} + \frac{1}{E_2}\right)\right]^{1/3} \tag{3.18}$$

where F is the force on the probe tip, r the tip radius, E_1 Young's modulus for the probe material, and E_2 Young's modulus for the material being measured.

Rather than attempting to calculate R_{sp} from Eqs. (3.16) and (3.18), the usual procedure is carefully to construct a tip, plot the measured R_{sp} for several known resistivities of homogeneous material, and prepare a calibration curve. Experience has shown that reproducibility of a given tip increases with use up to perhaps 1,000 measurements, after which it remains constant until damaged. In order to shorten the "burn-in" cycle, various probe treatments such as a short abrasion with a sandblaster have been used with some success.

The calibration curve varies with the probe pressure and the rate of pressure loading but ordinarily is more reproducible with increasing pressure. However, loading should be low enough to prevent fracture and minimize plastic flow of either the semiconductor or probe tip. The mean pressure P over the contact area described by Eq. (3.18) is given by

$$P = 0.42F^{1/3}\left[r\left(\frac{1}{E_1} + \frac{1}{E_2}\right)\right]^{2/3} \tag{3.19}$$

and plastic flow can be expected when P is about 0.4 of the indentation hardness.[81]

Even though resistivity itself is independent of orientation in cubic materials, spreading-resistance calibration curves are observed to vary with orientation.[86] The surface finish can also affect the calibration curve; so care should be taken to ensure that the same kind of surface is used for all measurements which use a given calibration curve. It has also been tacitly assumed that the resistance of the probe is negligible and that the metal-semiconductor barrier resistance is very small. The latter may not necessarily be true, but if the barrier is reproducible, the calibration curve will accommodate it.

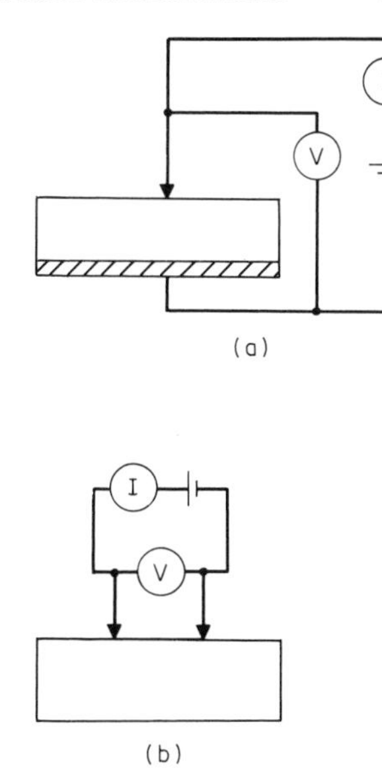

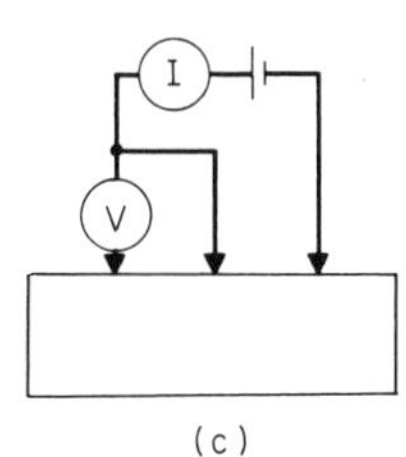

Fig. 3.15. Various spreading-resistance probe arrangements. (c)

The actual contacting configuration may take several forms. It can be a single movable probe for one contact, and a broad area contact for the current-return path. It can be two closely spaced points so that the measured resistance is actually the sum of two spreading resistances and hence has twice the problem of reproducibility of a single probe, but has the advantage of being compact and self-contained. As a compromise, current in and current out can each be by point-contact probe, but an additional voltage probe can be added so that only the voltage drop across one is measured. These are all illustrated in Fig. 3.15.

Most of the potential drop of a point contact occurs within a distance $3a$ from the point.[86] Thus, resistance readings will be nearly independent of what lies at depths greater than $3a$, and conversely, if the sample changes concentration within that distance, or if the whole sample is thinner than $3a$, readings must be treated accordingly. It is possible to make corrections for arbitrary concentration profiles below the surface, but the mathematics are quite tedious.[92] However, if reasonably accurate results are to be obtained when profiling near p-n junctions or abrupt steps such as at an epitaxial film–substrate interface, such corrections are a necessity. In addition to mathematical corrections, some experimental caution must also be exercised when examining thin layers because with heavy probe loading, the top layer may be so physically damaged and deformed that the next layer down is the one which contributes to the spreading resistance. For silicon, layers of less than 2 μm usually begin to require less probe pressure than would normally be used for good reproducibility.

3.6 NONCONTACTING RESISTIVITY METHODS

Noncontacting resistivity methods* can be broken into several categories: (1) those in which the sample is capacitance-coupled to a resistance-measuring apparatus,[93-100] (2) those which measure the additional losses of an inductance when it is coupled to the sample, [101-108] (3) those which measure the decay of eddy currents induced in the sample (most applicable in the 10^{-5} to 10^{-10} Ω-cm range and hence not ordinarily used for semiconductors),[109] (4) those which measure the amount of shielding the sample affords,[110] (5) microwave measurements in which the sample perturbs the transmission or reflection characteristics of a wave guide or cavity,[111-128] and (6) those which depend on a force interaction between the sample and its environment.[129,130] These measurements can be made without inducing probe damage and, if properly done, will have no chance of adding contamination even at high temperatures where contacting probes could react with the semiconductor. In addition, the effect of high-resistivity grain boundaries can be minimized. Contact-resistance effects are removed, though variations in sample size can change the coupling between sample and measuring circuit. However, despite these potential advantages and the considerable investigation in the early 1960s none of them have gained acceptance.

The resistance measurements required for the first two methods are usually done at megahertz frequencies. Bridges, Q meters, and amplitude of oscillation have all been reported, but the wide variety of rf configurations available has by no means been exhausted. The choice appears to be primarily one of availability and personal preference, though for a given resistivity and frequency range some will perform better than others.

Bridges can be direct-reading in series or parallel equivalent resistance, but Q-meter readings must be converted into the equivalent resistance. With Q meters, very low impedance samples should be put in series with the inductance, and high ones in parallel. However, with commercial equipment, the exposed terminals may allow only parallel connections. In that case, in order to keep the Q high enough for oscillation, a small capacitance may be added in series with low-resistivity samples (for example, by increasing the plate-to-sample gap), though sensitivity may suffer. Regardless of whether a bridge or a Q meter is used, the problem finally is one of relating resistance change to sample resistivity. For the specific case of cylindrical and spherical samples centered within a cylindrical solenoid, exact expresssions have been derived,[101] but for other geometries, experimentally determined calibration curves must largely be used, though some corrections for slice thickness have been calculated.[131]

Instead of using either a bridge or a Q meter, the sample can be coupled to an oscillator-resonant circuit which, if appropriately designed, will oscillate with an amplitude which depends on the resistivity. Depending on the frequency and the conductivity of the sample, the energy will penetrate different depths into the sample, so that a sample which is infinitely thick for one resistivity range may not be at another.[132,133]

There are several methods by which microwaves can be used for resistivity measurement.[111-128] However, before any extensive program to use microwave resistivity measurements is embarked upon, carrier behavior in that frequency range should

*There are also noncontacting optical methods for directly determining carrier concentration. They are described a little later in this chapter.

be investigated, since resistivity may not be frequency-independent.[134] Silicon and germanium can be dependably measured up to at least a few gigahertz. The percent power transmitted through a tight-fitting sample inserted in a waveguide can be calculated as a function of resistivity. Measurements will then give the resistivity without any calibration requirements, but the precision shaping of the sample is a major disadvantage. Fitted samples with a high-conductivity backplating can also be used. In this arrangement the frequency is adjusted until the waveguide section with the sample is a quarter wavelength, and then the standing-wave ratio is measured. The change in Q of a resonant cavity can be measured with the sample mounted inside as a post. To remove the necessity of accurate cutting, slices, large slabs, or whole crystals (if they have one smooth surface) can be placed against the open end of a waveguide or a hole in the resonant cavity and the standing-wave ratio or change in Q measured. If both the phase and the amplitude of the reflected wave are determined, the resistivity of thin layers (e.g., epitaxial films) can be measured if the substrate resistivity is known.[128]

3.7 CARRIER CONCENTRATION

There are several methods that measure either carrier concentration or ionized-impurity concentration rather than resistivity. Most of them can, if desired, be calibrated in terms of resistivity, but in any event, for well-known materials conversion from one to the other is a simple matter.

Hall Coefficient. The Hall coefficient may be used to determine the carrier concentration. For details, see Chap. 5.

Thermoelectric Probe.[135,136] The voltage developed between a hot probe contacting the surface to be measured and large-area cold contact can be used as an indication of carrier concentration, but the sensitivity is very poor.

Thermal Rebalance.[137] In general, the temperature vs. resistivity curve exhibits a peak in resistivity, and the sharpness of the peak is a function of impurity concentration.* For those materials such as silicon and germanium which have a maximum resistivity not far above room temperature, it is possible to measure the resistance of a sample at room temperature, increase the temperature to the point where the resistance has gone through its peak and decreased to its original value, and determine the carrier concentration from the observed ΔT. Thus, the resistivity is determined without any knowledge of the geometry of the sample.† It is necessary, however, for the contacts to be ohmic and have a much smaller temperature coefficient than the semiconductor. The major limitations of this method are the requirement for an ohmic contact and the fact that it is a destructive test.

Three-Point Probe.[138-145] Diode breakdown voltage has been used to evaluate material "quality" since the work on germanium and silicon in the early 1940s. If good diodes can be constructed, breakdown voltage can be related to carrier concentration through suitable curves.‡ However, even a poor diode formed by a point

*If there is appreciable compensation, interpretation is difficult because the width depends in a complex manner on total impurity concentration.

†In a similar vein, the slope of the *R-T* curve near room temperature has been used as a measure of carrier concentration in tellurium. (See H. H. Hall, "Task A-1, Development of a Low Impedance Pressure Gage," Physics Department, University of New Hampshire, NOrd 10358 *Rep.* 33, 1951.)

‡S. M. Sze and J. C. Irvin, Impurity Levels in GaAs, Ge, and Silicon at 300°K, *Solid State Electron.*, **11**:599–602 (1968).

contact on a polished surface, combined with an empirical curve, can give reasonable reproducibility. As long as the point is slightly farther away from a conducting or high-recombination surface boundary than the width of the depletion layer at breakdown, the reading is independent of boundary effects. Because of this it is suited for measuring thin high-resistivity layers on low-resistivity substrates. If the space charge reaches a boundary before avalanche begins, large currents will flow and the unit will appear in breakdown. Thus, instead of a single-line calibration curve, multivalues occur, as shown in Fig. 3.16*a*. In the simplest form, a curve tracer or oscilloscope can be used to observe the break in the *I–V* curve as voltage across the sample is increased (Fig. 3.16). The forward drop across the broad area contact can affect reproducibility and is eliminated by using an additional voltage probe shown in Fig. 3.16*b* ("three-point probe").

To prevent local heating and possible damage to the semiconductor when breakdown occurs, some form of current limiting and pulsing should be used. Either a series of short pulses of increasing amplitude[139] or a single sawtooth pulse can be used with circuitry which detects the point of maximum voltage and automatically stops current flow and reads the maximum voltage. When using calibration curves, the fact that the "breakdown voltage" will depend on the probe material, probe radius, probe loading, probe spacing, pulse length, and surface preparation should be kept in mind, and care should be taken not to change any of those parameters without a corresponding change in calibration.

Optical Methods. Both the optical-absorption coefficient and index of refraction of semiconductors are dependent on the free-electron or hole concentration. It is therefore possible in principle to determine the carrier concentration optically.

Such a method has the potential advantage of being nondestructive as well as being able to respond to the properties near the surface, independent of what might be below it (and thus it is useful for evaluating epitaxial layers). Its disadvantage is that the range of concentrations that can be examined is usually very limited and the equipment rather complex.

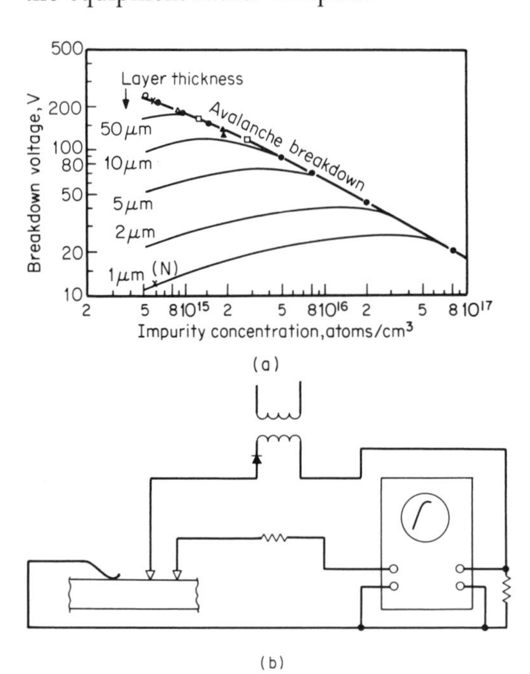

Fig. 3.16. Three-point calibration curve and circuit. (*From Allen, Clevenger, and Gupta.*[139])

For wavelengths corresponding to energies less than the bandgap of the semiconductor, the absorption coefficient α is reasonably sensitive to resistivity (carrier concentration).[146,147] For shorter wavelengths, the total absorption coefficient is ordinarily orders of magnitude greater than the absorption due to carriers. In the long-wavelength region, as the carrier concentration decreases, α drops to the point where, unless the sample is very thick, reflective losses predominate. At that point, small variations in surface finish will completely overshadow any effect of concentration change. The wavelength used is not critical, but the free-carrier losses increase as the wavelength; so a longer wavelength may be of some advantage. Care should of course be taken to ensure that the wavelength chosen does not coincide with an absorption line due to some other phenomenon. An example would be the 9-μm oxygen band in silicon. The total absorption is a measure of the total number of carriers in the path, and can be used as an alternate to conventional sheet resistance measurements for evaluating diffused layers in slices provided both surfaces are optically finished.[202]

From a convenience standpoint, reflectivity measurements are much more desirable, since they require just one polished surface and the only thickness limitation is that it be greater than a few $(1/\alpha)$s. In the same wavelength range, the reflectivity R is related to impurity concentration only through changes in the absorption coefficient and is relatively insensitive. Typically, R will be approximately 0.1 percent higher for an absorption coefficient of 1,000 per centimeter than for an α of zero. Very small changes in reflectivity may be detected by using the method of attenuated total reflection.[156] When the light wave travels from a medium of higher to one of lower refractive index, it will be reflected unless the angle of incidence ϕ is greater than the critical angle ϕ_c, in which case it will be totally reflected. When ϕ is very close to ϕ_c, the amplitude of the reflected wave is quite sensitive to the index of refraction of the second medium. Thus, otherwise indiscernible changes in the index can produce readily measured changes in reflectivity. Rather than fixing the angle very close to critical and measuring changes in reflectivity, the reflectivity can be measured as a function of angle. In that case the critical angle changes as the absorption coefficient of the sample changes.[157]

Plasma resonance [148-155] produces a minimum in the reflectivity vs. wavelength curve that shifts appreciably with free-carrier concentration in the higher concentration ranges (typically 5×10^{18} to 5×10^{20} atoms/cm^3 for silicon) and is readily measurable. It should be possible to calculate the minimum for any given concentration; however, the minimum is usually experimentally measured and calibration curves constructed.[154] When applied to circumstances where there are large concentration gradients perpendicular to the surface, additional interpretation difficulties arise, since the radiation will penetrate to depths of about $1/\alpha$ and in a diffused layer, for example, the concentration could change orders of magnitude in that depth. It is possible, however, by measuring the sheet resistance and assuming a known profile (e.g., gaussian or complementary error function), to predict the surface concentration of diffused layers.[155] When the layers are very thin, or when a uniform sample of low carrier concentration is examined, the reflectivity minimum is found to be very broad. Without some additional signal processing the accuracy of locating the exact minimum is therefore greatly reduced. Multiple scans combined with multichannel digital integration have been used to reduce trace noise, and electronic differentiation of the trace can help pinpoint the minimum more closely.[155] In some cases, the depth of the minimum can be used to estimate compensation.[203] Also, the envelope of the *IR* interference pattern often used for measuring epitaxial layer thickness (see Chap. 6) will exhibit a minimum due to the plasma resonance of the substrate and can thus be used to check its concentration.[204]

Instead of trying to measure the effect of carrier absorption on reflectivity, it is also possible to measure the optical constants n and k directly by use of an ellipsometer. The wavelength of light used must be in the range where carrier effects will be observable, and this usually implies infrared. For example, silicon measurements have been in the neighborhood of 50 μm. Unfortunately, instrumentation problems associated with long-wavelength ellipsometers are fairly severe, and accuracy is not exceptional.[158]

Capacitance-Voltage Measurements. The differential capacitance of a depletion layer, whether formed by Schottky barrier, p-n junction, or MIS structure, can be used to determine the majority carrier-concentration profile as the space-charge-region width is changed by a reverse bias voltage. The general expression is

$$\frac{1}{N_1(x_1)} + \frac{1}{N_2(x_2)} = \frac{q\varepsilon\varepsilon_0(dC/dV)}{A^2C^3} \tag{3.20}$$

where N_1 and N_2 are the net majority carrier concentrations on each side of the junction at positions x_1 and x_2, which are the respective distances that the space-charge region has moved out from the junction owing to the applied voltage V. C is the capacitance measured by a small-signal high-frequency capacitance bridge, A is the junction area, q the electronic charge, and $\varepsilon\varepsilon_0$ the dielectric constant. All the terms in the expression can be measured except N_1 and N_2. For $N_2 \gg N_1$ (i.e., an abrupt junction), the $1/N_2$ term can be neglected and N_1 determined. Such C-V measurements are widely used and have many variations for special circumstances. They are, however, more device-oriented and as such are not described in detail. References 159 to 163 and those contained therein are suggested for background reading before these measurements are pursued.

Deep Level Impurities. The methods described have been most applicable to shallow fully ionized impurities which contribute to conductivity. However, there are also electrical measuring techniques which can be used to determine the concentration of some deep and intermediate level impurities. They all involve measuring either the capacitance or current of a p-n junction as a function of temperature. The procedure requires packaged diodes and the capability of detecting capacitance changes of a few tenths of a picofarad. Again, such methods are more device oriented and will not be discussed.[205-208]

3.8 RESISTIVITY PROFILING

It is often desirable to determine the impurity profile normal to the surface of a slice after some operation (e.g., epitaxial deposition, diffusion, or ion implantation) or to measure resistivity fluctuations laterally across the surface. Each of these requirements usually necessitates a finer resolution than is normally used and thus requires special procedures and/or equipment. Various direct methods are summarized in Tables 3.6 and 3.7 as well as some chemical and electrochemical techniques. The latter depend on differences of etching or plating speed with impurity concentration and are most useful in looking at variations across a large surface area. It is difficult to determine the actual magnitude or resistivity in this manner,

Table 3.6. Methods of Profiling Normal to Surface

Method	Description	Value measured	Limitations
1. Use of variable-width four-point probe	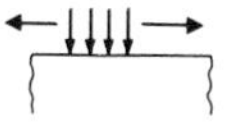	$\rho(z)$	Sensitivity poor, interpretation difficult
2. Four-point probe combined with beveling	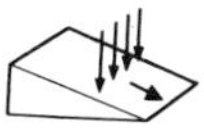	$\rho(z)$	Destructive, assumes uniform resistivity in x direction
3. Four-point probe combined with successive layer removal	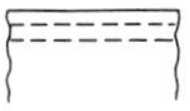	$\rho(z)$	Destructive, requires careful measurement of material removal; resistivity must remain constant or increase with depth
4. Differential resistance of diode in breakdown		$\rho(z)$	Must be sure breakdown occurs when profile desired. Requires pulsed current.
5. Same as 2 and 3 except using spreading resistance	—	$\rho(z)$	
6. Capacitance–voltage		$N(z)$	Requires diode, depth limited unless multiple diodes are combined with steps or bevels
7. Depth of junction combined with different resistivities	 ρ_1 ρ_1 ρ_1	$N(z)$	Destructive, requires generating same profile in several samples
8. Double beveling	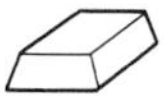	$N(z)$	Destructive, requires a diffusion, which may change distribution
9. Neutron activation combined with successive layer removal		$N(z)$	Impurity must be amenable to neutron activation; destructive, requires careful measurement of material removal
10. Radio tracer combined with successive layer removal		$N(z)$	Impurity must be radioactive; destructive, requires careful measurement of material removal
11. Ion microprobe	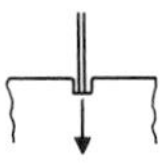	$N(z)$	Difficult to determine depth of hole and hence z
12. Selective staining, plating, and etching	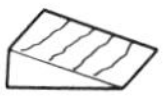	$N(z)$	Magnitude of variation unknown

Table 3.7. Methods of Profiling Laterally

Method	Description	Value measured	Limitations
Four-point probe	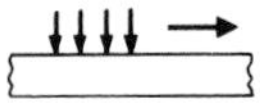	$\rho(x)$	Resolution limited by probe spacing
Spreading resistance	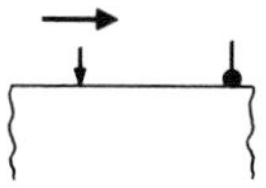	$\rho(x)$	Probably the best method listed
Two-point probe	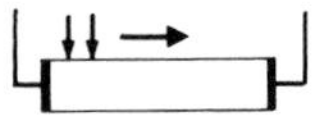	$\rho(x)$	Requires end contacts and known sample geometry. Resolution limited by probe spacing
Moving voltage probe	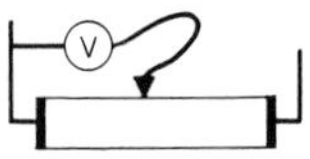 V	$\rho(x)$	Requires end contacts and known sample geometry
Moving electron beam	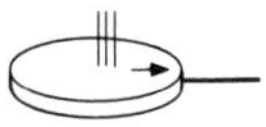		Interpretation difficult, requires E-beam source
Moving-spot photovoltage	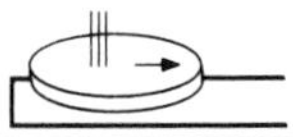	$N(x)$	Interpretation difficult
X-ray microprobe		$N(x)$	Sensitivity poor
Selective staining, plating, and etching	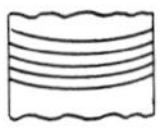	$N(x)$	Magnitude of variation unknown
Multiple MIS capacitors		$N(x)$	Resolution limited by photolithography

but spatial variations become clearly visible. Sirtl, copper, or anodic etching have all been used for delineation.

Normal to the Surface. When the samples are very thick relative to normal probe spacing and the resistivity variations are slowly varying, changes in spacing, either by increasing the S of the equal-spaced system or by using some configuration with one probe movable, can be used to estimate resistivity vs. depth. Interpretation is difficult, resolution is poor, and the method is seldom used, though it does find application in geophysical prospecting.[164,165]

More commonly, the four-point probe is combined with material removal, by either angle section or successive layer removal. The method of interpretation is basically the same in either case, but each has certain practical advantages. Historically, parallel sections were removed by mechanical polishing, using a carefully constructed jig. These may have varying degrees of complexity[166-171] but if the thickness is determined independently, e.g., by weighing, simple equipment similar to that described in Chap. 7 may be used. A lapping block with three diamond stops on compound screws used with conventional lapping equipment gives good control, but initial alignment of the plane of the stops with the semiconductor surface is difficult. With such arrangements, it is difficult to maintain parallelism, but by deliberately allowing polishing to produce a wedge-shaped section and making measurements along the incline, the necessity for parallelism is eliminated.[172] The angle does not need to be ground to some preset value, because it can be measured afterward. However, in order to obtain good resolution, very small angles, e.g., 0.1 to 0.2°, are required. If the total thickness to be examined is greater than the sample length L, a sequential process of incremental angle lapping and probing may be employed. The incremental steps should be chosen so that some overlap of sheet resistance from step to step is possible.

When parallel layers are desired, it is more reproducible to oxidize the surface and then remove the oxide. Where applicable, e.g., silicon, anodic oxidation is preferred[173] so that possible high-temperature redistribution of impurities does not occur.

If the impurity profile is one of rapidly increasing surface concentration away from the surface, the removal of a layer will affect the average resistivity of the remaining material very little and sensitivity will be very poor. For this reason it is better to profile from high to low concentration even if it means measuring from the back of a slice.

Interpretation of data for either angle or parallel sectioning is as follows: The probe spacing to total layer thickness ratio is assumed large enough that sheet-resistance values are measured, in which case the sheet resistance R_s of the ith layer removed is given by

$$R_s = \frac{R_i R_{i+1}}{R_{i+1} - R_i} \tag{3.21}$$

where R_i is the sheet value before removal of the ith layer and R_{i+1} is the value after removal of the layer. If Δx is the thickness of the ith layer,[174]

$$\rho(x) = \frac{R_i R_{i+1}}{R_{i+1} - R_i} \Delta x \tag{3.22}$$

The computed ρ is very sensitive to small instrumentation errors in R_i and R_{i+1} and on x_i and x_{i+1}. To minimize these effects, instead of using Eq. (3.22), make use of the relation

$$\sigma(x) = \frac{dG}{dx} \tag{3.23}$$

where G is the sheet conductance of the layer.[175] The logarithm of the sheet conductance will have some functional dependence of depth such that

$$F(x) = \ln G(x) \tag{3.24}$$

Differentiation with respect to x gives

$$\frac{dF}{dx} = \frac{(1/G)dG}{dx} \tag{3.25}$$

so that

$$\frac{1}{\sigma(x)} = \rho(x) = \frac{dx}{dG} = \frac{1}{G\,dF/dx} \tag{3.26}$$

Thus log G vs. x can first be plotted and smoothed as desired (either by machine or manually),[172,176] the slope taken at any desired value of x, and ρ determined all without resorting to the use of small differences in either R_s or x. As an alternate experimental procedure, contacts can be made on opposite ends of a rectangular sample by using a "wrap-around" diffusion and the sheet resistance determined by conventional I–V measurements.[177]

Either bevel or parallel sectioning can be combined with spreading-resistance measurements, but different interpretation is required. The main contribution to spreading resistance will be in a layer less than one-fifth the diameter of the contact area and virtually all from a depth less than 1.5 diameters. Thus, until the tip gets within 1.5 diameters of an abrupt discontinuity, it will be measuring an average value for a thin layer of fixed width, independent of the total sample thickness. For close spacing, corrections have been computed.[86,87,209]

Voltage-capacitance measurements are ideal for making profile measurements, but the depth is limited by the width of the space charge at avalanche. In order to extend the range, steps may be etched into the surface and diodes made on each step,[2] or a bevel much like that used for four-point profiling will serve the same purpose.[178] Instead of determining N directly from the C–V measurements, it is also possible to measure the resistance of the bulk semiconductor remaining between the space-charge boundary and the back contact.[179,180] As the voltage across the junction is increased, the space charge moves closer to the back and has the same electrical effect as physically removing part of the layer. Two ways of measuring the resistance have been used. One is to employ extra contacts and a lateral current flow; the other is to determine the series resistance from Q measurements of the capacitor. The differential resistance of the device after avalanche breakdown can also be used for extending the profile depth, provided avalanche really occurs where the profile is desired.[210,211]

Probably the most sensitive and widest-range method of all, but one that is applicable only if the impurity profile to be measured can be simultaneously introduced into several separate samples, is the measurement of the p-n-junction depth in materials of differing resistivity.[174] The depth of the junction may be determined by staining or other techniques (Chap. 7), and the concentration at that depth is equal to the background concentration of that particular sample. For silicon it is not unreasonable to measure a profile which ranges from 10^{12} to 10^{19} atoms/cm^3.

Double beveling, combined with diffusion of a known profile and an appropriate stain, can be used to give a pictorial view of the concentration variation.[181] Consider Fig. 3.17, in which it is desired to know the profile normal to surface A. First make a diffusion into the B face (which has been produced by beveling at a low angle θ_1) such that its concentration vs. depth is known and its type is opposite that already in the sample. Next, bevel the sample again as shown by face C and stain to delineate the junction. For any given depth z_j below the original surface given by $z_j = x \sin \theta_2$, the concentration is approximately equal to that of the diffusion at a depth of $\bar{x} \sin \theta_2$.

If the impurity is amenable to neutron-activation analysis, the sample can be irradiated and successive parallel layers removed from surface. Counting can then be done on either the layers removed or the remaining material. If the impurity itself is radioactive, similar removal and counting techniques can be used. In either

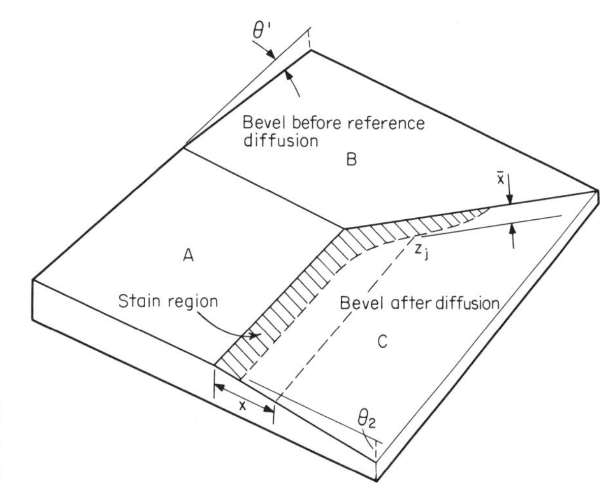

Fig. 3.17. Double-bevel method of displaying impurity concentration. (*Adapted from Sils et al.*[181])

case, of course, handling, sectioning, etc., must be done in a work area which provides protection for the operator. Because of this restriction, plus the requirements for irradiation and counting, these two methods are seldom used for routine analysis, but they are exceedingly useful for laboratory investigations. X-ray microprobe fluorescence can be used if combined with a bevel but is not particularly sensitive and is most applicable for high concentrations of impurities.

Lateral Profiling. The simplest method of lateral profiling is just to step four-point probes across the surface. Resolution is limited by probe spacing, and edge corrections should be used as required. If finer resolution is desired, spreading resistance can be used. It has proved very successful in finding inhomogeneities in silicon,[82,182] and in fact when combined with beveling, can be used to construct three-dimensional impurity profiles.[212] If fine resolution is not required and there is concern over probe damage, very small metal contacts with expanded pads for probing can be used.[213] When samples of uniform cross section are available, either a two-point or a moving-voltage probe can be moved across the surface. Similarly, electron-beam scans are applicable.[53,183-189] (Also see Sec. 3.3.) The photovoltage or photoconductivity change generated by a moving spot of light can also be used,[190-201] and while difficult, interpretation of the signals for circular slices as well as for filaments is possible and gives reasonable correlation with probe measurements.[201] It is interesting to note that the resolution limit is given not necessarily by the light-spot size but, depending on which is larger, by either it or the carrier-diffusion length.

REFERENCES

1. H. Frank, Influence of Damaged Surface Layer on Resistivity and Mobility of Thin Semiconductor Sheets, *Solid State Electron.*, **9**:609–614 (1966).
2. Earl Alexander, S. Watelski, Ken Bean, W. R. Runyan, and Murray Bullis, "Development of Epitaxial Structures for Radiation Resistant Silicon Solar Cells," *NASA* Report CR461, Clearing House for Federal Scientific and Technical Information, May 1966.
3. Edward N. Clark and Robert L. Hopkins, Electrical Conductivity of Mechanically Disturbed Germanium Surfaces, *Phys. Rev.*, **91**:1566 (1953).
4. W. J. Patrick, Measurement of Resistivity and Mobility in Silicon Epitaxial Layers on a Control Wafer, *Solid State Electron.*, **9**:203–211(1966).
5. George G. Harmon and Theodore Higier, Some Properties of Dirty Contacts on

Semiconductors and Resistivity Measurements by a Two-Terminal Method, *J. Appl. Phys.,* **33**:2198–2206 (1962).

6. Lydon J. Swartzendruber, "Calculations for Comparing Two-Point and Four-Point Probe Resistivity Measurements on Rectangular Bar-shaped Semiconductor Samples," National Bureau of Standards, *Tech. Note* 241, 1964; J. Marshall Reber, Potential Distribution in a Rectangular Semiconductor Bar for Use with Four-Point Probe Measurements, *Solid State Electron.,* **7**:525–529 (1964); Serg Jandl, K. D. Usadel, and Gaston Fischer, Resistivity Measurements with Samples in the Form of a Double Cross, *Rev. Sci. Instr.,* **45**:1479–1480 (1974).
7. M. G. Buehler, J. M. David, R. L. Mattis, W. E. Phillips, and W. R. Thurber, Planar Test Structures for Characterizing Impurities in Silicon, Ab. No. 171, Spring Meeting, *E.C.S. Extended Abstracts,* **75-1** (1975).
8. E. Earleywine, L. P. Hilton, and D. Townley, Measuring the Properties of Semiconductor Grade Materials, *SCP and Solid State Tech.,* **8**:17–28 (October 1965).
9. F. Padovani and G. Valant, Resistivity Characterization of Semiconductor Crystal Ingots, *J. Electrochem. Soc.,* **120**:585–587 (1973).
10. L. B. Valdes, Resistivity Measurements on Germanium for Transistors, *Proc. IRE,* **42**:420–427 (1954).
11. Helen H. Gegenwarth, Correction Factors for the Four-Point Probe Resistivity Measurements on Cylindrical Semiconductors, *Solid State Electron.,* **11**:787–789 (1968).
12. Sadayuki Murashima, Hitoshi Kanamori, and Fumio Ishibashi, Correction Devisors for the Four-Point Probe Resistivity Measurement of Cylindrical Semiconductors, *Japan. J. Appl. Phys.,* **9**:58–67 (1970); Sadayuki Murashima and Fumio Ishibashi, Correction Devisors for the Four-Point Probe Resistivity Measurements on Cylindrical Semiconductors, II, *Japan. J. Appl. Phys.,* **9**:1340–1346 (1970).
13. Erik B. Hansen, On the Influence of Shape and Variations of the Sample on Four-Point Measurements, *Appl. Sci. Res.,* **8B**:93–104 (1960).
14. A. Marcus and J. J. Oberly, Four-Probe Resistivity Measurements on Rectangular Semiconductor Filaments, *IRE Trans. Electron Devices,* **ED-3**:161–162 (1956).
15. G. Knight, Measurement of Semiconductor Parameters in Lloyd P. Hunter (ed.), "Handbook of Semiconductor Electronics," McGraw-Hill Book Company, New York, 1956.
16. J. N. Hummel, A Theoretical Study of Apparent Resistivity in Surface Potential Methods, AIME, *Tech. Pub.* 418, 1931.
17. A. Uhlir, Jr., The Potentials of Infinite Systems of Sources and Numerical Solutions of Problems in Semiconductor Engineering, *Bell System Tech. J.,* **34**:105 (1955).
18. Method of Test for Resistivity of Silicon Using Four-Point Probes, Standard Test Method F 43, American Society for Testing and Materials, Philadelphia, 1968.
19. R. Rymaszewski, Relationship between the Correction Factor of the Four-Point Probe Value and the Selection of Potential and Current Electrodes, *J. Sci. Instr.,* (2) **2**:170–174 (1969); P. J. Severin, Four-Point-Probe Resistivity Measurements on Silicon Heterotype Epitaxial Layers with Altered Probe Order, *Philips Res. Rept.* **26**:279–297 (1971).
20. F. M. Smits, Measurement of Sheet Resistivities with the Four-Point Probe, *Bell System Tech. J.,* **37**:711–718 (1958).
21. M. A. Logan, An AC Bridge for Semiconductor Resistivity Measurement Using a Four-Point Probe, *Bell System Tech. J.,* **40**:885 (1961).
22. J. F. Combs and M. P. Albert, Diameter Correction Factors for the Resistivity Measurement of Semiconductor Slices, *Semicond. Prod.,* **6**:26–27 (Feb., 1963).
23. M. P. Albert and J. F. Combs, Correction Factors for Radial Resistivity Gradient Evaluations of Semiconductor Slices, *IEEE Trans. Electron Devices,* **ED-11**:148–151 (1964).
24. Lydon J. Swartzendruber, "Correction Factor Tables for Four-Point Probe Resistivity Measurements on Thin, Circular Semiconductor Samples," National Bureau of Standards, *Tech. Note* 199, 1964.

25. Lydon J. Swartzendruber, Four-Point Probe Measurement of Non-Uniformities in Semiconductor Sheet Resistivity, *Solid State Electron.,* **7**:413–422 (1964).
26. D. E. Vaughan, Four-Point Probe Resistivity Measurements on Small Circular Specimens, *Brit. J. Appl. Phys.,* **12**:414–416 (1961).
27. M. A. Logan, Sheet Resistivity Measurements on Rectangular Surfaces—General Solutions for Four-Point Probe Conversion Factors, *Bell System Tech. J.,* **46**:2277–2322 (1967).
28. P. M. Hall and J. T. Koo, Sheet Resistance Measurements, *J. Appl. Phys.,* **38**:3112–3116 (1967).
29. S. B. Catalano, Correction Factors for Square-Array and Rectangular-Array Four-Point Probe near Conducting or Non-Conducting Boundaries, *IEEE Trans. Electron Devices,* **ED-10**:185–188 (1963).
30. M. G. Buehler, A Hall Four-Point Probe on Thin Plates, *Solid State Electron.,* **10**:801–812 (1967).
31. Frank Keywell and George Dorosheski, Measurements of the Sheet Resistivity of a Square Wafer with a Four-Point Probe, *Rev. Sci. Instr.,* **31**:833–837 (1960).
32. A. Mircea, Semiconductor Sheet Resistivity Measurements on Square Samples, *J. Sci. Instr.,* **41**:679–681 (1964).
33. M. G. Buehler and G. L. Pearson, Magnetoconductive Correction Factors for an Isotropic Hall Plate with Point Sources, *Solid State Electron.,* **9**:395–407 (1966).
34. L. J. van der Pauw, A Method of Measuring Specific Resistivity and Hall Effect of Discs of Arbitrary Shape, *Philips Res. Repts.,* **13**:1–9 (1958).
35. L. J. van der Pauw, A Method of Measuring the Resistivity and Hall Coefficient on Lamellae of Arbitrary Shape, *Philips Tech. Rev.,* **20**:220–224 (1959); Ronald Chwang, B. S. Smith and C. R. Crowell, Contact Size Effects on the van der Pauw Method for Resistivity and Hall Coefficient Measurements, *Solid State Electron.,* **17**:1217–1227 (1974); Patrick M. Hemenger, Measurement of High Resistivity Semiconductors Using the van der Pauw Method, *Rev. Sci. Instr.,* **44**:698–700 (1973).
36. S. Amer, van der Pauw's Method of Measuring Resistivities on Lamellae of Non-Uniform Resistivity, *Solid State Electron.,* **6**:141–145 (1963); W. L. V. Price, Electric Potential and Current Distribution in a Rectangular Sample of Anisotropic Material With Applications to the Measurement of the Principal Resistivities by an Extension of van der Pauw's Method, *Solid State Electron.,* **16**:753–762 (1973).
37. P. A. Schumann, Jr., and E. E. Gardner, Four-Point Probe Evaluation of Silicon n/n+ and p/p+ Structures, *Trans. Met. Soc. AIME,* **233**:602 (1965).
38. P. A. Schumann, Jr., and J. F. Hallenbeck, Jr., A Novel Four-Point Probe for Epitaxial and Bulk Semiconductor Resistivity Measurements, *J. Electrochem. Soc.,* **110**:538 (1963).
39. P. J. Olshefski, Constant Current Generator Measures Semiconductor Resistance, *Electronics,* **34**:63 (Nov. 24, 1961).
40. A. L. Barry and W. D. Edwards, Circuit to Facilitate the Measurement by the Four Point Probe Method of the Resistivity of Silicon in the Range 0.002 to 10,000 Ohm-Cm, *J. Sci. Instr.,* **39**:119–121 (1962).
41. H. G. Rudenberg, Resistivity Measuring Techniques in Semiconductors, *Semicond. Prod.,* **2**:28–34 (September 1959).
42. C. C. Allen and W. R. Runyan, An AC Silicon Resistivity Meter, *Rev. Sci. Instr.,* **32**:824–828 (1961).
43. Data Sheet, J. A. Radley Research Institute, Reading, England.
44. T. M. Dauphinee and E. Mooser, Apparatus for Measuring Resistivity and Hall Coefficient of Semiconductors, *Rev. Sci. Instr.,* **26**:660–664 (1955).
45. W. D. Edwards, Resistivity Measuring Circuit Using Chopped Direct Current, *J. Sci. Instr.,* **42**:432–434 (1965).
46. G. Fischer, D. Greig, and E. Mooser, Apparatus for the Measurement of Galvanomagnetic Effects in High Resistance Semiconductors, *Rev. Sci. Instr.,* **32**:842–846 (1961).

47. J. H. Fermor and A. Kjekshus, Servocontrolled Measuring Bridge for Semiconductors of High Resistivity, *Rev. Sci. Instr.,* **36:**763–766 (1965).
48. L. S. Swartzendruber, F. H. Ulmer, and J. A. Coleman, Direct Reading Instrument for Silicon and Germanium Four-Probe Resistivity Measurements, *Rev. Sci. Instr.,* **39:**1855–1863 (1968).
49. John F. Hallenback, Jr., and Daniel Piscitally, Precision Four-Point Probe Resistivity Instrumentation in Rolf R. Haberecht and Edward L. Kern (eds.), "Semiconductor Silicon," The Electrochemical Society, New York, 1969.
50. E. H. Putley, "The Hall Effect and Semiconductor Physics," Dover Publications, Inc., New York, 1968.
51. L. G. Sapugin and V. M. Ivko, Circuit for Measuring and Recording the Resistance of Semiconductors in Coordinates $\operatorname{Log} R = f(1/T)$, *Soviet Phys. Solid State,* **2:**1346–1351 (1961).
52. Luther Davis, Jr., Lawrence G. Rubin, and W. D. Straub, Rapid Determination of Some Electrical Properties of Semiconductors, *IRE Prof. Group Electron Devices,* **ED1:**34–40 (1954).
53. Chusuke Manakata and Hiroshi Watanabe, Measurement of Resistance by Means of Electron Beam—III, *Japan. J. Appl. Phys.,* **8:**1307–1309 (1969).
54. H. H. Lehner, Probing Techniques for Measuring the Potential Distribution in Semiconductors, *Rev. Sci. Instr.,* **38:**699–700 (1967).
55. Pierre Wagrez and Albert Soffa, A New Technique for Measuring the Resistivity of Individual Semiconductor Dice, *Semicond. Prod.,* **6:**23–25 (February 1963).
56. V. A. Dorin and M. M. Kazlov, Probe of Silicon Carbide for Testing Semiconductor Materials, *Zavodsk. Lab.,* **30:**206 (1964).
57. C. L. Paulnack, Technique for Two and Four Point Resistivity Measurements in GaAs, *Rev. Sci. Instr.,* **35:**1715–1717 (1964).
58. H. Frank and Samy Abdel Azim, Measurement of Diffusion Profile of Zn in n-type GaAs by a Spreading Resistance Technique, *Solid State Electron.,* **10:**727–728 (1967).
59. R. H. Creamer, The Measurement of the Electrical Resistivity of Silicon, *Brit. J. Appl. Phys.,* **7:**149 (1956).
60. Alex Cybriwsky, Wide Temperature Range Four Point Probe Device for Measuring Electrical Resistivity, *Rev. Sci. Instr.,* **37:**961–962 (1966).
61. R. A. Cooper and E. Lerner, Four-Point Mercury Contact Probe for Electrical Resistivity Measurement of Thin Films, *Rev. Sci. Instr.,* **39:**1207–1208 (1968); P. J. Severin and H. Bulle, Four-Point Probe Measurements on N-Type Silicon with Mercury Probes, *J. Electrochem. Soc.,* **22:**133–137 (1975); P. J. Severin and H. Bulle, Spreading Resistance Measurements on N-type Silicon Using Mercury Probes, *J. Electrochem. Soc.,* **22:**137–142 (1975); P. J. Severin and G. Poodt, Capacitance-Voltage Measurements with a Mercury-Silicon Diode, *J. Electrochem. Soc.,* **119:**1384–1389 (1972).
62. Gerald Abowitz and Emil Arnold, Simple Mercury Drop Electrode for MOS Measurements, *Rev. Sci. Instr.,* **38:**564–565 (1967).
63. D. W. F. James and R. G. Jones, On the Four-Probe Method of Resistivity Measurement, *J. Sci. Instr.,* **42:**283 (1965).
64. D. B. Gasson, A Four-Point Probe Apparatus for Measuring Resistivity, *J. Sci. Instr.,* **33:**85 (1956).
65. C. L. Paulnack and N. J. Chaplin, Minimal Maintenance Probe for Precise Resistivity Measurement of Semiconductors, *Rev. Sci. Instr.,* **33:**873–874 (1962).
66. John K. Kennedy, Four-Point Probe for Measuring the Resistivity of Small Samples, *Rev. Sci. Instr.,* **33:**773–774 (1962).
67. Alexander L. MacDonald, Julius Soled, and Carl A. Stearns, Four-Probe Instrument for Resistivity Measurement of Germanium and Silicon, *Rev. Sci. Instr.,* **24:**884–885 (1953).
68. A & M Fell, Ltd., "Resistivity Measurement Literature."

69. P. P. Clerx, Mechanical Aspects of Testing Resistivity of Semiconductor Materials and Diffused Layers, *Solid State Tech.*, **12**:16 (June 1969).
70. R. W. Germann and D. B. Rogers, Four-Probe Device for Accurate Measurement of Temperature Dependence of Electrical Resistivity on Small Irregular Shaped Single Crystals with Parallel Sides, *Rev. Sci. Instr.*, **37**:273–274 (1966).
71. Dr. C. Goodman, Standard Telephone Laboratory, private communication, 1963.
72. D. Dew-Hughes, A. H. Jones, and G. E. Brock, Improved Automatic Four-Point Probe, *Rev. Sci. Instr.*, **30**:920–922 (1959).
73. J. C. Brice and A. A. Stride, A Continuous Reading Four-Point Resistivity Probe, *Solid State Electron.*, **1**:245 (1960).
74. R. Hall, Minimizing Errors of Four-Point Probe Measurements on Circular Wafers, *J. Sci. Instr.*, **44**:53–54 (1967).
75. J. K. Hargreaves, and D. Millard, The Accuracy of Four-Probe Resistivity Measurements on Silicon, *Brit. J. Appl. Phys.*, **13**:231–234 (1962).
76. W. Murray Bullis, "Standard Measurements of the Resistivity of Silicon by the Four-Probe Method," National Bureau of Standards, *Rept.* 9666, December 1967.
77. Donald R. Zrudsky, Harry D. Bush, and John R. Fassett, Four-Point Sheet Resistivity Techniques, *Rev. Sci. Instr.*, **37**:885–890 (1966).
78. W. M. Bullis, F. H. Brewer, C. D. Kolstad, and L. J. Swartzendruber, Temperature Coefficient of Resistivity of Silicon and Germanium near Room Temperature, *Solid State Electron.*, **11**:639–646 (1968).
79. L. H. Garrison, Proper Current Input for True Resistivity Measurements of Single Crystal Silicon of Both n- and p-Type and Single Crystal Slices, *Solid State Tech.*, **9**:47–50 (1966).
80. David D. Valley, Probe Shows Silicon Resistivity Accurately, *Electronics*, **34**:70–73 (Mar. 31, 1961).
81. R. G. Mazur and D. H. Dickey, A Spreading Resistance Technique for Resistivity Measurements on Silicon, *J. Electrochem. Soc.*, **113**:255–259 (1966).
82. R. G. Mazur, Resistivity Inhomogeneities in Silicon Crystals, *J. Electrochem. Soc.*, **114**:255–259 (1967).
83. J. M. Adley, M. R. Poponiak, C. P. Schneider, P. A. Schumann, Jr., and A. H. Tong, The Design of a Probe for the Measurement of the Spreading Resistance of Semiconductors, in Rolf R. Haberecht and Edward L. Kern (eds.), "Semiconductor Silicon," The Electrochemical Society, New York, 1969.
84. D. C. Gupta and J. Y. Chan, A Semiautomatic Spreading Resistance Probe, *Rev. Sci. Instr.*, **41**:176–179 (1970).
85. W. A. Keenan, P. A. Schumann, Jr., A. H. Tong, and R. P. Phillips, A Model for the Metal-Semiconductor Contact in the Spreading Resistance Probe, in Bertram Schwartz (ed.), "Ohmic Contacts to Semiconductors," The Electrochemical Society, New York, 1969.
86. E. E. Gardner, P. A. Schumann, and E. F. Gorey, Resistivity Profiles and Thickness Measurements on Multilayered Semiconductor Structures by the Spreading Resistance Technique, in B. Schwartz and N. Schwartz (eds.), "Measurement Techniques for Thin Films," The Electrochemical Society, New York, 1967; P. J. Severin, Measurement of Resistivity of Silicon by the Spreading Resistance Method, *Solid State Electron.*, **14**:247–255 (1971); P. J. Severin, Measurement of the Resistivity and Thickness of a Heterotype Epitaxially Grown Silicon Layer with the Spreading-Resistance Method, *Philips Res. Rept.*, **26**:359–372 (1971).
87. Dinesh C. Gupta, Determination of Mobility and Its Profile in n/n^+ Silicon Epitaxial Layers, *J. Electrochem. Soc.*, **116**:670–671 (1969).
88. R. Holm, "Electric Contacts Handbook," 3d ed., Springer-Verlag OHG, Berlin, 1958.
89. Henry C. Torrey and Charles A. Whitmer, "Crystal Rectifiers," McGraw-Hill Book Company, New York, 1948.

90. F. P. Bowden and D. Tabor, "The Friction and Lubrication of Solids," Clarendon Press, Oxford, 1950.
91. F. L. Jones, "Physics of Electrical Contacts," Oxford University Press, London, 1957.
92. P. A. Schumann, Jr., and E. E. Gardner, Spreading Resistance Correction Factors, *Solid State Electron.,* **12**:371–375 (1969).
93. Von Wolfgang Keller, Measurement of Resistivity of Semiconductors with High Frequencies, *Z. Angew. Phys.,* **11**:346–350 (1959).
94. V. G. Sidyakin and E. T. Skorik, Measuring the Resistance of Semiconductors at High Frequencies, *Instr. Exptl. Tech.,* 326–329 (1960).
95. Paul J. Olshefski, A Contactless Method for Measuring Resistivity of Silicon, *Semicond. Prod.,* **4**:34–36 (1961).
96. I. R. Weingarten and M. Rothberg, Radio Frequency Carrier and Capacitative Coupling Procedures for Resistivity and Lifetime Measurements on Silicon, *J. Electrochem. Soc.,* **108**:167–171 (1961).
97. S. T. Kynev, M. K. Sheinkman, I. B. Shul'ga, and V. D. Fursenko, Contactless Method for Measuring the Parameters of Certain Semiconductors, *Instr. Exptl. Tech.,* 376–381 (1962).
98. Jun-Ichi Nishizawa, Yuzo Yamoguchi, Naotoshi Shoji, and Yoshio Tominga, Application of Siemens Method to Measure the Resistivity and Lifetime of Small Slices of Silicon, in Marvin S. Brooks and John K. Kennedy (eds.), "Ultrapurification of Semiconductor Materials," The Macmillan Company, New York, 1962.
99. C. A. Bryant and J. B. Gunn, Noncontact Technique for the Local Measurement of Semiconductor Resistivity, *Rev. Sci. Instr.,* **36**:1614–1617 (1965).
100. Nobuo Miyamoto and Jun-Ichi Nishizawa, Contactless Measurements of Resistivity of Slices of Semiconductor Materials, *Rev. Sci. Instr.,* **38**:360–367 (1967).
101. H. K. Henisch and J. Zucker, Contactless Method for the Estimation of Resistivity and Lifetime of Semiconductors, *Rev. Sci. Instr.,* **27**:409–410 (1956).
102. R. C. Powell and A. L. Rasmussen, A Radio Frequency Permittimeter, *IRE Trans. Instr.,* **I-9**:179–184 (1960).
103. J. C. Brice and P. Moore, Contactless Resistivity Meter for Semiconductors, *J. Sci. Instr.,* **38**:307 (1961).
104. J. E. Zimmerman, Measurement of Electrical Resistivity of Bulk Metals, *Rev. Sci. Instr.,* **32**:402–405 (1961).
105. S. J. Yosim, L. F. Grantham, E. B. Luchsinger, and R. Wike, Electrodeless Determination of Electrical Conductivities of Melts at Elevated Temperatures, *Rev. Sci. Instr.,* **34**:994–996 (1963).
106. T. O. Poehler and W. Liben, Induction Measurement of Semiconductor and Thin Film Resistivity, *Proc. IEEE,* **52**:731–732 (1964).
107. R. W. Haisty, Electrodeless Measurements of Resistivities over a Very Wide Range, *Rev. Sci. Instr.,* **38**:262–265 (1967).
108. R. W. Haisty, Measurement of High Resistivities by the Electrodeless Falling Sample Method, *Rev. Sci. Instr.,* **39**:778 (1968).
109. J. Le Page, A. Bernalte, and D. A. Linholm, Analysis of Resistivity Measurements by the Eddy Current Decay Method, *Rev. Sci. Instr.,* **39**:1019–1026 (1968).
110. A. C. Brown, A Non-contacting Resistivity Meter, *J. Sci. Instr.,* **41**:472–473 (1964).
111. T. S. Benedict and W. Shockley, Microwave Observations of the Collision Frequency of Electrons in Germanium, *Phys. Rev.,* **89**:1152–1153 (1953).
112. H. Hsieh, S. M. Goldey, and S. C. Brown, A Resonant Cavity Study of Semiconductors, *J. Appl. Phys.,* **25**:302–307 (1954).
113. J. G. Linhart, I. M. Templeton, and R. Dunsmuir, A Microwave Resonant Cavity Method for Measuring the Resistivity of Semiconducting Materials, *Brit. J. Appl. Phys.,* **7**:36–38 (1956).
114. T. Kohane and M. H. Siruetz, Measurement of Microwave Resistivity by Eddy Current Loss in Spheres, *Rev. Sci. Instr.,* **30**:1059–1060 (1959).

115. G. L. Allerton and J. R. Seifert, Resistivity Measurements of Semiconductors at 9000 MC, *IRE Trans. Instr.,* **I-9**:175–179 (1960).
116. T. Kohane, The Measurement of Microwave Resistivity by Eddy Current Loss in Small Spheres, *IRE Trans. Instr.,* **I-9**:184–186 (1960).
117. K. S. Champlin and R. R. Krongard, The Measurement of Conductivity and Permittivity of Semiconductor Spheres by an Extension of the Cavity Perturbation Method, *IRE Trans. Microwave Theory Tech.,* **MTT-9**:545–551 (1961).
118. H. Jacobs, F. A. Brand, J. D. Meindl, M. Benanti, and R. Benjamin, Electrodeless Measurement of Semiconductor Resistivity at Microwave Frequencies, *Proc. IRE,* **49**:928–932 (1961).
119. K. S. Champlin and D. B. Armstrong, Explicit Forms for the Conductivity and Permittivity of Bulk Semiconductors in Wave Guides, *Proc. IRE,* **50**:232 (1962).
120. B. R. Nag and S. K. Roy, Microwave Measurements of Conductivity and Dielectric Constant of Semiconductors, *Proc. IRE,* **50**:2515–2516 (1962).
121. J. N. Bhar, Microwave Techniques in the Study of Semiconductors, *Proc. IEEE,* **51**:1623–1631 (1963).
122. J. Lindmayer and M. Kutsko, Reflection of Microwaves from Semiconductors, *Solid State Electron.,* **6**:377–382 (1963).
123. B. R. Nag, S. K. Roy, and C. K. Chatterjee, Microwave Measurement of Conductivity and Dielectric Constant of Semiconductors, *Proc. IEEE,* **51**:962 (1963).
124. M. W. Gunn, The Microwave Measurement of the Complex Permittivity of Semiconductors, *Proc. IEEE,* **52**:185 (1964).
125. D. A. Holmes, D. L. Feucht, and H. Jacobs, Microwave Interaction with a Semiconductor Post, *Solid State Electron.,* **7**:267–273 (1964).
126. J. A. Naber and D. P. Snowden, Application of Microwave Reflection Technique to the Measurement of Transient and Quiescent Electrical Conduction of Silicon, *Rev. Sci. Instr.,* **40**:1137–1141 (1969).
127. Keith S. Champlin, Gary H. Glover, and John D. Holm, Bulk Microwave Conductivity of Semiconductors Determined from TE_{01}^{0} Reflectivity of Boule Surface, *IEEE Trans. Instr. Meas.,* **IM-18**:105–110 (1969).
128. M. R. E. Bichara and S. P. R. Poitevin, Resistivity Measurement of Semiconducting Epitaxial Layers by the Reflection of a Hyperfrequency Electromagnetic Wave, *IEEE Trans. Instr. Meas.,* **IM-13**:323–328 (1964).
129. T. Ogawa, Measurement of the Electrical Conductivity and Dielectric Constant without Contacting Electrodes, *J. Appl. Phys.,* **32**:583–592 (1961).
130. M. W. Ozelton and J. R. Wilson, A Rotating Field Apparatus for Determining Resistivities of Reactive Liquid Metals and Alloys at High Temperatures, *J. Sci. Instr.,* **43**:359–363 (1966).
131. A. C. Brown, Thickness Corrections to Resistivity Measurement by Q-Meter, *J. Sci. Instr.,* **41**:335–336 (1964).
132. Edward C. Jordan and Keith G. Balmain, "Electromagnetic Waves and Radiating Systems," Prentice-Hall, Inc., Englewood Cliffs, N.J., 1968.
133. A. H. Frei and M. J. D. Strutt, Skin Effect in Semiconductors, *Proc. IRE,* **48**:1272–1277 (1960).
134. O. Sandus, Comments on Electrodeless Measurements of Semiconductor Resistivity at Microwave Frequencies, *Proc. IRE,* **50**:473–474 (1962).
135. C. C. Allen and E. G. Bylander, Evaluation Techniques for, and Electrical Properties of Silicon Epitaxial Films, in John B. Schroeder (ed.), "Metallurgy of Semiconductor Materials," Interscience Publishers, a division of John Wiley & Sons, Inc., New York, 1963.
136. E. Batifol and G. Duraffourg, *J. Phys. Radium,* **24**(Suppl. to No. 11):207A (1960).
137. J. R. Biard and Stacy B. Watelski, Evaluation of Germanium Epitaxial Films, *J. Electrochem. Soc.,* **109**:705–709 (1962).
138. John Brownson, A Three-Point Probe Method for Electrical Characterization of Epi-

taxial Films, Electrochemical Society Meeting, May 1962; *J. Electrochem. Soc.,* **111**:919–924 (1964).

139. C. C. Allen, L. H. Clevenger, and D. C. Gupta, A Point Contact Method of Evaluating Epitaxial Layer Resistivity, *J. Electrochem. Soc.,* **113**:508–510 (1966).
140. E. E. Gardner, J. F. Hallenback, Jr., and P. A. Schumann, Jr., Comparison of Resistivity Measurement Techniques on Epitaxial Silicon, *Solid State Electron.,* **6**:311–313 (1963).
141. E. E. Gardner and P. A. Schumann, Jr., Measurement of Resistivity of Silicon Epitaxial Layers by the Three-Point Probe Technique, *Solid State Electron.,* **8**:165–174 (1965).
142. P. A. Schumann, Jr., M. R. Poponiak, J. F. Hallenback, Jr., and C. P. Schneider, Measurement Electronics for the Three-Point Probe, *Solid State Tech.,* **11**:32–36 (November 1968).
143. P. A. Schumann, Jr., A Theoretical Model of the Three-Point Probe Breakdown Technique, *J. Electrochem. Soc.,* **115**:1197–1203 (1968).
144. M. H. Norwood, Three-Point Probe Calibration for GaAs, *J. Electrochem. Soc.,* **112**:875 (1965).
145. H. Frank, Contribution to the Measurement of Resistivity of nn^+ and pp^+ Epitaxial Layers by the Punch through Method (in German), *Phys. Stat. Solidi,* **18**:401–414 (1966).
146. W. G. Spitzer and H. Y. Fan, Infrared Absorption in n-Type Silicon, *Phys. Rev.,* **108**:268–271 (1957).
147. N. J. Harrick, Semiconductor Type and Local Doping Determination through the Use of Infrared Radiation, *Solid State Electron.,* **1**:234–244 (1960).
148. W. G. Spitzer and H. Y. Fan, Determination of Optical Constants and Carrier Effective Mass of Semiconductors, *Phys. Rev.,* **106**:882–890 (1957).
149. David F. Edwards and Paul D. Maker, Quantitative Measurement of Semiconductor Homogeneity from Plasma Edge, *J. Appl. Phys.,* **33**:2466–2468 (1962).
150. I. Kudman, A Non-destructive Measurement of Carrier Concentration in Heavily Doped Semiconductor Materials and Its Application to Thin Surface Layers, *J. Appl. Phys.,* **34**:1826–1827 (1963).
151. L. A. Murray, J. J. Rivera and P. A. Hoss, Infrared Reflectivity of Heavily Doped Low-Mobility Semiconductors, *J. Appl. Phys.,* **37**:4743–4745 (1966).
152. E. E. Gardner, W. Kappallo, and C. R. Gorden, Measurement of Diffused Semiconductor Surface Concentrations by Infrared Plasma Reflection, *Appl. Phys. Lett.,* **9**:432–434 (1966).
153. P. A. Schumann and R. P. Phillips, Comparison of Classical Approximations to Free Carrier Absorption in Semiconductors, *Solid State Electron.,* **10**:943–948 (1967).
154. P. A. Schumann, Jr., Plasma Resonance Calibration Curves for Silicon, Germanium and Gallium Arsenide, *Solid State Tech.,* **13**:50–52 (1970).
155. Toshio Abe and Yoshio Nishi, Non-destructive Measurement of Surface Concentrations and Junction Depths of Diffused Layers, *Japan. J. Appl. Phys.,* **7**:397–403 (1968).
156. T. G. R. Rawlins, Measurement of the Resistivity of Epitaxial Vapor Grown Films of Silicon by an Infrared Technique, *J. Electrochem. Soc.,* **111**:810–814 (1964).
157. Dinesh C. Gupta, Non-destructive Determination of Carrier Concentration in Epitaxial Silicon Using a Total Internal Reflection Technique, *Solid State Electron.,* **13**:543–552 (1970).
158. Charlie E. Jones and A. Ray Hilton, Optical Properties by Far Infrared Ellipsometry, *J. Electrochem. Soc.,* **115**:106–107 (1968).
159. J. Hilibrand and R. D. Gold, Determination of the Impurity Distribution in Junction Diodes from Capacitance-Voltage Measurements, *RCA Rev.,* **21**:245–252 (1960).
160. C. Jund and R. Poirer, Carrier Concentration and Minority Carrier Lifetime Measurement in Semiconductor Epitaxial Layers by the MOS Capacitance Method, *Solid State Electron.,* **9**:315–319 (1966).
161. M. G. Buehler, Peripheral and Diffused Layer Effects on Doping Profiles, *IEEE Trans. Electron Devices,* **ED-19**:1171–1178 (1972).

162. K. H. Zaininger and F. P. Heiman, The C-V Technique as an Analytical Tool—Part I, *Solid State Tech.,* **13**:49-55 (May 1970).
163. K. H. Zaininger and F. P. Heiman, the C-V Technique as an Analytical Tool—Part II, *Solid State Tech.,* **13**:46-55 (June 1970).
164. J. J. Jakosky, Exploration Geophysics, *Times-Mirror Press,* Los Angeles, 1940.
165. E. W. Carpenter and G. M. Habberzam, A Tri-potential Method of Resistivity Prospecting, *Geophysics,* **21**:455-469 (1956).
166. Harry Letaw, Jr., Lawrence M. Slifkin, and William M. Portnoy, A Precision Grinding Machine for Diffusion Studies, *Rev. Sci. Instr.,* **25**:865-868 (1954).
167. L. M. L. J. Leblans and M. L. Verheijke, A Precision Grinding Machine for Tracer Diffusion Studies, *Philips Tech. Rev.,* **25**:191-194 (1963-1964).
168. B. Goldstein, Precision Lapping Device, *Rev. Sci. Instr.,* **28**:289-290 (1957).
169. H. W. Schamp, D. A. Oakes, and N. M. Reed, Grinder for Sectioning Solid Diffusion Specimens, *Rev. Sci. Instr.,* **30**:1028-1031 (1959).
170. H. J. de Bruin, and Ralph L. Clark, Precision Lapping Device for the Determination of Diffusion Coefficients in Ceramic Materials, *Rev. Sci. Instr.,* **35**:227-228 (1964).
171. W. C. Dunlap, Jr., Diffusion of Impurities in Germanium, *Phys. Rev.,* **94**:1531-1540 (1954).
172. Stacy B. Watelski, W. R. Runyan, and R. C. Wackwitz, A Concentration Gradient Profiling Method, *J. Electrochem. Soc.,* **112**:1051-1053 (1965).
173. B. McDonald and F. C. Collins, Anodic Sectioning of Diffused Silicon p-n Junctions, *Bull. Am. Phys. Soc.,* **6**:106 (March 1961).
174. Howard Reiss and C. S. Fuller, Diffusion Processes in Germanium and Silicon, in N. B. Hannay (ed.), "Semiconductors," Chap. 6, Reinhold Publishing Corporation, New York, 1960.
175. Don L. Kendall, Texas Instruments Inc., unpublished work. See also Refs. 172 and 176.
176. R. A. Evans and R. P. Donovan, Alternative Relationship for Converting Incremental Sheet Resistivity Measurements into Profiles of Impurity Concentration, *Solid State Electron.,* **10**:155-157 (1967).
177. M. F. Lamorte, Calculation of Concentration Profiles and Surface Concentration from Sheet-Conductance Measurements of Diffused Layers, *Solid State Electron.,* **1**:164-171 (1960).
178. T. E. McGahon and W. H. Hackett, Jr., A Technique for Determining p-n Junction Doping Profiles and Its Application to GaP, *Rev. Sci. Instr.,* **41**:1182-1183 (1970).
179. H. Kressel and M. A. Klein, Determination of Epitaxial-Layer Impurity Profiles by Means of Microwave Diode Measurements, *Solid State Electron.,* **6**:309-310 (1963).
180. D. Pomerantz, Measuring Mobility and Density of Charge Carriers near a p-n Junction, *J. Electrochem. Soc.,* **112**:196-200 (1965).
181. V. Sils, R. W. Berkstrosser, P. Wang, and T. A. Longo, Doping Profile Determination in Epitaxial Silicon Films, *Electrochem. Tech.,* **2**:138-143 (1964).
182. D. C. Gupta, J. Y. Chan, and P. Wang, Observations on Imperfections in Silicon Material Using the Spreading Resistance Probe, *J. Electrochem. Soc.,* **117**:1611-1612 (1970).
183. Chusuke Munakata, Detection of Resistivity Striations in a Ge Crystal with an Electron Beam, *Japan. J. Appl. Phys.,* **5**:336 (1966).
184. Hiroshi Watanabe and Chusuke Munakata, Measurement of Resistance by Means of Electron Beam—I, *Japan. J. Appl. Phys.,* **4**:250-258 (1965).
185. Chusuke Munakata, Bulk Electron Voltaic Effect, *Japan. J. Appl. Phys.,* **4**:697 (1965).
186. Chusuke Munakata and Hiroshi Watanabe, Measurement of Resistance by Means of Electron Beam—II, *Japan. J. Appl. Phys.,* **5**:1157-1160 (1966).
187. Chusuke Munakata, An Electron Beam Method of Measuring Resistivity Distribution in Semiconductors, *Japan. J. Appl. Phys.,* **6**:963-971 (1967).
188. Chusuke Munakata, An Application of Beta Conductivity to Measurement of Resistivity Distribution, *J. Sci. Instr.,* (2) **1**:639-642 (1968).

189. J. Tauc, Generation of an EMF in Semiconductors with Nonequilibrium Current Carrier Concentrations, *Revs. Mod. Phys.,* **29**:308–324 (1957).
190. J. Oroshnik, and A. Many, Evaluation of the Homogeneity of Germanium Single Crystals by Photovoltaic Scanning, *J. Electrochem. Soc.,* **106**:360–362 (1959).
191. C. D. Cox, Bulk Photoeffects in Inhomogeneous Semiconductors, *Can. J. Phys.,* **38**:1328–1342 (1960).
192. J. Oroshnik and A. Many, Quantitative Photovoltaic Evaluation of the Resistivity Homogeneity of Germanium Single Crystals, *Solid State Electron.,* **1**:46–53 (1960).
193. E. I. Rashba and V. A. Romanov, A Photoelectric Method of Observing Inhomogeneity with Depth in Semiconductors, *Soviet Phys. Solid State,* **2**:2393–2396 (1961).
194. D. T. Kokorev and N. F. Kovtonyuk, Analysis of the Homogeneity of Semiconductor Materials by Using the Method of the Volume Photo-EMF, *Instr. Exptl. Tech.,* **28**:382–387 (1962).
195. K. Thiessen and H. Hornung, Measurement of Inhomogeneous Distribution of Recombination Centers in Germanium by Means of Photoconductive and Photomagnetoelectric Effects (in German), *Phys. Stat. Solidi,* **2**:1158–1164 (1962).
196. I. A. Baev and E. G. Valyashko, Study of the Homogeneity of Semiconductor Crystals with the Use of a Moving Light Probe, *Soviet Phys. Solid State,* **6**:1357–1361 (1964).
197. M. I. Iglitsyn, D. I. Levinson, and V. U. Chernopisskii, Checking the Uniformity of Monocrystalline Germanium by the Single Probe Method, *Ind. Lab.,* **30**:251–254 (1964).
198. M. I. Iglitsyn, A. A. Meier, O. V. Karagioz, D. I. Levinzon, and A. V. Ivanov, A Single-Probe Method of Measuring the Resistivity of Semiconductors with Alternating Current, *Ind. Lab.,* **31**:1355–1357 (1965).
199. M. G. Mil'vidshii, S. P. Grishina, and A. V. Berkova, Detecting Inhomogeneities in Silicon Single Crystals with Infrared Transillumination, *Ind. Lab.,* **31**:586–588 (1965).
200. I. A. Baev and E. G. Valyashko, An Investigation of the Distribution of Inhomogeneous Regions in Semiconductors, *Soviet Phys. Solid State,* **7**:2093–2099 (1966).
201. W. Murray Bullis (ed.), "Methods of Measurement for Semiconductor Materials, Process Control, and Devices," National Bureau of Standards, *Tech. Note* 488, July 1969.
202. A. H. Tong, Nondestructive and Contactless Sheet Resistance and Junction Depth Measurements by Infrared Transmission through the Diffusion, in Howard R. Huff and Ronald R. Burgess (eds.), "Semiconductor Silicon 1973," pp. 596–605, Electrochemical Society, Princeton, 1973.
203. J. K. Kung and W. G. Spitzer, Infrared Reflectivity and Free Carrier Absorption of Si-doped, N-type GaAs, *J. Electrochem. Soc.,* **121**:1482–1487 (1974).
204. Alvin H. Tong, Paul A. Schumann, Jr., and William A. Keenan, Epitaxial Substrate Carrier Concentration Measurement by the Infrared Interference Envelope (IRIE) Technique, *J. Electrochem. Soc.,* **119**:1381–1384 (1972).
205. M. G. Buehler, Impurity Centers in PN Junctions Determined from Shifts in the Thermally Stimulated Current and Capacitance Response with Heating Rate, *Solid State Electron.,* **13**:69–79 (1972).
206. Martin G. Buehler, Thermally Stimulated Measurements: The Characterization of Defects in Silicon p-n Junctions, in Howard R. Huff and Ronald R. Burgess (eds.), "Semiconductor Silicon 1973," pp. 549–560, Electrochemical Society, Princeton, 1973.
207. D. V. Lang, Fast Capacitance Transient Apparatus: Application to ZnO and O Centers in GaP p-n Junctions, *J. Appl. Phys.,* **45**:3014 (1974).
208. D. V. Lang, Deep-Level Transient Spectroscopy: A New Method to Characterize Traps in Semiconductors, *J. Appl. Phys.,* **45**:3023–3032 (1974).
209. Bernard L. Morris, Some Device Applications of Spreading Resistance Measurements on Epitaxial Silicon, *J. Electrochem. Soc.,* **121**:422–426 (1974).
210. G. H. Glover and W. Tantraporn, Doping Profile Measurements from Avalanche Space-Charge Resistance: A New Technique, *J. Appl. Phys.,* **46**:867–874 (1975).

211. G. H. Glover and W. Tantraporn, R-I Profiling: A New Technique for Measuring Semiconductor Doping Profiles, *Appl. Phys. Lett.,* **25**:348–349 (1974).
212. P. A. Schumann, Jr., C. P. Schneider, and L. A. Pietrogallo, Three-Dimensional Impurity Profile of an Epitaxial Layer, *Solid State Tech.,* **16**:54–55 (Sept., 1973).
213. J. Burtscher, H. W. Dorendorf, and J. Krausse, Electrical Measurement of Resistivity Fluctuations Associated with Striations in Silicon Crystals, *IEEE Trans. Electron. Devices,* **ED-20**:702–708 (1973).

4

Lifetime

4.1 INTRODUCTION

When the equilibrium number of holes and electrons is disturbed by introducing additional carriers, the *lifetime* is a measure of the time required for the excess carriers to recombine. Their behavior in an infinite block of material with no complications* is described by

$$-\frac{\partial p}{\partial t} = \frac{\Delta p}{\tau} \qquad -\frac{\partial n}{\partial t} = \frac{\Delta n}{\tau} \tag{4.1}$$

where n, p = number of carriers present at time t
Δn, Δp = excess number of carriers
τ = lifetime

or

$$\Delta n = Ae^{-t/\tau} \qquad \Delta p = Be^{-t/\tau} \tag{4.2}$$

where A and B are constants.

Should the block not be infinite, the carriers which diffuse to the surface will generally combine at a rate quite different from the one in the bulk. The most common way of characterizing such surface recombination is to define a surface-recombination velocity s through the equations

$$J_n = qs\Delta n \qquad J_p = qs\Delta p \tag{4.3}$$

or

$$D_n\frac{\partial n}{\partial x} = s\Delta n \qquad D_p\frac{\partial p}{\partial x} = s\Delta p \tag{4.4}$$

where J is the perpendicular component of the current density flowing to or from the surface. When the appropriate equations are solved for some specific geometry

*No trapping and negligible diffusion.

rather than the infinite block described by Eq. (4.1), Eq. (4.4) can be used as one of the boundary conditions and will give an effective lifetime lower than that which would be observed in the bulk unless $s = 0$. Some specific cases are discussed in a later section.

Excess holes can directly recombine with conduction-band electrons, and excess electrons can directly recombine with valence-band holes, in which case the lifetimes for holes and electrons will be equal. If traps are present, there are two limiting cases. Should the captured carrier reside in the trap only for a time short compared with the lifetime before combining with an opposite type of carrier, both hole and electron lifetime will still be equal, and the trap is ordinarily referred to as a *recombination center*. However, if upon trapping, the carrier resides there for a longer time, the process is referred to as *trapping*, and lifetime will usually vary with type and the method of measurement. The details of the various trapping models that give rise to the differing lifetimes will not be considered in any detail. Occasionally, the method used will measure the generation rate of carriers rather than their recombination rate. Further, depending on the approach and sample geometry, the measured value may be determined by either bulk or surface-recombination processes, and depending on the injection level, the value of τ can theoretically vary by a factor of 2. Therefore, before lifetimes measured by various methods are compared, some care should be taken to ensure that there is in fact some justification for the comparison.

Literally dozens of approaches have been reported, although they can in general be grouped in two categories. One uses a single conductivity-test sample wnich may be as simple as a rectangular specimen of millimeter dimensions to observe the decay of carriers directly. The other uses p-n junctions or metal-oxide-semiconductor (MOS) capacitors, usually in the form of finished devices, and attempts to correlate some device parameter with lifetime. Normally the material manufacturers strive for long lifetime, whereas many finished devices require either the lowest possible lifetime or at least some controlled intermediate value. When lifetime measurements are to be used for initial material characterization and/or process control, those methods which require the fewest additional fabrication steps are the most desirable since any additional steps not only slow down the evaluation process but may also materially change the lifetime. However, since the lifetime in finished devices is very process-sensitive and is generally so important to device performance, considerable effort has been expended in methods to interpret device performance in terms of lifetime.

As the technology of a given material matures and the lifetime of the average material produced becomes more than adequate for its intended use, fewer lifetime measurements are required of the starting material. Indeed the majority of silicon crystals grown are never checked for lifetime. Similarly, as the device processing becomes more fixed, there is less inclination to determine actual lifetime in the devices. What is usually done is to monitor the device parameter of interest and relate it directly to some process variation which affects lifetime. For example, in making fast diodes, instead of measuring recovery times, calculating lifetime, relating that lifetime value to a specific concentration of recombination centers, and then further considering the relation between the number of centers and the mode of introduction of that recombination center, curves of diode recovery time are plotted directly against gold-diffusion temperature and used for process control and optimization.

Table 4.1 summarizes some forty variations of lifetime-measurement techniques that will be discussed. The final choice of a particular method or variation depends primarily on the magnitude of the lifetime, the sample size, and whether the sample can only be a piece of bulk material or a completed device is available. The usual working range for several of these methods is shown in Table 4.2. It should be remembered, however, that with special emphasis and attention to a particular case, nearly any method can be extended in either direction. But the photoconductive decay (PC) method is so simple it is used whenever possible, so there has been little inclination to extend other methods into the longer-lifetime region. The one exception has been that as longer and longer lifetimes have been achieved after various processing stages, the PC method sometimes becomes size-limited (for example, with slices). In such circumstances surface photovoltage, which allows approximately a 4 to 1 reduction in thickness for a given lifetime measurement, becomes very attractive even though it is quite tedious. For cases where an oxide at least a few hundred angstroms thick is present (e.g., after the first step in bipolar and MOS processing) the MOS capacitor will allow measurements over a rather broad range of lifetimes and is not necessarily destructive. At the other end of the scale, diode recovery measurements are favored whenever possible, again because of simplicity. However, in keeping with the theme of this volume, which is devoted to bulk measurements, device-related methods are not discussed, although references are given in Table 4.1.

4.2 PHOTOCONDUCTIVE DECAY

In this method the excess carriers are generated by irradiating the sample with light of short enough wavelength to produce hole-electron pairs. The conductivity σ of the sample is directly proportional to the number of carriers, and the conductivity change is proportional to the number of excess carriers, or

$$\Delta\sigma = Ce^{-t/\tau} \tag{4.5}$$

Thus monitoring the conductivity after the light is removed allows the lifetime to be determined.

Detection of Decay. The simplest method of observing the carrier decay is by use of the circuit of Fig. 4.1. If current is supplied by a constant-current source, then it can be shown[96] that the time constant for the voltage decay is quite close to the effective carrier time constant and is given by

$$\tau_{\text{eff}} = \tau_{\text{volt}}\left(1 - \frac{\Delta V}{V_0}\right) \tag{4.6}$$

if ΔV is small compared with the dark IR drop V_0 developed by the constant current across the portion of the sample illuminated by the chopped light. Actual measurement of the time constant is by means of an oscilloscopic display. Then the variable sweep can be used to match a curve drawn on the face. Other methods include using a fixed sweep and measuring the time for an amplitude reduction to $1/e$ of the value at the start of the time measurement, by generating a second curve from a photocell and a variable but known RC time constant and matching it to the photodecay curve by using a dual-trace oscilloscope, or by putting one signal on the x plates and the other on the y plates.[10,19,33,37] The latter has the added advantage of visually displaying difficulties arising when conditions such as excessive light

Table 4.1. Literature Guide to Methods of Measuring Carrier Lifetime

Method	Reference
Diffusion-length methods:	
Traveling spot	1, 11, 12, 13, 28, 71, 103, 125, 140, 153, 154
Flying spot	7, 30, 125
Dark spot	81
Drift field	2, 12, 14, 16, 20, 32
Emitter-point efficiency	12
Conductivity-decay methods:	
Photoconductive decay	
Direct observation of resistivity	4, 12, 21, 33, 37, 53, 55, 57 80, 94, 96, 107, 116, 134, 171
Q changes	107, 114, 139, 167
Microwave absorption	66, 69, 75, 76, 88, 97, 111, 118, 121
Spreading resistance	20
Eddy-current losses	165
Infrared-absorption coefficient	26, 47, 62, 133
Pulse decay	
Direct observation of resistivity	3, 10, 40, 41
Microwave absorption	66, 69, 88
High-energy-radiation-generated carrier decay	31, 117, 135, 137
Conductivity-modulation methods:	
Photoconductivity	
Steady state	18, 25, 46, 82, 136, 146
Modulated source	19, 45, 49, 52, 89, 102, 156, 160, 180
Infrared detection, steady state	26
Infrared detection, modulated source	62
Q changes	27
Microwave absorption	75
Eddy-current losses	116
Pulse injection—spreading resistance	20, 40, 41
Device methods:	
Diode open-circuit voltage decay	15, 17, 23, 98, 112, 145, 169, 181
Diode reverse-current behavior	6, 8, 9, 35, 39, 73, 101, 113, 120, 127, 128, 132, 141, 145, 150, 155, 166, 172, 173, 174, 177, 185, 189
Diffusion capacitance	108, 123
Junction photocurrent or photovoltage	50, 68, 83, 84, 91, 99, 115, 131, 142, 159, 187, 193
I-V	130, 190
Stored charge	127, 129, 144, 148
Current-distortion effects	113, 161, 183, 188
Base transport	24, 51, 60, 95, 126, 145, 151
Collector response	145, 158
Offset voltage	
MOS capacitance method:	152, 163, 164, 175, 176, 180, 182, 184
Photomagnetoelectric effect	34, 58, 64, 72, 74, 77, 100, 104, 105, 109, 122, 124, 162, 179, 192
Surface photovoltage:	
Steady state	92, 179, 192
Decay	42
Miscellaneous methods:	
Photoluminescence	147
Cathodoluminescence	168
Suhl and related effects	5, 44
Electroluminescence	93
Charge-collection efficiency	61, 149
Noise	78, 116
Lifetime definitions and interpretation	22, 29, 36, 48, 54, 56, 59, 70, 85, 86, 90, 138, 157

Adapted in part from W. Murray Bullis, "Measurement of Carrier Lifetime in Semiconductors—An Annotated Bibliography Covering the Period 1949–1967," National Bureau of Standards, *Tech. Rept.* AFML-TR-68-108, 1968.

Table 4.2. Range and Limitations of Lifetime Measuring Methods

Method	Lower limitation	Lifetime, s (10^{-9}, 10^{-6}, 10^{-3})	Upper limitation	Misc. limitation
Diode recovery	Pulsing instrumentation		Device geometry	
PME–PC	Inadequate signal to noise			Complex equipment very sensitive to surface effects
Surface photovoltage	Inadequate signal to noise		Sample size	Time-consuming measurement
Photodecay	Inadequate signal to noise, lack of rapid cutoff light source.	*	Sample size	
MOS capacitance	Doping level to n_i ratio		Surface recombination	

Adapted from Joseph Horak, "Minority Carrier Lifetime Measurements on Silicon Material for Use in Electron Irradiation Studies," AF Contract F19628-67-0043, *Scientific Report,* May 1968.

*May be extended into this range by using high-energy electron excitation.

intensity preclude an exponential decay. If the semiconductor signal is exponential and both amplifiers are matched in phase, a straight line will appear. If a complex Lissajous figure appears, the data should be suspected.[37] An excessive electric field may cause some of the carriers to be transported to the contacts before normal recombination can occur. This will give rise to a reduced value of measured lifetime, and usually a nonexponential decay as well.[80]

The use of microwave absorption for detecting the excess carriers and their subsequent decay is quite appealing because there is then no necessity for making ohmic contacts to the specimen. To that end considerable effort has been expended, but in general the added complexity of the microwave equipment overshadows contacting difficulties. It is first necessary to establish that the microwave detector signal is indeed proportional to the number of excess carriers. For a simple post inserted into a waveguide, theory would indicate that only under very limited conditions is it true,[69,88,97] although good agreement between microwave and contact measurements has been reported.[66,75,76] By completely filling the cavity with the sample, the limitations are considerably reduced and the calculations simplified, but with the added complexity of requiring a carefully sized sample.[121] To keep from shaping the sample at all, coupling to odd-shaped pieces via an open waveguide or horn has been used. Again, there are complications of interpretation, as well as more susceptibility to errors caused by reflections and a reduced signal-to-noise ratio.[111,118]

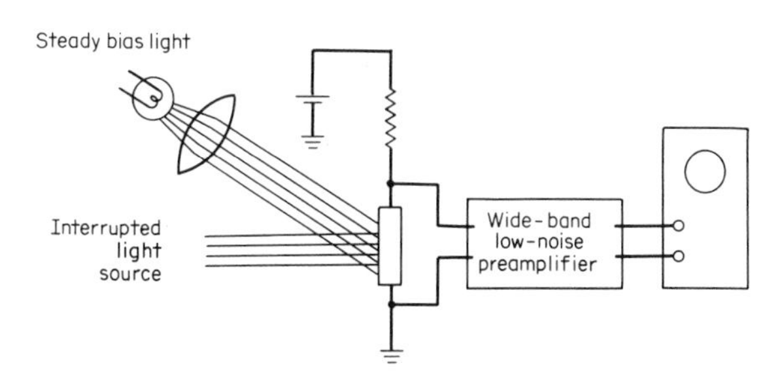

Fig. 4.1. Photodecay equipment.

The same advantage of no direct electrical contact can be realized by using lower frequencies and capacitance coupling. The actual circuitry can vary considerably and may range from rf bridges[107] to Q meters.[27] Coupling to rods, e.g., crystals in plastic bags, can be conveniently accomplished by having wide metal bands which clamp around the rod.[107] Slices, however, can best be laid over two adjacent flat-plate electrodes covered by a thin plastic or mica spacer.[114,167] Eddy currents can also be used by inserting the sample into an rf coil and observing the change of oscillator loading after a light pulse. A "cage" to short out the axial component of the field will make the power absorbed vs. conductivity linear over a much broader range of $\sigma/\omega\varepsilon$ than would otherwise be possible[165] (σ = conductivity, ε = dielectric constant).

An increase in the number of free carriers will increase the long-wavelength optical absorption. Thus by monitoring the absorption coefficient, the decay of carriers can be followed, again without direct contact.[26] The straightforward approach requires an infrared detector with a response time shorter than that of the lifetime being measured. However, because of the nonlinear relationship between the transmitted intensity and the number of excess carriers, if the light generating the carriers is chopped, the average value of transmitted intensity will depend on the chopping frequency. Thus very slow detectors can be used,[62,133] but sensitivity is not very good.

Light Source. The simplest sort of light source is the combination of an incandescent bulb and a rotating-blade mechanical chopper. It does, however, have the disadvantage of producing a rather slow light-pulse fall time for any reasonable disk rotational speed. Despite this handicap, it has been widely used for measuring the lifetime of germanium. Fall times can be decreased by using a combination of rotating mirror and a slit so that a larger angular velocity of the beam relative to the limiting aperture can be obtained without either excessive speed or large wheel diameter,[21] as shown in Fig. 4.2*a*. This approach has been carried to the extreme by using a mirror-sample separation of several feet and a many-faceted mirror rotated tens of thousands of revolutions per minute.[79,87] The mirror velocity can be reduced (or the pulse made more rapid) by using multiple reflections,[38,94] as shown in Fig. 4.2*b* and *c*, but eventually the limit is reached, not by the inability to increase relative velocity further but by the decrease in light intensity available at the sample. One distinct advantage that mechanically modulated systems have over others is that the spectral distribution of the source can be chosen independently. Such an independent choice is very important when studying narrow-bandgap materials where the wavelength might be several micrometers. Modulation may also be effected by the use of a Kerr cell, but electrical noise associated with the high-voltage pulses required to operate the cell is difficult to shield against. The natural decay of arcs in air,[63,65] hydrogen,[43] or xenon[21] has also been used to produce short light pulses. Xenon flash tubes are particularly convenient, and an ordinary commercial strobe light can be used. Air arcs were the light source in most of the early silicon lifetime-test sets. Neither of them can be used to measure lifetimes of less than 2 or 3 μs, however, since that is the order of the decay time of the ionized gas. Filters are sometimes used to remove short wavelengths and thereby reduce surface effects. For silicon a thin slice of silicon or GaAs may be used as the filter.

Laser pulses can be obtained with very fast fall times and thus can be used for light sources.[198] Modulation of the defocused beam of a high-intensity short-persistence cathode-ray tube is useful for moderate lifetimes,[19] and for certain selected wavelengths, light-emitting diodes can be used.[156]

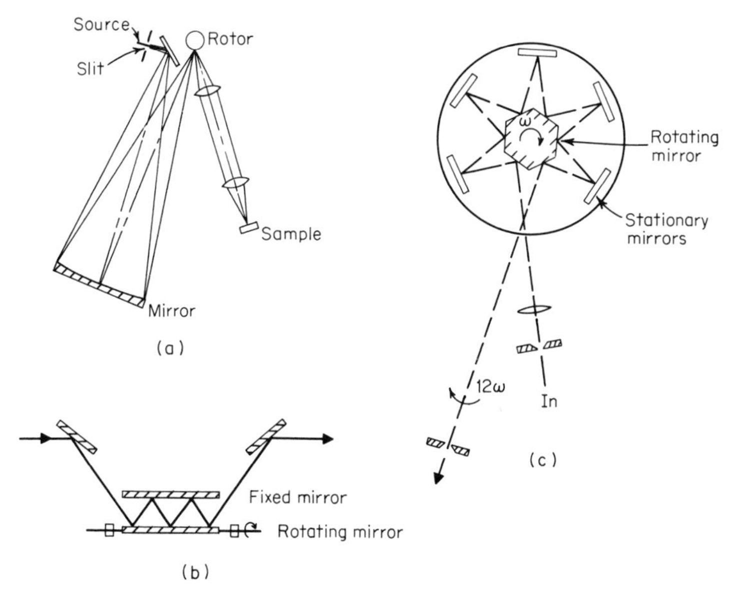

Fig. 4.2. Mechanical methods of modulating light. (*a*) Single rotating mirror. (*Adapted from Williams.*[87]) (*b*) Multiple reflections from a rotating mirror combined with a fixed mirror. (*Adapted from Holeman and Hilsum.*[94]) (*c*) Multiple reflections from a rotating mirror combined with a fixed mirror. (*Adapted from Garbuny, Vogl, and Hansen.*[38])

If the absorption coefficient is very high, the light may be predominantly absorbed near the surface and give rise to excess carrier densities comparable with the doping carrier density. If this happens, the initial decay will not be a simple exponential and will be slower than expected. However, after enough time has elapsed for some of the carriers either to diffuse into the bulk or to recombine, satisfactory measurements can be made.[32]

Boundary Effects. Because surface recombination can seriously affect interpretation of photodecay curves, it is recommended practice either to calculate the surface contribution or else to use very large samples.[21] In any event, light with wavelengths near the band edge (and thus with relatively low absorption coefficients) should be used to ensure that the carriers are generated in the body of the semiconductor and not all at the surface. For those materials which have a pronounced dependence of τ on the ratio of excess carriers to equilibrium number of majority carriers, high concentrations near the surface will give a longer measured lifetime and may partially counteract the effect of the excess surface recombination. Nevertheless, both circumstances should be avoided if possible.

The use of large samples and sandblasted surfaces is the procedure recommended by the IRE standards.[96] The exact size of sample required is somewhat subjective, and varies with the diffusion coefficient of the minority carriers. Figure 4.3 gives a suggested minimum sample dimension as a function of lifetime to be measured for various diffusion coefficients. For convenience, typical high-resistivity D values are also tabulated.

If the dimensions are not as large as indicated in Fig. 4.3, the bulk lifetime can be obtained by applying a suitable correction factor to the measured value. If the

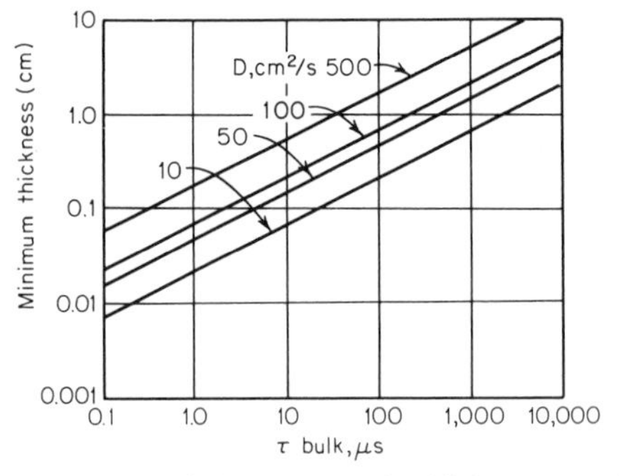

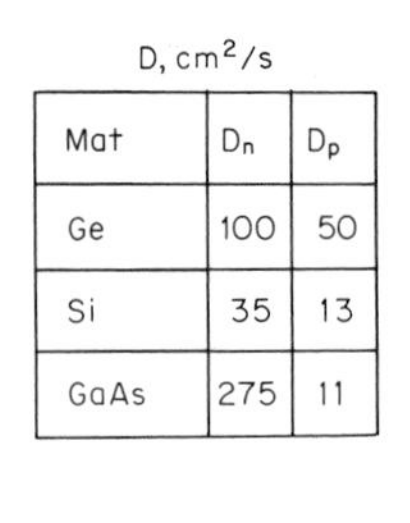

D, cm²/s

Mat	D_n	D_p
Ge	100	50
Si	35	13
GaAs	275	11

Fig. 4.3. Minimum sample thickness required for photodecay lifetime measurement. (*Adapted in part from Stevenson and Keyes.*[21])

sample is rectangular and is sandblasted so that the surface-recombination term is very large,[21]

$$\frac{1}{\tau} = \frac{1}{\tau_{\text{eff}}} - \pi^2 D\left(\frac{1}{A^2} + \frac{1}{B^2} + \frac{1}{C^2}\right) \tag{4.7}$$

where D is an ambipolar diffusion coefficient given by

$$D = \frac{n + p}{(n/D_p) + (p/D_n)} \tag{4.8}$$

and A, B, and C are the rectangular dimensions. For reasonably heavily doped material, D reduces to D_n for p-type and D_p for n-type. In the event the surface recombination is not extremely large, the correction equations are much more complex, and the reader is referred to Refs. 21, 53, 55, and 57. Corrections have also been calculated for circular cross sections uniformly illuminated over the whole surface by either penetrating or nonpenetrating light.[53] For radius a and length l, limiting values are[53]

$$\frac{1}{\tau} = \frac{1}{\tau_{\text{eff}}} - \frac{2s}{a} - \frac{\pi^2 D}{l^2} \qquad \left(s < \frac{D}{a}\right) \tag{4.9}$$

$$\frac{1}{\tau} = \frac{1}{\tau_{\text{eff}}} - D\left[\left(\frac{2.4s}{D + as}\right)^2 + \frac{\pi^2}{l^2}\right] \qquad \left(s > \frac{D}{a}\right) \tag{4.10}$$

For $s = 0$ over the surface the only effect is that of the end contacts, which are still assumed to have $s = \infty$. When l is very large, $\tau = \tau_{\text{eff}}$. When $s \to \infty$, Eq. (4.10) reduces to [21]

$$\frac{1}{\tau} = \frac{1}{\tau_{\text{eff}}} - \pi^2 D\left(\frac{9}{16a^2} + \frac{1}{l^2}\right)$$

and is the right-circular-cylinder counterpart of Eq. (4.7).

The complete solution for the decay is actually a series of exponentials with different time constants. Their sum is not a simple exponential, but the higher-order terms decay rapidly and leave the single exponential as described by Eqs. (4.7), (4.9), and (4.10). Figure 4.4 shows the effect to be expected, and in principle it should be possible to determine both s and τ from a measurement of τ_{eff}.[53] To prevent misinterpretation when using Eqs. (4.7), (4.9), or (4.10), IRE standards recommend

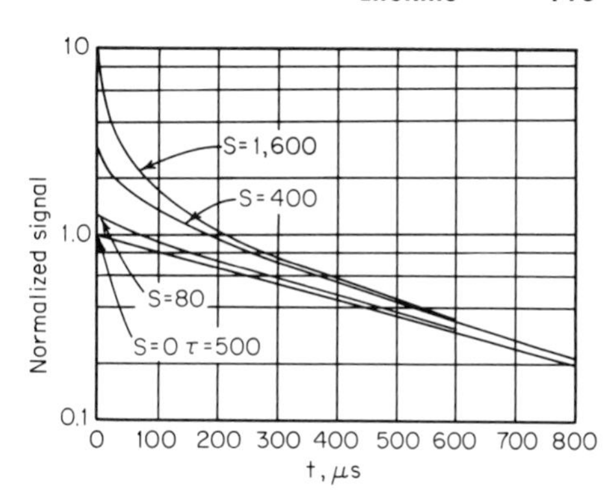

Fig. 4.4. Calculated effect of surface recombinations on photoconductive decay of a cylindrical ingot. The radius was assumed to be 0.5 cm, the length infinite, and $D = 40\ \text{cm}^2/\text{s}$. *(Adapted from McKelvey.*[53]*)*

that no measurement of the decay slope be made until the signal has decayed to less than 60 percent of its peak value if half or less of the width of the sample is exposed to illumination. If more than half of the width is illuminated, no measurement should be made until the signal is less than 25 percent of the peak value.

A rather complex alternate approach has been proposed which is based on the fact that a forward-biased p-n junction should present a low recombination surface. Then thin samples, e.g., slices with a skin of opposite type, could be contacted so that the inner region behaves as a standard photodecay sample and the junctions formed by the skin are forward-biased so that surface recombination is virtually eliminated. For such an arrangement the light must be penetrating enough to generate carriers in the central core.[171] As another alternative, light can enter radially from the edge of a relatively thick slice and the photo current collected by a contact completely covering one side and a centered contact of reduced diameter on the other. The contacts will then only collect carriers which have recombined well into the slice.[199]

If trapping is present, a long tail on the decay curve will be observed. Shining a continuous background light on the sample[12,16] or heating it to about 70°C should remove the effect and allow useful measurements to be made. Germanium usually can be measured with a minimum of trapping difficulties, whereas with silicon, lifetimes an order of magnitude too long may be obtained because of trapping.

4.3 ALTERNATE INJECTION METHODS

Contact Injection. Instead of generating carriers by light as just described, it is possible, particularly with long-lifetime samples, to inject sufficient carriers from the end contacts to change the sample conductance measurably. Injection can then be abruptly stopped and the conductance decay observed exactly as before.[3] If the injection pulse is not long compared with the lifetime, the voltage ΔV will not be exponential, but that can be corrected by plotting time vs. $\log f(\Delta V)$, where $f(\Delta V)$ is given in Ref. 3. By injecting from a point contact and then sampling the spreading resistance as a function of time, the lifetime in very localized areas can be studied.[20]

High-Energy Particles. A pulsed high-energy electron source may be used to generate the carriers. Van de Graaff accelerators can be pulsed with fall times of the order of 5×10^{-9} s and thus can be used to measure lifetimes of the order of 10^{-8} s.[31]

Flash X-Ray. Very short x-ray pulses can be generated and used instead of light. They have the advantage of supplying more carriers than can normally be produced by a high-speed light flash, but the high-voltage pulses required to produce the x-rays (up to 150 kV) make shielding more difficult.

4.4 PHOTOCONDUCTIVITY MODULATION

Sinusoidal light excitation can be used instead of pulses and the phase angle θ between the photoconductivity of the sample and the impinging light interpreted in terms of τ_{eff}. τ_{eff} is found through the expression

$$\tan \theta = \omega \tau_{eff}$$

As before τ_{eff} will depend on the geometry and the relative value of τ, s, and surface-barrier capacitance.[19] Experimentally, the modulated light falls on both the sample and a detector known to introduce negligible phase shift between the incident light and the electrical signal. For relatively low frequencies the signal from the photocell can be processed through a parallel RC network and compared with the signal from the sample. If amplitudes and the RC product are adjusted, a null will occur for $RC = \tau_{eff}$. When the process can be characterized by a single time constant, i.e., sL/D is small, the RC time constant will be independent of frequency. Alternative arrangements have included the use of the signal driving the x and y plates of an oscilloscope to match the RC network phase shift against that of the sample,[52] and the use of a vector voltmeter.[186] Kerr cells have been used to provide modulation frequencies of up to 4 MHz for measuring InSb lifetimes in the 10^{-8} s range.[108] For even higher frequencies, a Michelson interferometer with one mirror oscillating[156] or a modulated laser[186] can be used.

An estimate of τ should be made and the modulation frequency ω chosen so that $\omega\tau \leq 0.3$. For maximum sensitivity the voltage across the sample should be as high as possible without overheating the sample or sweeping out too many carriers. The latter can be checked by gradually increasing the field until τ_{eff} begins to decrease, and then reducing it slightly. Correction curves are available if it is desirable to operate with somewhat higher fields.[49] With no trapping, white light, a thickness w greater than ten times the diffusion length, and a lapped surface,

$$\tau \simeq 1.2\, \tau_{eff} \tag{4.11}$$

The actual numerical relation will vary somewhat depending on the absorption coefficient α, and the choice of 1.2 is a compromise that covers both Si and Ge. Should the thickness be such that it is not large compared with L, the use of correction curves will still allow τ to be found.[89]

One interesting variation of this method allows measurement of radiative recombination lifetime independently of other processes by using a monochrometer tuned to the recombination radiation wavelength between a photodetector and the sample. The phase of the modulation of this signal rather than that from the photoconductivity is then compared with that of the incident light.[186]

4.5 PHOTOCONDUCTIVITY

Instead of following the transient behavior of carriers after they are generated, the increase in steady-state conductivity can be examined. In general, the change in resistance ΔR is given by[46]

$$\Delta R = \rho^2 lq(\mu_n + \mu_p)\frac{\Delta p}{A} \tag{4.12}$$

where ρ is the dark resistivity of the sample, l the length of sample irradiated, and A the sample cross section. When using a constant-current source, the voltage change ΔV will be proportional to ΔR. The number of excess carriers Δp and Δn will depend on lifetime, light intensity and absorption coefficient, surface-recombination velocity, and sample dimensions. The exact expression is somewhat intractable, but provided the sample is thick enough to absorb all the light, it can be expressed as

$$\Delta p = \phi\beta(1 - R)f(\alpha, L, s, D, \text{and sample dimensions}) \tag{4.13}$$

where ϕ is the incident-light intensity, β is the efficiency of converting photons to hole-electron pairs (usually considered 1), R is the reflection coefficient, and L is the carrier diffusion length, given by $L = \sqrt{D\tau}$. By making various assumptions regarding the relative magnitude of the various parameters, considerable simplification can be achieved. The problem is to pick simplifications that are physically realizable and also amenable to experimental interpretation.

Spatial variations in lifetime can in principle be measured by using a relatively small spot of light[136] or an electron beam[153] and scanning it over the surface. Since only a small region is illuminated, the sensitivity is reduced; so careful signal processing is required.

Experimentally none of the ΔV readings actually have to be taken with steady illumination. As long as the frequency is relatively low, the light may be chopped and ac signal processing used. To simplify instrumentation, bridges have been constructed using a photoresistor in one of the balancing arms so that light intensity and current variations are both compensated.[102]

4.6 SURFACE PHOTOVOLTAGE

When light shines on the surface of a semiconductor without an intentional junction, a surface photovoltage will occur which is similar to the voltage developed at a p-n junction. By capacitively coupling to the surface, the voltage can be detected without direct contact and can be used to measure the carrier diffusion length.[42,92] It is a rather unique method in that surface recombination does not influence the measured value of lifetime, and it can be used even if there is extensive majority-carrier and moderate minority-carrier trapping.[181] There are restrictions in the ratio of Δp, Δn to n and p that can be tolerated, but in general, they are less severe than in some of the other methods. The sample thickness must be greater than $\sim 4L$. (If used, for example, on epitaxial layers, if the layer is greater than $4L$ there are no problems; if it is less[192] than $\sim 0.5L$, the substrate diffusion length is measured.) The expression for the equilibrium numbers of carriers is similar to that used for photoconductivity, except that now only those carriers generated within approximately one diffusion length of the surface will contribute to the photovoltage and

$$\Delta p = \frac{\beta\phi(1 - R)}{D/L + s}\frac{\alpha L}{1 + \alpha L} \tag{4.14}$$

In deriving this expression, the value of α was not allowed to go to an extreme, since the surface voltage V_s is measured as a function of α. V_s will be some function of Δp which actually need not be known in order to determine L. That is, $V_s = f(\Delta p)$, or conversely, $\Delta p = F(V_s)$. From this, the light intensity ϕ can be expressed in terms of V_s. Thus

$$\phi = F(V_s)M\left(1 + \frac{1}{\alpha L}\right) \tag{4.15}$$

where
$$M \equiv \frac{s + D/L}{\beta(1 - R)}$$

For a given sample, M should be a constant as long as the wavelength is not changed enough to affect either β or R appreciably. Experimentally, if α is changed by varying the wavelength and ϕ is always adjusted to give the same value of V_s, $F(V_s)$ should also remain constant. Under those conditions a plot of ϕ vs. $1/\alpha$ extrapolated to $\phi = 0$ gives L. That is,

$$\phi = 0 = 1 + \frac{1}{\alpha L} \tag{4.16}$$

The experimental arrangement is shown in Fig. 4.5*a*, and a typical plot in Fig. 4.5*b*. The wave analyzer is used only as a narrow-band detector. Other instrumentation such as a lock-in amplifier can be used instead. Either a monochrometer

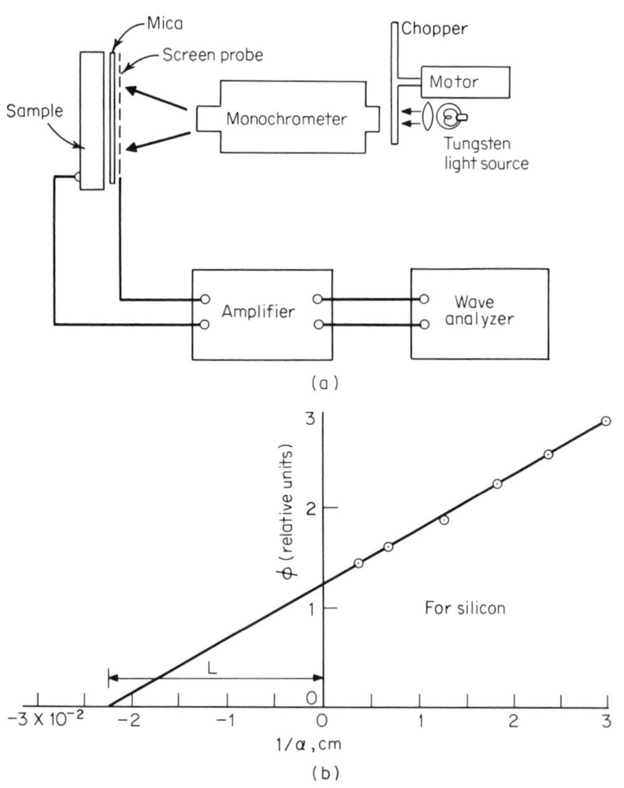

Fig. 4.5. Use of surface photovoltage to determine the diffusion length L from I_0 and α. (*From Joseph Horak, "Minority Carrier Lifetime Measurements on Silicon Material for Use in Electron Irradiation Studies," AF Contract F19628-67-0043, Scientific Report, May 1968.*)

or a series of interference filters can be used for changing the wavelength. Filters have the advantage of providing large-area illumination inexpensively. The (sample-screen) combination must be free of vibration so that no capacitor modulation at the chopping frequency occurs. The pickup-capacitor plate must allow light through it and can be either a fine wire mesh or a semitransparent evaporated-metal film. Some care must be given to the value of α used in the calculations, since for the case of silicon at least, it appears to depend on surface treatment and may vary enough to alter the shape of the ϕ, $1/\alpha$ curve substantially.[192]

4.7 PEM EFFECT

The photoelectromagnetic (PEM) effect is the Hall effect of a photoinduced diffusion current. That is, the light-generated hole-electron pairs will, under the influence of the magnetic field as shown in Fig. 4.6, separate and cause either a short-circuit current to flow or an open-circuit voltage to develop. Both the short-circuit current and the open-circuit voltage are proportional to $\sqrt{L}$ and so can be used to measure L and hence τ.[64] The photoconductance and photovoltage depend on L rather than the square root, and hence as τ decreases, those effects will become immeasurably small more rapidly than the PEM effect. For that reason it was used very early to study short-lifetime semiconductors. As with the other methods discussed, the geometry can be chosen so that the surface recombination is dominant. If it is desired to study high values of s, the PEM effect is very useful.[58]

The simplest theory gives

$$I_{sc}(\text{per unit width}) = \phi\beta qLB(\mu_n + \mu_p) \tag{4.17}$$

where B is the magnetic induction. When the thickness is not many times L, the surface recombination not zero, and the light absorbed at varying depths, the expressions become increasingly complicated. However, if both the zero-magnetic-field photoconductance ΔG and the PEM short-circuit current are measured with the same light intensity, the ratio of ΔG to I_{sc} removes the need to know either ϕ, β, or s. Under these conditions, and assuming $L \ll$ thickness,

$$\tau = \frac{B^2 D(\Delta G)^2}{(I_{sc})^2} \tag{4.18}$$

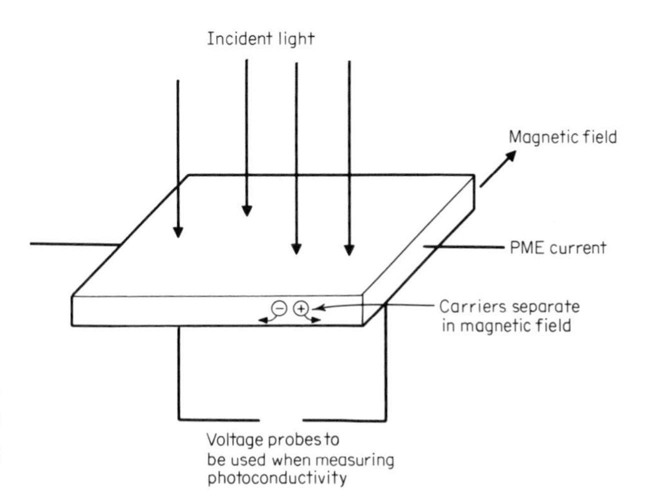

Fig. 4.6. The photoelectromagnetic (also referred to as PME) method for measuring lifetime.

4.8 DIFFUSION-LENGTH MEASUREMENTS

In several of the preceding methods, L was deduced by measuring current or voltage as some parameter such as the depth of light absorption was changed. It is also possible to use geometries such that the separation of the light source and the detector of carriers can be varied at will. For example, carriers can be generated by shining light on the surface and collected some distance away by a suitable biased point-contact electrode[1] or built-in p-n junction if one is available.* The carrier motion is described by the continuity equation, and for the particular geometry used, the equation must be solved for current collected as a function of distance between the collector and point of generation.[4] For a line of light falling on a large expanse (Fig. 4.7), the solution is in terms of a zero-order Hankel function, provided that the geometrical restrictions $5d_0 < r < \frac{1}{4}z$ are followed.[4] (See Fig. 4.7 for a definition of terms.) For large r/L the solution reduces to an exponential, as it does for most other geometries. That is, there will usually be a region that can be approximated by $I = I_0 \exp(-r/L)$. Therefore, if response vs. distance is plotted on semilog paper, the diffusion length is given by the slope of the line.

These measurements are quite insensitive to trapping, but surface recombination can cause errors much as it does for photodecay. If the surface-recombination velocity s is small compared with the diffusion velocity $\sqrt{D/\tau}$, negligible error will occur. Thus, for materials like Ge, which tend to have long lifetimes and low-recombination-velocity surfaces, there is little trouble. Si, however, usually has a lower lifetime and much higher surface recombination. One approach is to calculate curves for various L, s combinations and choose the best fit.[1,103] Other details may be found in Ref. 22. A better way perhaps is to treat the surface to produce either a low or at least a reproducible value of s. Etching is often used for low s, and lapping or sandblasting for very high s. Surface treatments to give heavy surface accumulation should reduce surface recombination and have been suggested for silicon.[28] Inversion layers can cause serious error, since they may give rise to rather large signals far from the collector. These measurements normally give only the diffusion length L, so that to find τ, a determination of the diffusion coefficient is necessary. However, by moving the spot at a uniform velocity great enough to be an appreciable fraction of $\sqrt{D/\tau}$, both D and τ can be estimated from the asymmetry of the response.[7] An alternate approach is to use chopped light and measure both the variation of signal intensity with distance from the source and the phase difference between the light source and the signal.[13]

*The general method does not depend on the mode of detection, and indeed the infrared absorption of the free carriers has been used to make diffusion-length measurements in Ge.[47]

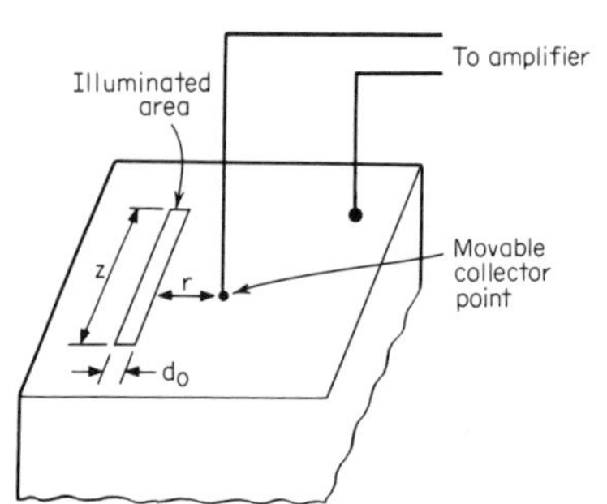

Fig. 4.7. Geometry for diffusion-length measurements.

For materials such as Si, Ge, and GaAs, an ordinary incandescent bulb and conventional optics are satisfactory, but for materials like InSb much longer wavelength and special optics (e.g., reflective) may be necessary. An additional advantage of reflective optics is that the spot size can be determined using visible light and will not change as the wavelength is increased. In order to simplify instrumentation, chopped light and ac amplifiers are ordinarily used. However, unless the chopping frequency ω is such that $\omega^2\tau^2 \ll 1$, the measurement value will not be the lifetime but something between it and a limiting value of $2/\omega$.[13] For lifetimes of 100 μs, chopping frequencies of 100 Hz are appropriate.

An electron beam can also be used to generate the carriers. It has the advantage of providing a very small spot size and allows the measurement of very short diffusion lengths. If done in conjunction with a scanning electron microscope, the area near the junction can first be viewed in the normal way and the exact area of interest chosen.[140]

Microwave detection of the carriers can be used for materials with long diffusion lengths.[75,118] The procedure is complicated and offers no advantage unless there is no way to establish electrical contact to the sample.

By using a p-n junction as the detector and a bevel across it as shown in Fig. 4.8, the distance from source to detector can be varied by moving a light spot (or electron beam[191]) down the incline. For $x_j > 2L$, $\alpha L > 3$, and $\alpha D \gg s$, the short-circuit current is given by

$$I_{sc} = \frac{A \exp(-x_j/L)}{1 + sL/D} \tag{4.19}$$

where A is a constant involving the various optical parameters. By keeping the light intensity and s constant and plotting I_{sc} vs. x_j, L can be determined from [131,142]

$$L = \frac{dx_j}{d(\ln I_{sc})} \tag{4.20}$$

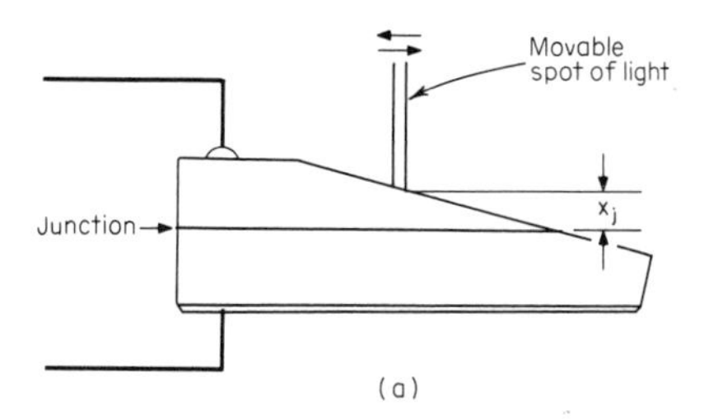

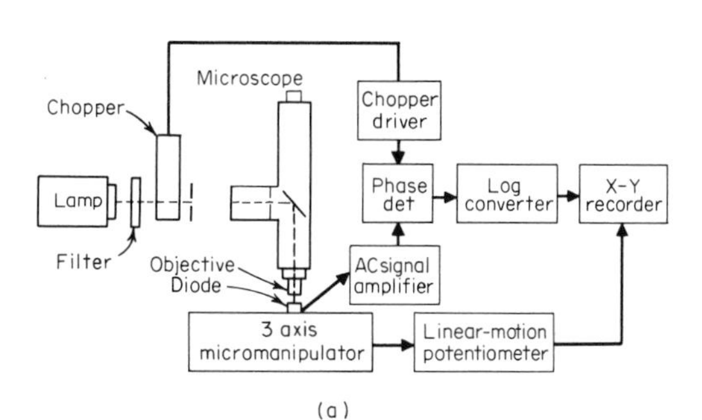

Fig. 4.8. Details of bevel probing. (*Adapted from Ashley and Biard.*[194])

In terms of the distance x along the incline which makes an angle θ with the plane of the junction

$$L = \frac{\sin\theta(dx)}{d(\ln I_{sc})} \tag{4.21}$$

If there are two discrete layers of the same type but of different L's (e.g., n^+, n) above the junction, the response curve will show a break when the light spot crosses the boundary.[142] The equipment necessary is as shown in Fig. 4.8*b*. A metallurgical microscope can be used to provide small spot size* for wavelengths out to about 1 μm and is usually the basis for such equipment. Additionally, a light chopper and phase-sensitive detector or other narrow-band amplifier are necessary in order to provide the required signal-to-noise ratio over the two to three decades of I_{sc} that should be plotted.

The wavelength can be chosen to ensure that α is sufficiently large for Eq. (19) to be valid. In some cases, e.g., GaP, this may require ultraviolet.[197]

4.9 DRIFT

Historically, this is the oldest of the methods discussed.[2] A pulse of minority carriers is injected into a thin rod of material and swept along the rod by an applied field. Some distance down the rod the carriers are collected. From the decrease in amplitude during the time of travel, the lifetime may be calculated. A rectangular pulse originally injected will gradually take a bell distribution and spread more and more as time increases. Thus, pulse amplitude is no longer a measure of the number of surviving carriers. For a very short pulse that area will be proportional to

$$\text{Pulse height } \sqrt{t} \tag{4.22}$$

or

$$\text{Pulse height } \sim t^{-1/2} \exp \frac{-t}{\tau} \tag{4.23}$$

However, for long pulses and the short times used, the $t^{1/2}$ dependence is small and usually neglected in comparison with the exponential.

If there are no traps, the envelope of the detected carrier should be symmetrical. However, trapping causes a tail to occur (Fig. 4.9), much like that observed in photodecay.[16] The collector may also be nonlinear in that the collector current is not always proportional to the minority-carrier concentration over the range of concentrations likely to be encountered. In particular, efficiency is usually degraded at very low levels. Thus, either pulse amplitudes must be rather large, or as has been suggested for silicon, the collector may be continually flooded with light.[12]

Rather than moving the collector point or changing the sweep field in order to

*When L is very short, the size and light distribution in the spot may affect the calculations. For a discussion, see Ref. 200.

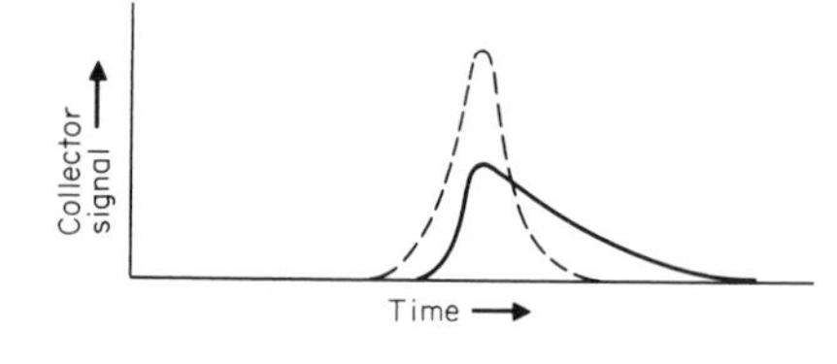

Fig. 4.9. Collector signal shape as a function of time. Dotted line without trapping. Solid line with trapping. (*Adapted from Hornbeck and Haynes.*[16])

vary the time between injection of the carriers and their collection, a collector voltage pulse can be used. That is, the carriers can be allowed to diffuse for a time, and then the drift field applied to sweep them to the collector for detection. A plot of collector response vs. delay time will then vary as $\exp(-t/\tau)$.[14,20]

4.10 MISCELLANEOUS METHODS

Charge-Collection Efficiency. For very high resistivity materials, if carrier pulses are generated by high-energy particles such as alpha particles and the sample is used as a counter, the pulse amplitude is given by

$$A = A_0 \frac{\mu\tau E}{d}\left[1 - \exp\frac{-d}{\mu\tau E}\right] \tag{4.24}$$

where E is the electric field, d the counter (sample) thickness, which must be large compared with the particle range, μ the carrier mobility, and τ the lifetime. A plot of A vs. E will enable the $\mu\tau$ product to be determined so that if μ is independently known, τ is available.[61,149]

Noise. Noise in semiconductors can sometimes be related to the lifetime when the generation-recombination process is the predominant noise generator. In that case

$$\tau = \frac{V_n^2(R + R_L)^4\, n(AL)}{4V^2R^2R_L^2\Delta f} \tag{4.25}$$

where V_n is the noise voltage, R the sample resistance, R_L the series load resistance, n the carrier density, AL the sample volume, V the applied voltage, and Δf the bandwidth used to measure the noise (the center frequency assumed to be $\ll 1/\tau$).[78,106]

Electroluminescence. For wide-bandgap semiconductors the excitation frequency at which the electroluminescence intensity is reduced by one-third is approximately equal to $1/\tau$.[93]

REFERENCES

1. F. S. Goucher, Measurement of Hole Diffusion in n-Type Germanium, *Phys. Rev.*, **81**:475 (1951).
2. J. R. Haynes and W. Shockley, The Mobility and Life of Injected Holes and Electrons in Germanium, *Phys. Rev.*, **81**:835–843 (1951).
3. D. Navon, R. Bray, and H. Y. Fan, Lifetime of Injected Carriers in Germanium, *Proc. IRE*, **40**:1343–1347 (1952).
4. L. B. Valdes, Measurement of Minority Carrier Lifetime in Germanium, *Proc. IRE*, **40**:1420–1423 (1952).
5. L. P. Hunter, E. J. Huibregtse, and R. L. Anderson, Current Carrier Lifetimes Deduced from Hall Coefficient and Resistivity Measurements, *Phys. Rev.*, **91**:1315–1320 (1953).
6. E. M. Pell, Recombination Rate in Germanium by Observation of Pulsed Reverse Characteristic, *Phys. Rev.*, **90**:278–279 (1953).
7. G. Adam, A Flying Light Spot Method for Simultaneous Determination of Lifetime and Mobility of Injected Current Carriers, *Physica*, **20**:1037–1041 (1954).
8. R. H. Kingston, Switching Time in Junction Diodes and Junction Transistors, *Proc. IRE*, **42**:829–834 (1954).
9. B. Lax and S. F. Neustadter, Transient Response of a p-n Junction, *J. Appl. Phys.*, **25**:1148–1154 (1954).

10. A. Many, Measurement of Minority Carrier Lifetime and Contact Injection Ratio on Transistor Materials, *Proc. Phys. Soc.,* **B67**:9–17 (1954).
11. D. G. Avery and D. P. Jenkins, Measurements of Diffusion Length in Indium Antimonide, *J. Electron.,* **1**:145–151 (1955).
12. J. B. Arthur, W. Bardsley, A. F. Gibson, and C. A. Hogarth, On the Measurement of Minority Carrier Lifetime in n-Type Silicon, *Proc. Phys. Soc.,* **B68**:121–129 (1955).
13. D. G. Avery and J. B. Gunn, The Use of a Modulated Light Spot in Semiconductor Measurements, *Proc. Phys. Soc.,* **B68**:918–921 (1955).
14. N. F. Durrant, Measurement of Minority Carrier Lifetimes in Semiconductors, *Proc. Phys. Soc.,* **B68**:562–563 (1955).
15. B. R. Gossick, On the Transient Behavior of Semiconductor Rectifiers, *J. Appl. Phys.,* **26**:1356–1365 (1955).
16. J. A. Hornbeck and J. R. Haynes, Trapping of Minority Carriers in Silicon. I. p-Type Silicon, *Phys. Rev.,* **97**:311–321 (1955).
17. S. R. Lederhandler and L. J. Giacoletto, Measurement of Minority Carrier Lifetime and Surface Effects in Junction Devices, *Proc. IRE,* **43**:477–483 (1955).
18. Sumner Mayburg, dc Photoconductivity Technique for the Determination of Lifetime of Minority Carriers in Silicon Filaments, *Rev. Sci. Instr.,* **26**:616–617 (1955).
19. B. H. Schultz, Analysis of the Decay of Photoconductance in Germanium, *Philips Res. Rept.,* **10**:337–348 (1955).
20. W. G. Spitzer, T. E. Firle, M. Cutler, R. G. Schulman, and M. Becker, Measurement of the Lifetime on Minority Carriers in Germanium, *J. Appl. Phys.,* **26**:414–417 (1955).
21. D. T. Stevenson and R. J. Keyes, Measurement of Carrier Lifetimes in Germanium and Silicon, *J. Appl. Phys.,* **26**:190–195 (1955).
22. W. van Roosebroeck, Injected Current Carrier Transport in a Semi-infinite Semiconductor and the Determination of Lifetimes and Surface Recombination Velocities, *J. Appl. Phys.,* **26**:380–391 (1955).
23. H. L. Armstrong, On Open Circuit Transient Effects in Point Contact Rectifier, *J. Appl. Phys.,* **27**:420–421 (1956).
24. D. M. Evans, Measurements on Alloy-Type Germanium Transistors and Their Relation to Theory, *J. Electron.,* **1**:461–476 (1955).
25. K. D. Glinchuk, E. G. Miseliuk, and E. I. Rashba: Measurement of Rate of Recombination of Carriers by Conductivity Modulation, *Soviet Phys. Tech. Phys.,* **1**:2521–2528 (1956).
26. N. J. Harrick, Lifetime Measurements of Excess Carriers in Semiconductors, *J. Appl. Phys.,* **27**:1439–1442 (1956).
27. H. K. Henisch and J. Zucker, Contactless Method for the Estimation of Resistivity and Lifetime of Semiconductors, *Rev. Sci. Instr.,* **27**:409–410 (1956).
28. C. A. Hogarth, On the Measurement of Minority Carrier Lifetimes in Silicon, *Proc. Phys. Soc.,* **B69**:791–795 (1956).
29. S. G. Kalashnikov, Recombination of Electrons and Holes when Various Types of Traps Are Present, *Soviet Phys. Tech. Phys.,* **1**:237–247 (1956).
30. O. V. Sorokin, Measurement of the Lifetime, Diffusion Coefficient, and Rate of Surface Recombination for Non-equilibrium Current Carriers in a Thin Semiconductor Specimen, *Soviet Phys. Tech. Phys.,* **1**:2390–2396 (1956).
31. G. K. Wertheim and W. M. Augustyniak, Measurement of Short Carrier Lifetimes, *Rev. Sci. Instr.,* **27**:1062–1064 (1956).
32. R. L. Watters and G. W. Ludwig, Measurement of Minority Carrier Lifetime in Silicon, *J. Appl. Phys.,* **27**:489–497 (1956).
33. H. L. Armstrong, Comparator Method for Optical Lifetime Measurements on Semiconductors, *Rev. Sci. Instr.,* **28**:202 (1957).
34. T. M. Buck and F. S. McKim, Measurements on the PME Effect in Germanium, *Phys. Rev.,* **106**:904–909 (1957).

35. M. Byczkowski and J. R. Madigan, Minority Carrier Lifetime in p-n Junction Devices, *J. Appl. Phys.*, **28**:878–881 (1957).
36. D. H. Clarke, Semiconductor Lifetime as a Function of Recombination State Density, *J. Electron. Control,* **3**:375–386 (1957).
37. A. R. Engler and C. J. Kevane, Direct Reading Minority Carrier Lifetime Measuring Apparatus, *Rev. Sci. Instr.*, **28**:548–551 (1957).
38. M. Garbuny, T. P. Vogl, and J. R. Hansen, Method for the Generation of Very Fast Light Pulses, *Rev. Sci. Instr.*, **28**:826–827 (1957).
39. J. C. Henderson and J. R. Tillman, Minority Carrier Storage in Semi-conductor Diodes, *Proc. IEE,* **104B**:318–332 (1957).
40. M. I. Iglitsyn, Iu. A. Kontsevoi, Measuring the Lifetime of Charge Carriers in Semiconductors, *Soviet Phys. Tech. Phys.*, **2**:1306–1315 (1957).
41. M. I. Iglitsyn, Iu. A. Kontsevoi, and V. D. Kudin, Measurement of Lifetime in Monocrystalline Silicon, *Soviet Phys. Tech. Phys.*, **2**:1316–1321 (1957).
42. E. O. Johnson, Measurement of Minority Carrier Lifetimes with the Surface Photovoltage, *J. Appl. Phys.*, **28**:1349–1353 (1957).
43. J. H. Malmberg, MillimicroSecond Duration Light Source, *Rev. Sci. Instr.*, **28**:1027–1029 (1957).
44. O. V. Sorokin, A Method for Measuring the Volume Lifetime and the Diffusion Coefficient of Current Carriers by Measuring the Resistance of a Semiconductor in a Magnetic Field, *Soviet Phys. Tech. Phys.*, **2**:2572–2574 (1957).
45. L. J. van der Pauw, Analysis of the Photoconductance in Silicon, *Philips Res. Rept.*, **12**:364–376 (1957).
46. H. M. Bath and M. Cutler, Measurement of Surface Recombination Velocity in Silicon by Steady-State Photoconductance, *J. Phys. Chem. Solids,* **5**:171–179 (1958).
47. Lennart Huldt and Torsten Staflin, Infrared Absorption of Photogenerated Free Carriers in Germanium, *Phys. Rev. Lett.*, **1**:236–237 (1958).
48. G. Bemski, Recombination in Semiconductors, *Proc. IRE,* **46**:990–1004 (1958).
49. N. B. Grover and E. Harnik, Sweep-out Effects in the Phase Shift Method of Carrier Lifetime Measurements, *Proc. Phys, Soc. London,* **72**:267–269 (1958).
50. R. Gremmelmaier, Irradiation on p-n Junctions with Gamma Rays: A Method for Measuring Diffusion Length, *Proc. IRE,* **46**:1045–1049 (1958).
51. F. J. Hyde, Some Measurements on Commercial Transistors and Their Relation to Theory, *Proc. IEE,* **105B**:45–52 (1958).
52. E. Harnik, A. Many, and N. B. Grover, Phase Shift Method of Carrier Lifetime Measurements in Semiconductors, *Rev. Sci. Instr.*, **29**:889–891 (1958).
53. J. P. McKelvey, Volume and Surface Recombination of Injected Carriers in Cylindrical Semiconductor Ingots, *IRE Trans. Electron Devices,* **ED-5**:260–264 (1958).
54. A. Many and R. Bray, Lifetime of Excess Carriers in Semiconductors, in "Progress in Semiconductors," vol. 3, pp. 117–151, John Wiley & Sons, Inc., New York, 1958.
55. B. K. Ridley, Measurement of Lifetime by the Photoconductive Decay Method, *J. Electron. Control,* **5**:549–558 (1958).
56. D. J. Sandiford, Temperature Dependence of Carrier Lifetime in Silicon, *Proc. Phys. Soc.*, **71**:1002–1006 (1958).
57. A. C. Sim, A Note on Surface Recombination Velocity and Photoconductive Decays, *J. Electron. Control,* **5**:251–255 (1958).
58. W. Van Roosbroeck and T. M. Buck, Methods for Determining Volume Lifetimes and Surface Recombination Velocities, in F. J. Biondi (ed.), "Transistor Technology," vol. 3, pp. 309–324, D. Van Nostrand Company, Inc., Princeton, N.J. 1958.
59. R. N. Zitter, Role of Traps in the Photoelectromagnetic and Photoconductive Effects, *Phys. Rev.*, **112**:852–855 (1958).
60. M. B. Das and A. R. Boothroyd, On the Determination of the Minority Carrier Lifetime in the Base Region of Transistors, *J. Electron. Control,* **7**:534–539 (1959).

61. W. D. Davis, Lifetimes and Capture Cross Sections in Gold-doped Silicon, *Phys. Rev.*, **114**:1006–1008 (1959).
62. L. Huldt, Optical Method for Determining Carrier Lifetimes in Semiconductors, *Phys. Rev. Lett.*, **2**:3–5 (1959).
63. Quentin A. Kerns, Frederick A. Kirsten, and Gerald C. Cox, Generator of Nanosecond Light Pulses for Phototube Testing, *Rev. Sci. Instr.*, **30**:31–36 (1959).
64. T. S. Moss, "Optical Properties of Semiconductors," Academic Press, Inc., New York, 1959.
65. G. Porter and E. R. Wooding, Simple Light Source of about 10 mμs Duration, *J. Sci. Instr.*, **36**:147 (1959).
66. A. P. Ramsa, H. Jacobs, and F. A. Brand, Microwave Techniques in Measurement of Lifetime in Germanium, *J. Appl. Phys.*, **30**:1054–1060 (1959).
67. A. C. Sim, A Note on the Use of Filters in Photoconductivity Decay Measurements, *Proc. IEE,* **106B**(Suppl. 15):308–310 (1959).
68. M. Waldner, Measurement of Minority Carrier Diffusion Length and Lifetime by Means of the Photovoltaic Effect, *Proc. IRE,* **47**:1004–1005 (1959).
69. H. A. Atwater, Microwave Measurement of Semiconductor Carrier Lifetimes, *J. Appl. Phys.*, **31**:938–939 (1960).
70. J. S. Blakemore and K. C. Nomura, Influence of Transverse Modes on Photoconductive Decay in Filaments, *J. Appl. Phys.*, **31**:753–761 (1960).
71. Z. Bodo', Determination of Minority Carrier Lifetime in Semiconductors, in "Solid-State Physics Electronics Telecommunications," pp. 194–198, Academic Press, Inc., New York, 1960.
72. B. Ya Moizhes, Elimination of Edge Effect during the Measurement of the Photomagnetic EMF in Semiconductor, *Soviet Phys. Solid State,* **1**:1135–1138 (1960).
73. A. E. Bakanowski and J. H. Forster, Electrical Properties of Gold-doped Diffused Silicon Computer Diodes, *Bell System Tech. J.*, **39**:87–104 (1960).
74. C. Hilsum and B. Holeman, "Carrier Lifetime in GaAs," *Proceedings of the International Conference on Semiconductor Physics,* Prague, 1960.
75. H. Jacobs, A. P. Ramsa, and F. A. Brand, Further Considerations of Bulk Lifetime Measurement with a Microwave Electrodeless Technique, *Proc. IRE,* **48**:229–233 (1960).
76. R. D. Larrabee, Measurement of Semiconductor Properties through Microwave Absorption, *RCA Rev.*, **21**:124–129 (1960).
77. E. G. Landsberg, The Influence of Geometrical Dimensions of Samples in Measurements of Diffusion Length Using the Photomagnetic Method, *Soviet Phys. Solid State,* **2**:777–781 (1960).
78. S. Okazaki and H. Oki, Measurement of Lifetime in Germanium from Noise, *Phys. Rev.,* **118**:1023–1024 (1960).
79. C. G. Peattie, W. J. Odom, and E. D. Jackson, An Ultra-short Lifetime Apparatus, *Proc. IEE,* **106B**(Suppl. 15):303–307 (1960).
80. B. K. Ridley, The Effect of an Electric Field on the Decay of Excess Carriers in Semiconductors, *Proc. Phys. Soc.*, **75**:157–161 (1960).
81. A. C. Sim, The Dark-Spot Method for Measuring Diffusion Constant and Length of Excess Charge Carriers in Semiconductors, *Proc. IEE,* **106B**(Suppl. 15):311–328 (1960).
82. V. K. Subashiev, V. A. Petrusevich, and G. B. Dubrovskii, Determination of Recombination Constants from the Spectral Response of Photoconductivity, *Soviet Phys. Solid State,* **2**:925–926 (1960).
83. V. K. Subashiev, Determination of Recombination Parameters from Spectral Response of a Photocell with a p-n Junction, *Soviet Phys. Solid State,* **2**:187–193 (1960).
84. V. K. Subashiev, G. B. Dubrovskii, and V. A. Petrusevich, Determination of the Recombination Constant and the Depth of a p-n Junction from the Spectral Characteristics of a Photocell, *Soviet Phys. Solid State,* **2**:1781–1782 (1961).

85. W. van Roosbroeck, Current Carrier Transport and Photoconductivity in Semiconductors with Trapping, *Phys. Rev.,* **119**:636–652 (1960).
86. W. van Roosbroeck, Theory of Current Carrier Transport and Photoconductivity in Semiconductors with Trapping, *Bell System Tech. J.,* **39**:515–614 (1960).
87. R. L. Williams, High Frequency Light Modulation, *J. Sci. Instr.,* **37**:205–208 (1960).
88. H. A. Atwater, Microwave Determination of Semiconductor Carrier Lifetimes, *Proc. IRE,* **49**:1440–1441 (1961).
89. S. V. Bogdanov and V. D. Kopylovskii, Concerning the Utilization of the Phase Method for Measuring Lifetime of Nonequilibrium Charge Carriers in Semiconductors, *Soviet Phys. Solid State,* **3**:674–679 (1961).
90. F. M. Berkovskii, S. M. Ryvkin, and N. B. Strokan, Effect of Trapping Levels on the Decay of Current through p-n Junctions, *Soviet Phys. Solid State,* **3**:169–172 (1961).
91. G. B. Dubrovskii, Determination of the Basic Parameters of Photovoltaic Cells with p-n Junctions with Respect to Their Spectral Characteristics, *Ind. Lab.,* **27**:1236–1239 (1962).
92. A. M. Goodman, A Method for the Measurement of Short Minority Carrier Diffusion Lengths in Semiconductors, *J. Appl. Phys.,* **32**:2550–2552 (1961).
93. G. G. Harman and R. L. Raybold, Measurement of Minority Carrier Lifetime in SiC by a Novel Electroluminescent Method, *J. Appl. Phys.,* **32**:1168–1169 (1961).
94. B. Holeman and C. Hilsum, Photoconductivity in Semi-insulating Gallium Arsenide, *J. Phys. Chem. Solids,* **22**:19–24 (1961).
95. F. J. Hyde and H. J. Roberts, Minority Carrier Lifetime in the Base Regions of Transistors, *J. Electron. Control,* **11**:35–46 (1961).
96. Institute of Radio Engineers, Measurement of Minority-Carrier Lifetime in Germanium and Silicon by the Method of Photoconductive Decay, *Proc. IRE,* **49**:1292–1299 (1961).
97. H. Jacobs, Microwave Determination of Semiconductor Carrier Lifetimes—Author's Reply, *Proc. IRE,* **49**:1441–1442 (1961).
98. W. H. Ko, The Reverse Transient Behavior of Semiconductor Junction Diodes, *IRE Trans. Electron Devices,* **ED-8**:123–131 (1961).
99. J. J. Loferski and J. J. Wysocki, Spectral Response of Photovoltaic Cells, *RCA Rev.,* **22**:38–56 (1961).
100. E. G. Landsberg, Photomagnetic Method of Measuring the Lifetime Electrons and Holes, *Ind. Lab.,* **27**:1226–1230 (1962).
101. M. A. Melehy and W. Shockley, Response of a p-n Junction to a Linearly Decreasing Current, *IRE Trans. Electron Devices,* **ED-8**:135–139 (1961).
102. A. A. Meier, E. A. Soldatov, and V. P. Sushkov, Some Methods of Measuring Lifetime of Nonequilibrium Charge Carriers Based on the Modulation of Photoconductivity, *Ind. Lab.,* **27**:1223–1226 (1962).
103. A. A. Meier, Measurement of the Local Diffusion Length and the Surface Recombination Rate with Respect to Two Points on Semiconductor Bars, *Ind. Lab.,* **27**:1230–1235 (1962).
104. V. A. Petrusevich, Photoelectric Methods of Studying Semiconductors, *Ind. Lab.,* **27**:1217–1223 (1962).
105. Yu. I. Ravich, Determining the Characteristic Parameters of Minority Carriers in Semiconductors from Measurements of Photoconductivity and the Photomagnetic Effect, *Soviet Phys. Solid State,* **3**:1162–1168 (1961).
106. T. P. Vogl, J. R. Hansen, and M. Garbuny, Photoconductive Time Constants and Related Characteristics of p-Type Gold Doped Germanium, *J. Opt. Soc. Am.,* **51**:70–75 (1961).
107. I. R. Weingarten and M. Rothberg, Radio-Frequency Carrier and Capacitive Coupling Procedures for Resistivity and Lifetime Measurements on Silicon, *J. Electrochem. Soc.,* **108**:167–171 (1961).
108. E. I. Aderovich, A. N. Gubkin, and B. D. Kopylovskii, Measurement of Short Lifetimes from Phase Characteristics of the Voltage Transmission Coefficient in a Circuit Containing a p-n Junction, *Soviet Phys. Solid State,* **4**:1359–1365 (1963).

109. A. R. Beattie, and R. W. Cunningham, Large-Signal Photomagnetoelectric Effect, *Phys. Rev.,* **125**:533–540 (1962).
110. S. C. Choo and E. L. Heasell, Technique for the Measurement of Short Carrier Lifetimes, *Rev. Sci. Instr.,* **33**:1331–1334 (1962).
111. S. Deb and B. R. Nag, Measurement of Lifetime of Carriers in Semiconductors through Microwave Reflection, *J. Appl. Phys.,* **33**:1604 (1962).
112. N. R. Howard, A Simple Pulse Generator for Semiconductor Diode Lifetime Measurements, *J. Sci. Instr.,* **39**:647–648 (1962).
113. S. M. Krakauer, Harmonic Generation, Rectification, and Lifetime Evaluation with the Step Recovery Diode, *Proc. IRE,* **50**:1665–1676 (1962).
114. J. Nishizawa, Y. Yamoguchi, N. Shoji, and Y. Tominaga, Application of Siemens Method to Measure the Resistivity and the Lifetime of Small Slices of Silicon, in "Ultrapurification of Semiconductor Materials," pp. 636–644, The Macmillan Company, New York, 1962.
115. W. Rosenzweig, Diffusion Length Measurement by Means of Ionizing Radiation, *Bell System Tech. J.,* **41**:1573–1588 (1962).
116. G. Suryan and G. Susila, Lifetime of Minority Carriers in Semiconductors: Experimental Methods, *J. Sci. Ind. Res.,* **21A**:235–246 (1962).
117. J. A. W. van der Does de Bye, Transient Recombination in Silicon Carbide, *Philips Res. Rept.,* **17**:419–430 (1962).
118. J. N. Bhar, Microwave Techniques in the Study of Semiconductors, *Proc. IEEE,* **51**:1623–1631 (1963).
119. L. W. Davies, The Use of P-L-N Structures in Investigation of Transient Recombination from High Injection Levels in Semiconductors, *Proc. IEEE,* **51**:1637–1642 (1963).
120. J. J. Dlubac, S. C. Lee, and M. A. Melehy, On the Measurement of Minority Carrier Lifetime in p-n Junctions, *Proc. IEEE,* **51**:501–502 (1963).
121. H. Jacobs, F. A. Brand, J. D. Meindl, S. Weitz, R. Benjamin, and D. A. Holmes, New Microwave Techniques in Surface Recombination and Lifetime Studies, *Proc. IEEE,* **51**:581–592 (1963).
122. V. K. Subashiev, Determination of Semiconductor Parameters from the Photomagnetic Effect and Photoconductivity, *Soviet Phys. Solid State,* **5**:405–406 (1963).
123. E. I. Adirovich, S. P. Lunezhev, and Z. P. Mironenkova, Measurement of 10^{-10} Sec Relaxation Times in p-n Junctions by the Phase Compensation Method, *Soviet Phys. Doklady,* **9**:999–1002 (1965).
124. A. R. Beattie and R. W. Cunningham, Large-Signal Photoconductive Effect, *J. Appl. Phys.,* **35**:353–359 (1964).
125. I. A. Baev, Measurements of the Minority Carrier Lifetime and Diffusion Coefficient in InSb by the Moving-Light-Spot Method, *Soviet Phys. Solid State,* **6**:217–221 (1964).
126. A. N. Daw and N. K. D. Chowdhury, On the Variation of the Lifetime of Minority Carriers in a Junction with the Level of Injection, *Solid State Electron.,* **7**:799–809 (1964).
127. H. J. Fink, Reverse Recovery Time Measurements of Epitaxial Silicon p-n Junctions at Low Temperatures, *Solid State Electron.,* **7**:823–831 (1964).
128. A. S. Grove and C. T. Sah, Simple Analytical Approximations to the Switching Times in Narrow-Base Diodes, *Solid State Electron.,* **7**:107–110 (1964).
129. A. Hoffman and K. Schuster, An Experimental Determination of the Carrier Lifetime in P-I-N Diodes from the Stored Carrier Charge, *Solid State Electron.,* **7**:717–724 (1964).
130. E. A. Ivanova, D. N. Nasledov, and B. V. Tsarenkov, Carrier Lifetime in the Space-Charge Layer of GaAs p-n Junctions, *Soviet Phys. Solid State,* **6**:604–606 (1964).
131. G. Jungk and H. Menniger, On the Measurement of the Diffusion Length of Minority Carriers in Semiconductors, *Phys. Stat. Solidi,* **5**:169–174 (1964).
132. H. J. Kuno, Analysis and Characterization of p-n Junction Diode Switching, *IEEE Trans. Electron Devices,* **ED-11**:8–14 (1964).
133. N. G. Nilsson, Determination of Carrier Lifetime, Diffusion Length, and Surface

Recombination Velocity in Semiconductors from Photo-excited Infrared Absorption, *Solid State Electron.*, **7**:455–463 (1964).
134. G. Susila, A Method for the Determination of Short Lifetime of Carriers in a Photoconductor from the Transient Photoresponse, *Indian J. Pure Appl. Phys.*, **2**:44–47 (1964).
135. V. A. J. van Lint, and J. W. Harrity, Carrier Lifetime Studies in High-Energy Irradiated Silicon, in "Radiation Damage in Semiconductors," pp. 417–423, Academic Press, Inc., New York, 1964.
136. I. A. Baev and E. G. Valyashko, An Investigation of the Distribution of Inhomogeneous Regions in Semiconductors, *Soviet Phys. Solid State,* **7**:2093–2099 (1966).
137. O. L. Curtis, Jr., and R. C. Wickenhiser, An Efficient Flash X-Ray for Minority Carrier Lifetime Measurements and Other Research Purposes, *Proc. IEEE,* **53**:1224–1225 (1965).
138. D. A. Evans and P. T. Landsberg, Theory of the Decay of Excess Carrier Concentrations in Semiconductors, *J. Phys. Chem. Solids,* **26**:315–327 (1965).
139. E. Earlywine, L. P. Hilton, and D. Townley, Measuring the Properties of Semiconductor Grade Materials, *Semicond. Prod. Solid State Tech.,* **8**:17–30 (1965).
140. H. Higuchi and H. Tamura, Measurement of the Lifetime of Minority Carriers in Semiconductors with a Scanning Electron Microscope, *Japan. J. Appl. Phys.,* **4**:316–317 (1965).
141. I. M. Muratov, On the Problem of Determining the Lifetime of Unbalanced Current Carriers, *Radio Eng. Electron. Phys.,* **10**:132–135 (1965).
142. M. H. Norwood and W. G. Hutchinson, Diffusion Lengths in Epitaxial Gallium Arsenide by Angle Lapped Junction Method, *Solid State Electron.,* **8**:807–811 (1965).
143. G. I. Suleiman, N. F. Kovtonyuk, and D. T. Kokorev, Automatic Apparatus for Recording the Lifetime Distribution of Non-Equilibrium Charge Carriers in Semiconductors, *Instr. Exptl. Tech.,* **8**:203–205 (1965).
144. K. Schuster, Determination of the Lifetime from the Stored Carrier Charge in Diffused p-s-n Rectifiers, *Solid State Electron.,* **8**:427–430 (1965).
145. K. Santha Kumari, and B. A. P. Tantry, Measurement of Lifetime of Minority Carriers in Junction Transistors, *Indian J. Pure Appl. Phys.,* **3**:380–384 (1965).
146. A. A. Vol'fson, S. M. Gorodetskii, and V. K. Subashiev, A Study of Photoconductivity in Heavily Doped p-Silicon, *Soviet Phys. Solid State,* **7**:53–57 (1965).
147. J. Vilms and W. E. Spicer, Quantum Efficiency and Radiative Lifetimes in p-Type Gallium Arsenide, *J. Appl. Phys.,* **36**:2815–2821 (1965).
148. P. W. C. Chilvers and K. Foster, Measurement of Lifetime and Transition Time in Charge Storage Diodes, *Electron. Lett.,* **2**:108–110 (1966).
149. J. A. Coleman and L. J. Swartzendruber, Effective Charge Carrier Lifetime in Silicon P-I-N Junction Detectors, *IEEE Trans. Nuclear Sci.,* **NS-13**:240–244 (1966).
150. L. A. Davidson, Simple Expression for Storage Time for Arbitrary Base Diode, *Solid State Electron.,* **9**:1145–1147 (1966).
151. M. Fournier and C. Lemyre, Measurement of the Effective Minority Carrier Lifetime in the Floating Region of a P-N-P-N Device, *IEEE Trans. Electron Devices,* **ED-13**:511–512 (1966).
152. C. Jund and R. Poirier, Carrier Concentration and Minority Carrier Lifetime Measurement in Semiconductor Epitaxial Layers by the MOS Capacitance Method, *Solid State Electron.,* **9**:315–319 (1966).
153. C. Munakata and H. Todokoro, A Method of Measuring Lifetime for Minority Carriers Induced by an Electron Beam in Germanium, *Japan. J. Appl. Phys.,* **5**:249 (1966).
154. C. Munakata, Measurement of Minority Carrier Lifetime with a Non-ohmic Contact and an Electron Beam, *Microelectron. Reliability,* **5**:267–270 (1966).
155. D. B. B. Owen and E. L. G. Wilkinson, Junction Recovery and Trapping in Silicon p^+n Junctions, *Intern. J. Electron.,* **20**:21–29 (1966).
156. C. M. Penchina and H. Levinstein, Measurement of Lifetimes in Photoconductors by Means of Optical Beating, *Infrared Phys.,* **6**:173–182 (1966).

157. B. G. Streetman, Carrier Recombination and Trapping Effects in Transient Photoconductive Decay Measurements, *J. Appl. Phys.*, **37**:3137–3144 (1966).
158. K. Santha Kumari, and B. A. P. Tantry, Rise and Fall Times of Transistors in Switching Operation with Finite Impedance, *Solid State Electron.*, **9**:730–733 (1966).
159. H. Y. Tada, Theoretical Analysis of Transient Solar Cell Response and Minority Carrier Lifetime, *J. Appl. Phys.*, **37**:4595–4596 (1966).
160. G. V. Zakhvatikin, O. V. Karagioz, V. N. Solomatin, and E. N. Gerasimov, Device for Measuring the Lifetime of Carriers in High-Resistance Silicon by Means of the Phase Method, *Instr. Exptl. Tech.*, **10**:444–448 (1966).
161. A. Bilotti, Measurement of the Effective Carrier Lifetime by a Distortion Technique, *Solid State Electron.*, **10**:445–448 (1967).
162. A. V. Dudenkova and V. V. Nikitin, Utilization of a Laser to Measure the Lifetime of Excess Carriers in Gallium Arsenide Single Crystals, *Soviet Phys. Solid State,* **9**:664–665 (1967).
163. F. P. Heiman, On the Determination of Minority Carrier Lifetime from the Transient Response of an MOS Capacitor, *IEEE Trans. Electron Devices,* **ED-14**:781–784 (1967).
164. S. R. Hofstein, Minority Carrier Lifetime Determination from Inversion Layer Transient Response, *IEEE Trans. Electron Devices,* **ED-14**:785–786 (1967).
165. R. M. Lichtenstein and H. J. Willard, Jr., Simple Contactless Method for Measuring Decay Time of Photoconductivity in Silicon, *Rev. Sci. Instr.,* **38**:133–134 (1967).
166. D. J. McNeill, Measurement of Minority Carrier Lifetime in Semiconductor Junction Diodes, *Am. J. Phys.*, **35**:282–283 (1967).
167. N. Miyamoto and J. Nishizawa, Contactless Measurement of Resistivity of Slices of Semiconductor Materials, *Rev. Sci. Instr.,* **38**:360–367 (1967).
168. D. B. Wittry and D. F. Kyser, Measurement of Diffusion Lengths in Direct-Gap Semiconductors by Electron-Beam Excitation, *J. Appl. Phys.*, **38**:375–382 (1967).
169. P. G. Wilson, Recombination in Silicon p-π-n Diodes, *Solid State Electron.*, **10**:145–154 (1967).
170. David K. Lynn, Charles S. Meyer, and Douglas J. Hamilton (eds.), "Analysis of Integrated Circuits," McGraw-Hill Book Company, New York, 1967.
171. R. J. Bassett and C. A. Hogarth, A Novel Method for the Direct Determination of the Small Signal Bulk Minority Carrier Lifetime in the Middle Region of a Semiconductor Sandwich, *Intern. J. Electron.*, **24**:301–316 (1968).
172. David P. Kennedy, Reverse Transient Characteristics of a p-n Junction Diode due to Minority Carrier Storage, *IRE Trans. Electron Devices,* **ED-9**:174–182 (1962).
173. J. L. Moll, S. Krakaver, and R. Shen, p-n Junction Charge-Storage Diodes, *Proc. IRE,* **50**:43–53 (1962).
174. Charles F. Fell and Wendell A. Johnson, Effective Lifetime, a Figure of Merit for Nanosecond Diodes, *IEEE Trans. Electron Devices,* **Ed-11**:306–308 (1964).
175. M. Zerbst, Relaxationseffekte an Halbleiter-Isolator-Grenzflächen, *Z. Angew. Phys.*, **22**:30–33 (1966).
176. Horst Preier, Different Mechanisms Affecting the Inversion Layer Transient Response, *IEEE Trans. Electron Devices,* **ED-15**:990–997 (1968).
177. R. J. Bassett, Observation on a Method of Determining the Carrier Lifetime in p^+-v-n^+ Diodes, *Solid State Electron.,* **12**:385–391 (1969).
178. Max Brousseau and Roland Schuttler, Use of Microwave Techniques for Measuring Carrier Lifetime and Mobility in Semiconductors, *Solid State Electron.,* **12**:417–423 (1969).
179. S. C. Choo and A. C. Sanderson, Bulk Trapping Effect on Carrier Diffusion Length as Determined by the Surface Photovoltage Method: Theory, *Solid State Electron.,* **13**:609–617 (1970).
180. P. Tománek, Measuring the Lifetime of Minority Carriers in MIS Structures, *Solid State Electron.,* **12**:301–303 (1969).

181. S. C. Choo and R. G. Mazur, Open Circuit Voltage Decay Behavior of Junction Devices, *Solid State Electron.,* **13:**553–564 (1970).
182. J. S. T. Huang, Bulk Lifetime Determination Using an MOS Capacitor, *Proc. IEEE,* **58:**1849–1850 (1970).
183. R. B. Renbeck and L. P. Hunter, Measurement of the Effective Carrier Lifetime by a Distortion Technique, *Solid State Electron.,* **13:**394–395 (1970).
184. D. K. Schroder and H. C. Nathanson, On the Separation of Bulk and Surface Components of Lifetime Using the Pulsed MOS Capacitor, *Solid State Electron.,* **13:**557–582 (1970).
185. Raymond H. Dean and Charles J. Nuese, A Refined Step-Recovery Technique for Measuring Minority Carrier Lifetimes and Related Parameters in Asymmetric p-n Junction Diodes, *IEEE Trans. Electron. Devices,* **ED-18:**151–158 (1971).
186. C. J. Hwang, An Improved Phase Shift Technique for Measuring Short Carrier Lifetime in Semiconductors, *Rev. Sci. Instr.,* **42:**1084–1086 (1971).
187. J. L. Lindström, Flash X-ray Irradiation of p-n Junctions: A Method to Measure Minority Carrier Lifetimes, Diffusion Constants and Generation Constants, *Solid State Electron.,* **14:**827–833 (1971).
188. R. Van Overstraeten, G. Declerck, and R. Mertens, Measurement of the Carrier Lifetime by an Impedance Technique, *Solid State Electron.,* **ED-14:**289–294 (1971).
189. W. Murray Bullis, National Bureau of Standards, *Tech. Note* 754, 1972.
190. P. U. Calzalari and S. Graffi, A Theoretical Investigation on the Generation Current in Silicon p-n Junctions under Reverse Bias, *Solid State Electron.,* **15:**1003–1011 (1972).
191. W. H. Hackett, Jr., Electron-Beam Excited Minority-Carrier Diffusion Profiles in Semiconductors, *J. Appl. Phys.,* **43:**1649–1654 (1972).
192. W. E. Phillips, Interpretation of Steady-State Surface Photovoltage Measurements in Epitaxial Semiconductor Layers, *Solid State Electron.,* **15:**1097–1102 (1972).
193. B. L. Smith and M. Abbott, Minority Carrier Diffusion Length in Liquid Epitaxial GaP, *Solid State Electron.,* **15:**361–370 (1972).
194. K. L. Ashley and James R. Biard, Optical Microprobe Response of GaAs, *IEEE Trans. Electron. Devices,* **ED-14:**429–432 (1967).
195. R. A. Logan and A. G. Chynoweth, Charge Multiplication in GaP p-n Junctions, *J. Appl. Phys.,* **33:**1649–1654 (1962).
196. J. L. Moll and S. A. Hamilton, Physical Modeling of the Step Recovery Diode for Pulse and Harmonic Generation Circuits, *Proc. IEEE,* **57:**1250–1257 (1969).
197. C. J. Hwang, S. E. Haszeko, and A. A. Bergh, UV Microprobe Technique for Measurement of Minority Carrier Diffusion Length in GaP p-n Junction Material, *J. Appl. Phys.,* **42:**5117–5119 (1971).
198. Jayaraman and C. H. Lee, Observation of Two-Photon Conductivity in GaAs with Nanosecond and Picosecond Light Pulses, *Appl. Phys. Lett.,* **20:**392–395 (1972).
199. Kathleen A. Carroll and Karl J. Casper, Separation of Surface and Bulk Minority Carrier Lifetimes in Silicon, *Rev. Sci. Instr.,* **45:**576–579 (1974).
200. Gaberiel Lengyel, Effect of Spot Size on the Determination of the Diffusion Length of Minority Carriers in p-n Junctions Using Scanned Light or Electron Beam Techniques, *Solid State Electron.,* **17:**510–511 (1974).

5

Mobility, Hall, and Type Measurements

5.1 MOBILITY

The carrier drift velocity v (much less than the actual carrier velocity, since the carrier traces out a much more lengthy, nearly random path) is proportional to the field $\mathbf{E}$ and is described by

$$\mathbf{v}_{\text{drift}} = \mu_d \mathbf{E} \tag{5.1}$$

where μ_d is the carrier drift mobility.[1] Alternately, μ may be defined and measured from

$$\mathbf{J} = \mu_c nq \mathbf{E} \tag{5.2}$$

where n is the carrier density. In this case, μ_c is referred to as the conductivity mobility and in principle should be the same as μ_d. The resistivity is given by $1/\mu_c nq$, so that a resistivity measurement coupled with a value of n will give the mobility for majority carriers.

The most common method of determining n, and the one most applicable over a wide range of materials and temperatures, is through the Hall coefficient* R, and as a matter of convenience, yet another mobility, the Hall mobility μ_H, is defined through

$$\mu_H = |R|\sigma \tag{5.3}$$

If the very simplest theory is assumed, μ_H is equal to $(3\pi/8)\mu_c$, but in fact, because of the complex behavior of most semiconductors, there may be appreciably more variation between the two.

Since μ relates two vectors, it is a second-rank tensor and in all but cubic crystals will be direction-dependent. Even then, if the current flow is restricted to very thin sheets, as, for example, in inversion layers, the mobility becomes anisotropic;[2] so measurements must comprehend this. For bulk majority carriers, Eqs. (5.2) and (5.3) are normally used. The mobility of inversion layers can be determined by measuring the characteristics of an MOS transistor which uses the inversion layer as its channel.[2]

*R_H is sometimes used instead of R to avoid confusion with resistance R. In this chapter resistance will be designated differently.

Such measurements must of necessity examine only inversion layers which produce working devices, but since those are the layers of most practical interest, such a restriction is not serious.

For minority carriers, the classical measurement of Haynes and Shockley[3] shown in Fig. 5.1 determines μ directly. The procedure is as follows. An electric field is established along a length of the semiconductor. A point-contact emitter is pulsed to inject minority carriers into the bar at point A. When the extra packet of carriers reach the collector at point B, they can be extracted and will show as a pulse on the oscilloscope. The mobility is given by

$$\mu_D = \frac{d}{Et} = \frac{d}{(V/L)t} \qquad \text{cm}^2/(\text{V-s}) \tag{5.4}$$

where V is the applied sweep voltage, t the time between pulse injection and its appearance at the collector, d the emitter-collector separation, and L the length of the bar.

As the arrival time increases, the pulse amplitude decreases, and the pulse width increases. The decrease in amplitude is primarily because of carrier recombination and can be used to deduce lifetime. The increase in width occurs because the carriers do not remain closely bunched but diffuse out from the high-concentration central region.[4] Note that if the emitter-collector spacing is too great, the carriers will all have recombined before reaching the collector. For Ge, whose lifetime may be in the millisecond range, spacings of up to 1 cm can be conveniently used. Materials with lower lifetimes and mobilities require commensurately lesser spacings. There are a number of experimental variations which may afford advantages in special circumstances. For example, instead of using the emitter contact of Fig. 5.1, partially injecting end contacts[5] or light flashes[6] can be used for carrier injection and the sweep field can be pulsed.[4,5]

Possible sources of error are conductivity modulation caused by the carrier injection, trapping, distortion of the field because of imperfect contacts, and injection level. By making measurements at several injected pulse heights and extrapolating to zero amplitude, the first can be minimized. When trapping is present, the value determined will be less than that actually displayed by an individual carrier during the time it is untrapped.[8,9] Shining a steady light on the surface may fill the traps and minimize their effect.[10] The greater the collector-emitter separation to area ratio and the greater the end contact–to–collector and end contact–to–emitter ratio, the

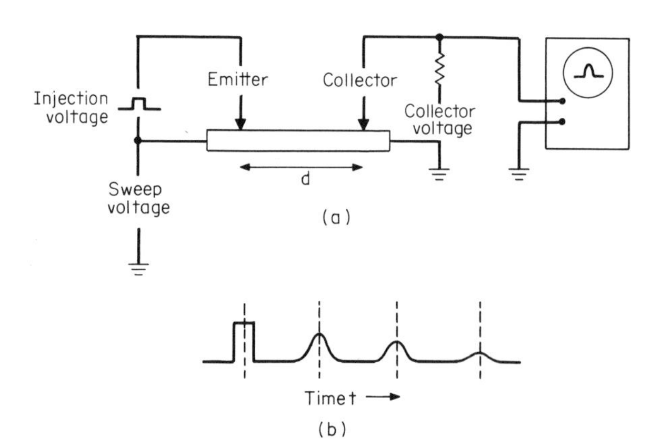

Fig. 5.1. Geometry for direct measurement of mobility (Haynes-Shockley experiment).

smaller the chance for field distortion to cause errors. If the injected carriers have a concentration which is an appreciable fraction of the majority-carrier concentration, the pulse moves at a rate determined by the ambipolar mobility[11] μ_a [see Eq. (4.8)].

There are also various device-related measurements that can be used to deduce mobility. Examples are transistor-base transit time (if diffusion-controlled), and the behavior of pulsed PIN diodes or particle detectors.[12,13] Unlike other methods, this one allows the mobility of both holes and electrons to be determined in the same sample. The method is also applicable to mobility measurements under high-field conditions, since until the carrier pulse is injected, the reverse-biased diode is dissipating very little power. The carrier pulse may be initiated by a burst of relatively low energy electrons (so they will have limited range in the semiconductor) or by a flash of high-intensity light.[14] Examples of its use include measurements of mobilities in Se and CdS.[15]

5.2 HALL EFFECT[1,16,17,ASTM F 76]

Hall measurements are widely used in the initial characterization of semiconductors to measure carrier concentration and/or carrier mobility. As a given semiconductor becomes more widely used and better understood, simpler techniques such as four-point resistivity measurements are usually substituted.

Unlike some of the measurements discussed, the Hall effect has been known for nearly 100 years, and has been used and written about ever since. When a magnetic field is applied at right angles to current flow, an electric field E_H is generated which is mutually perpendicular to the current and the magnetic field, and is directly proportional to the product of the current density and the magnetic induction. Thus

$$E_H = \frac{RIB}{A} \qquad V_H = \frac{RIB}{w} \tag{5.5}$$

where R is the Hall coefficient, I the current through the sample, A the sample cross section, w the thickness, and B the magnetic induction.* Figure 5.2 shows the geometry, sign convention, etc.

*There are actually several related galvanomagnetic effects, but the Hall effect is the one most used. Further, they are all tensors and may depend on crystal orientation. There may also be voltages generated for conditions other than E_R, H, and I mutually orthogonal. Sometimes these are also referred to as Hall voltages. One example is the planar Hall voltage, which is observed when the magnetic field lies in the plane $ABCD$ of Fig. 5.2.

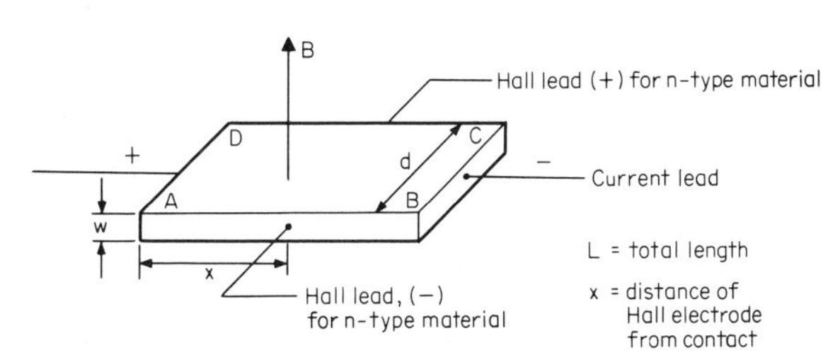

Fig. 5.2. Sign and dimension terminology for Hall bars.

Abbreviated Interpretation. Very simple theory predicts that R as defined in Eq. (5.5) is given by

$$R = \frac{-1}{nq}, \frac{1}{pq} \tag{5.6}$$

where q is the electronic charge and n or p the density of carriers. Thus the carrier type as well as concentration can also be determined from the Hall coefficient, since if the sign convention of Fig. 5.2 is followed, R is negative for n-type and positive for p-type. More elaborate theory predicts that

$$R = \frac{-r_n}{nq}, \frac{r_p}{pq} \tag{5.7}$$

where r is a constant, usually between 0.5 and 1.5, that depends on the specific details of conduction in a given material. For spherical energy bands,

$$r_n = r_p = \frac{3\pi}{8}$$

By making measurements at high enough magnetic field, r can be reduced to 1, regardless of the conduction mechanism.* In any event, the error introduced by determining n directly with no prior knowledge of r is not gross. This fact and the relative simplicity of Hall measurements have accounted for its widespread usage in the study of semiconductor materials.

For the case of mixed conduction (i.e., an appreciable number of both holes and electrons), the Hall coefficient has a contribution from each. For the small-field case

$$R = \frac{(1/q)(p - b^2 n)}{(bn + p)^2} \tag{5.8}$$

where b is the ratio of electron-to-hole conduction mobility. Examination of Eq. (5.8) shows that for n-type material, as the p/n ratio increases from zero, R becomes progressively smaller than would be predicted by supposing that only n-type carriers were present. For p-type material, as the n/p ratio increases, R will be zero when

$$b^2 n = p \tag{5.9}$$

and for $b^2 n > p$, R will change from + to − assuming $b > 1$, which it usually is.

From this behavior it is clear that a single measurement of R, without some background knowledge of the region in which the measurements were made, is very risky. Ordinarily, one should examine the material as a function of temperature. If the material is intrinsic, R will decrease as the temperature increases, as shown by curve a in Fig. 5.3. Its value is obtained by setting $n = p = n_i$ in Eq. (5.8) and is given by

$$R_i = \frac{(1/n_i q)(1 - b)}{1 + b} \tag{5.10}$$

If the material is extrinsic, the R will be independent of temperature, as shown by curve b of Fig. 5.3. The behavior of the transition from extrinsic to intrinsic is

*There may be an oscillation of the magnetic susceptibility of the material as a function of magnetic field. This is the de Haas–van Alphen effect and will lead to similar oscillations of the Hall voltage, magnetoresistance, etc. It is observed, however, only at temperatures below perhaps 10°K.

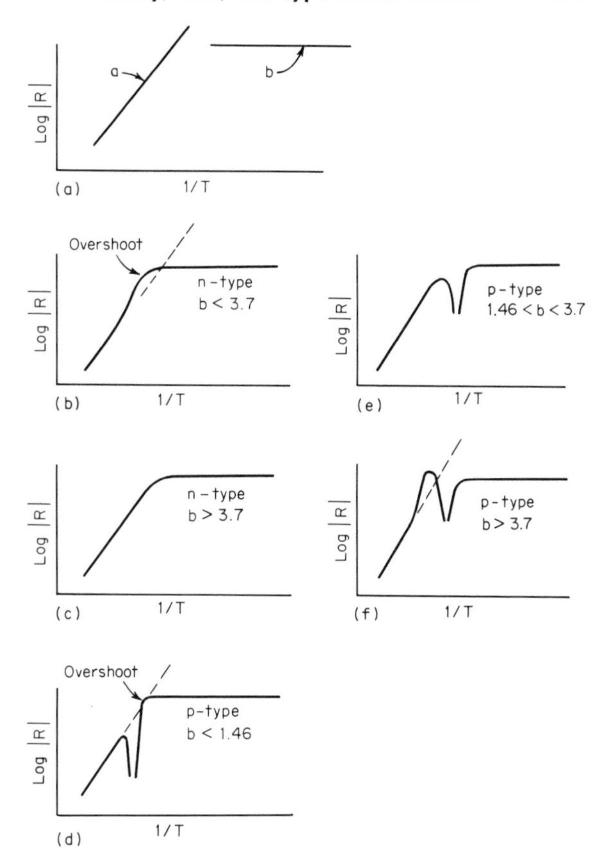

Fig. 5.3. Behavior of R in the transition region from extrinsic to intrinsic. (*Adapted from Putley*[16] *and Hunter.*[18])

illustrated in Fig. 5.3*b* to *f*[18]. If b is a function of temperature, it is possible to have Eq. (5.10) satisfied for more than one n/p ratio, in which case multiple $R = 0$ points are possible (tellurium shows two zero points). Another way to display the overshoot, as well as other relations, is to use the "Dunlap ellipse," which is obtained by combining Eq. (5.8) with the comparable expressions for conductivity and plotting it along with $R = \mu_H/\sigma$ on μ_H, $1/\sigma$ coordinates.[19] Because of the overshoot, an extrapolation or calculation of the intrinsic slope must be done only after the curve is extended far enough for the effect to become negligible.

One possible source of error in the use of Eq. (5.10) arises because of the finite and sometimes quite long carrier-diffusion length in semiconductors.[25] Equation (5.8) and those following were derived assuming zero lifetime. The carrier lifetime enters into the expression because the equilibrium carrier concentration is upset when the sample is in the magnetic field. For $p \gg n$ or the reverse there is little error regardless of the lifetime. When p is comparable with n, there is a difference of a factor of 2. If the ratio of the drift mobilities is different from the Hall mobility ratio, the difference may be somewhat more.[21] For cases intermediate between zero and infinite lifetime, if

$$S > (b + 1)\,\mu_p{}^2 B E_l$$

where S is the surface recombination velocity, μ_p the hole mobility, and E_l the longitudinal electric field, little error will occur.[22] Typically for Si and Ge, S need only be greater than a few hundred centimeters per second. Thus a surface which has had light sandblasting or grinding will generally be acceptable.

As Hall measurements in extrinsic specimens are extended to lower temperatures, the carriers from the doping impurities will begin to freeze out and R will no longer be temperature-independent. The number of carriers in that temperature region is given by[23]

$$p = \frac{K_A + N_D}{2}\left\{\left[1 + \frac{4K_A(N_A - N_D)}{(K_A + N_D)^2}\right]^{1/2} - 1\right\} \tag{5.11}$$

where N_A is the number of acceptor impurities and K_A is the equilibrium constant. If r is temperature-independent (which it may not be) R vs. $1/T$ can be converted directly to a $\log p$ or n vs. $1/T$ plot, which can then be interpreted in light of Eq. (5.11). If the excess impurity is predominantly of one atomic species, the slope of the curve will allow the activation energy and the degree of compensation to be determined.[16,23-26]

Calculations and Consistent Units. For either the meter-kilogram-second (mks) or centimeter-gram-second (cgs) system,

$$V_H = \frac{RBI}{w}$$

However, in the more commonly used practical units (Table 5.1),

Table 5.1. Units and Calculations

MKS SYSTEM	
Magnetic induction	B = tesla = 1 volt-s/meter^2
Electric field	E = volts/meter
Current	I = amperes
Current density	J = amperes/meter^2
Hall coefficient	R = meters^3/coulomb
PRACTICAL CGS UNITS	
Magnetic induction	B = gauss
Electric field	E = volts/cm
Current	I = amperes
Current density	J = amperes/cm^2
Hall coefficient	R = cm^3/coulomb
Carrier mobility	μ = cm^2/volt-s
Carrier diffusion coefficient	D = cm^2/s
ELECTROMAGNETIC CGS UNITS	
Magnetic induction	B = gauss
Electric field	E = abvolts/cm
Current	I = abamps
Current density	J = abamps/cm^2
CONVERSION FACTORS	
1 tesla	= 10^4 gauss = 1 weber/meter^2
1 volt	= 10^8 abvolts
1 amp	= 0.1 abamps
1 volt-amp-sec	= 1 watt-s = 1 joule = 10^7 ergs
k	= 1.38×10^{-16} erg/°K
q	= 1.6×10^{-19} coulombs/electron

$$V_H = \frac{10^8 RBI}{w}$$

where V_H is in volts, R in cubic centimeters per coulomb, B in gauss, I in amperes, and the thickness w in centimeters.

In calculations involving the effect of surface recombination, in practical units,

$$S > 10^{-8}\mu^2 BE_l(b + 1)$$

where B is in gauss, μ in square centimeters per volt-second, E_l in volts per centimeter, and S in centimeters per second.

The Hall angle is given by

$$\theta_H = \tan^{-1}\left(10^{-8}\frac{\mathrm{RB}}{\rho}\right)$$

where R is in cubic centimeters per coulomb, B in gauss, and ρ in ohm-centimeters.

There are a number of spurious voltages which will be included in the value read at the Hall terminals, but most of them can be eliminated by making a series of readings with the various combinations of current and magnetic field. In particular, with

$$\begin{array}{ll} +B, +I & V_{\text{meas}} \equiv V_1 = V_H + V_E + V_N + V_{RL} + V_M + V_T \\ -B, +I & V_{\text{meas}} \equiv V_2 = -V_H - V_E - V_N - V_{RL} + V_M + V_T \\ +B, -I & V_{\text{meas}} \equiv V_3 = -V_H - V_E + V_N + V_{RL} - V_M + V_T \\ -B, -I & V_{\text{meas}} \equiv V_4 = V_H + V_E - V_N - V_{RL} - V_M + V_T \end{array} \quad (5.12)$$

where V_H is the true Hall voltage, V_E the Ettingshausen voltage, V_N the Nernst voltage, V_{RL} the Righi-Leduc voltage, V_M the voltage generated because the Hall probes are not exactly electrically opposite each other, and V_T any thermoelectric voltage generated because of an externally imposed thermal gradient between the probes. By combining them,

$$\frac{V_1 + V_4 - V_2 - V_3}{4} = V_H + V_E \quad (5.13)$$

This is the established method of making dc Hall measurements and is the one recommended in ASTM F 76. As can be seen from Eq. (5.13) it does not separate out the Ettingshausen voltage, but that voltage is ordinarily quite small and can be neglected.

If alternating current of frequency ω is used,[27,28]

$$\begin{aligned} V_H &= \frac{RBI_0}{w} \sin \omega t \\ V_E &= A_1 \exp\left(\frac{-\omega}{\omega_0}\right) \sin \omega t \\ V_M &= A_2 \sin \omega t \\ &V_N, V_R, V_T \text{ are all dc} \end{aligned} \quad (5.14)$$

A_i are constants, and ω_0 is a time constant which depends on the properties of the material being measured. If the frequency is high enough, V_E can be reduced to as small a value as desired. By this means $V_H + V_M$ can be separated from all the terms without the time-consuming current and field reversals necessary to obtain the results described in Eq. (5.13). V_M is independent of the field except for the magnetoresistance effect changing the IR drop.

For the case of constant current and alternating field (an unlikely experimental choice),

$$V_H = \frac{RIB_0}{w} \sin \omega t$$

$$V_E, V_N, V_{RL} \sim \sin \omega t \tag{5.15}$$

$$V_M \sim \omega^2$$

In addition, because of the alternating magnetic field, an induced voltage V_I, proportional to dB/dt (i.e., $\omega \sin \omega t$), will be observed unless extreme care is taken in arranging leads, or unless a variable bucking coil is used.[97]

Should both alternating field and current be used, there will be a dc component of Hall voltage as well as one which occurs at double the frequency. Except for V_E, the others will all be of frequency ω. For different frequencies and specifically for the current frequency ω_1 greater than the field frequency ω_2 (usually the case since it is difficult to produce high-frequency alternating magnetic fields of the required magnitude),

$$V_H = \frac{R}{2w} I_0 B_0[\sin(\omega_1 - \omega_2)t + \sin(\omega_1 + \omega_2)t] \tag{5.16}$$

All the other voltages occur at different frequencies from V_H except for V_E, and it is attenuated by a factor dependent on the ratio of the ω's. If the Hall voltage is balanced against the voltage developed across a variable resistor z which has the sample current flowing through it, $R = zw/B$ and I need not be accurately measured. If the voltage between two contacts along the length of the sample is also measured in order to determine resistivity, and if both that measurement and the Hall-voltage measurement are made with the same sample current,

$$\mu_H \equiv \frac{R}{\rho} = \frac{V_H d_1}{V_\rho B d_1} \tag{5.17}$$

where d_1 is the separation of the resistivity contacts and V_ρ is the voltage measured. Again, the current need not actually be measured, other than to ensure that the longitudinal electric field is within bounds.

Experimental Procedures Based on ASTM F 76.*,† In making resistivity and Hall-effect measurements, spurious results can arise from a number of sources.

1. Photoconductive and photovoltaic effects can seriously influence the observed resistivity, particularly with nearly intrinsic material. Measurements should be made in a dark chamber unless experience shows that the material is insensitive to ambient illumination.
2. Minority-carrier injection can occur because of the electric field in the specimen. With material possessing high minority-carrier lifetime and high resistivity, such injection can result in a lowering of the resistivity for a distance of several millimeters along the bar. Carrier injection can be detected by repeating the measurements at lower applied voltages. In the absence of injection no increase in resistivity should be observed.

*This includes the more salient features of F 76. The reader may wish to consult the complete procedure, however.

3. Semiconductors have a significant temperature coefficient of resistivity. Thus the temperature of the specimen should be known and the current used should be small to avoid resistive heating. Resistive heating can be detected by a change in readings as a function of time starting immediately after the current is applied.
4. Spurious currents can be introduced in the testing circuit when the equipment is located near high-frequency generators.
5. High contact resistances may lead to spurious results.
6. Surface leakage can be a serious problem when measurements are made in the high-resistivity range. Surface effects can often be observed as a difference in measured value of resistivity or Hall coefficient when the surface condition of the specimen is changed.
7. In measuring high-resistivity samples, particular attention should be paid to possible leakage paths in other parts of the circuit such as switches, connectors, wires, and cables, which may shunt some of the current around the sample. Since high values of lead capacitance may lengthen the time required for making measurements on high-resistivity samples, connecting cable should be as short as practicable.
8. Inhomogeneities of the specimen impurity concentration or of the magnetic flux will cause the measurements to be inaccurate.
9. It is essential that in the case of parallelepiped or bridge-type specimens, measurements must be made on side contacts far enough removed from the end contacts that shorting effects can be neglected.
10. Thermomagnetic effects, with the exception of the Ettingshausen effect, and effects due to misalignment of the side contacts (in parallelepiped or bridge-type specimens) can be eliminated by suitable averaging of the measured Hall emfs. In general, the error due to the Ettingshausen effect is small and can be neglected, particularly if the sample is in good thermal contact with its surroundings.
11. For materials which are anistropic in Hall coefficient such as p-type germanium and n-type silicon, Hall measurements are affected by the orientation of current and magnetic field with respect to the crystal axes.
12. Spurious emfs which may occur in the measuring circuit, e.g., thermal emfs, can be detected by measuring the emf across the potential leads with no current flowing or with the potential leads shorted at the sample position.

Most measurements required in calculating the Hall effect should be measured to ± 1 percent. Flux uniformity should also be ± 1 percent over the sample area. Dewars inserted into the magnetic field should not alter it more than ± 1 percent. Current in the specimen should be restricted so that the associated electric field is less than 1 V/cm.

Samples should be single-crystal. They should be mounted in the Hall fixturing so that mechanical stress does not occur, either from clamping or from differential expansion if measurements are to be made at temperatures different from room. If the contact arrangement allows measurement of Hall voltage at more than one position on the sample, readings should agree within 10 percent. Otherwise inhomogeneity is a problem and the sample should be discarded.

Effect of Inhomogeneous Sample. The measurements defined by Eq. (5.5) are predicated on a sample with uniform properties. Actually, there are numerous cases in which it is desirable to make measurements on samples whose properties are inhomogeneous. These may be divided into four categories: those which are uniform in two dimensions, such as diffused layers;[29-33] those which are composed of a matrix of one property and inclusions of another;[34] those in which the material to be measured varies in a random manner so the best that can be done is to measure an average value;[35] and those which have radial gradients (e.g., coring in slices).[98,99] Specific directions may be found in the references indicated.

Effect of Sample Size, Shape, and Electrode Placement. Equation (5.5) assumed a rectangular sample with uniform end contacts, and the Hall electrodes several sample widths away from either contact. If the latter condition is not met, the contacts will partially short out the Hall voltage. In the event that only short, stubby samples are available, corrections[36-39] can be applied. However, the smaller the L/d ratio, the greater the sensitivity to errors in determining the sample dimensions, and the smaller the Hall voltage measured.

In the simple theory the Hall contacts are assumed to be infinitely small so that they do not distort the current flow. However, unless the Hall angle ($\tan^{-1} RH/\rho$) is very large, the error is still relatively small. Experimentally, contacting can be by sharp tips or very small (e.g., 1 mil) alloyed-wire contacts. More often, though, ears on the sample are used (Fig. 5.4). The ears serve two purposes. First, they allow a large area to be used for contacting without severely distorting the lines of current flow in the sample. Second, a contact made directly to the side of the bar will in general be noisier than one using an ear.[46,47] The multiplicity of contacts allows resistivity to be simply measured on the sample [required if the mobility is to be calculated by Eq. (5.6)] and affords redundancy in case one contact is broken.

If it is inconvenient to cut out rectangular bars, thin samples of arbitrary shape such as platelets or slices can be used if contacts are made at four places around the periphery as shown in Fig. 5.5.[40,41] If V_S/I_S is measured with and without a magnetic field,

$$R = \frac{(\Delta V/I)d}{B} \tag{5.18}$$

where ΔV is V_S measured with the field minus V_S measured without the field, and

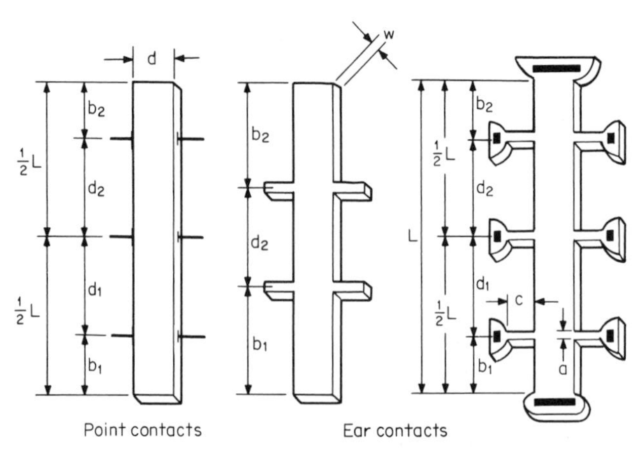

Fig. 5.4. Common Hall-bar configurations with ASTM F 76 suggested dimensions.

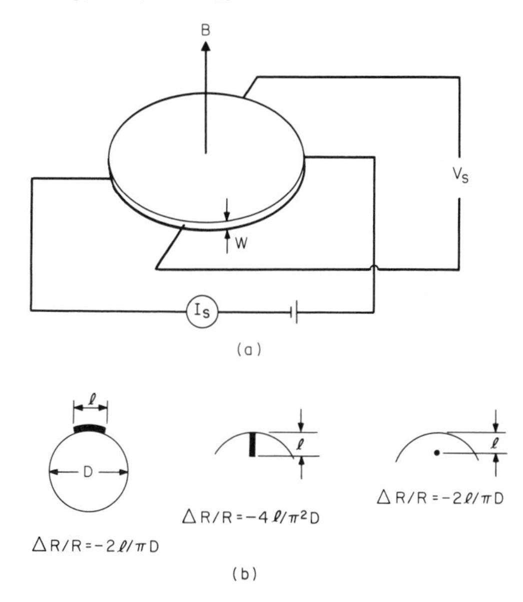

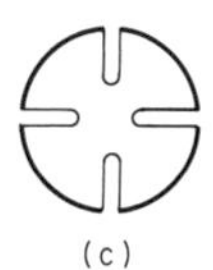

Fig. 5.5. Use of an arbitrarily shaped sample with randomly placed contacts for determining the Hall coefficient. (*b*) The error in the Hall coefficient if the probes are finite sized or misplaced. (*van der Pauw.*[41])

I is held constant for both measurements. The resistivity must also be measured on the same sample. The details of this procedure are discussed in Chap. 3.

A square array of contacts can be made inside but close to the perimeter of thin lamellae. Correction factors are available for rectangles, half-planes, quarter-planes, full, half-, and quarter-circles. Such an arrangement allows four-point Hall measurements much in the manner of four-point resistivity measurement. Such probes can be spaced very closely together and thus can be used when only a limited area is available.*[42-44] As the probe spacing to sample lateral dimensions become small (i.e., a sheet of infinite expanse), the method fails, but as a possible alternative, a hole can be cut in the sheet, and the probes placed around it or else the probes can be placed close to an edge.[43] The general concept can also be extended to make test structures on slices being used for device manufacture.

It is not necessary to have the Hall contacts exactly opposite each other, since a reversal of magnetic field will cancel out any initial unbalanced voltage. However, if the unbalanced voltage is much greater than the Hall voltage, error can be introduced, since V_H will be the small difference between two large numbers. Further, if electronic instrumentation is used, large unbalances may cause amplifier saturation. Several prebalancing schemes are available. One Hall contact may be used, along with a virtual contact generated by an external voltage divider. This method reduces the Hall voltage by a factor of 2 and may generate additional noise.

*Remember, however, that the theory is developed on the assumption that the probe spacing is large compared with the thickness of the sample.

Two contacts on one side plus a potentiometer between them may be used, but there will be some distortion of the current lines because of current flow between the two contacts. Current can be fed into corners of the sample, and the relative amount adjusted to provide balance at the Hall electrodes.[48]

Sample Preparation. Because of the complex shape of Hall bars they are most conveniently cut with an ultrasonic machine, although a fine-nozzle sandblaster or spark erosion can also be used. If the sample is very thin, it may be etched to shape. For epitaxial and diffused layers isolated from their substrates by p-n junctions, etching defined by standard photolithography techniques down through the junction to the substrate will suffice.[49] For built-in test structures in integrated-circuit slices p-n-junction sidewall isolation can be used. p-n-junction isolation places restrictions on the amount of current that can be passed through the sample without forward-biasing the junction and reducing isolation,[50] but fortunately Hall measurements are relatively insensitive to such leakage.

If the material has high carrier mobility and a long lifetime, the surface should be treated (e.g., sandblasting) to increase the surface-recombination velocity. Surface conduction or inversion layers can also radically change the measured value of the Hall coefficient. Materials such as high-resistivity silicon are particularly susceptible to surface conditions, and a series of treatments in boiling water[52] can change the calculated mobility by factors of 10.

Equipment. Fields in the few kilogauss range for good sensitive and laboratory-quality magnets are commonly used. The requirements for magnetic-field homogeneity are not severe; so no special care in pole-piece design such as is common in nuclear magnetic resonance (NMR) work is necessary. If extraordinarily inhomogeneous fields are used, corrections can be calculated.[53]

The Hall voltage can be measured by a simple potentiometer* and galvanometer, and the current flow by an ammeter. For high-resistivity samples the simple equipment fails for many reasons. Sensitivity will be reduced because of the low-impedance detector, and noise becomes more pronounced. Stray signals also cause more difficulty, and surface leakage across the sample holder may cause erroneous readings.[54] All leads need to be carefully shielded, and the fact that it is hard to ground one end of the sample and one terminal of the detector simultaneously makes external noise a real problem. By using only one Hall probe, the difficulty can be avoided, but then the sensitivity is reduced by a factor of 2.

The introduction first of vibrating-reed electrometers and then of high-input-impedance stabilized differential amplifiers has greatly simplified high-impedance instrumentation.[16,55,56] If a single-ended electrometer and both Hall probes are used, potentiometers can still be used to allow a low-impedance path to ground for both the sample and the detector.[57] More desirable is the use of a differential amplifier.[55] It will allow both sample and voltmeter to be grounded and will reject a large percentage (depending on the common-mode rejection ratio) of any extraneous noise which is simultaneously induced in both leads. Some additional improvement in signal to noise can often be obtained by integrating the signal over a several-minute interval.[55] If long leads are involved, the time constants associated with the lead capacitance–sample resistance may require inordinately long wait times between

*In Hall's first reported experiment [*Am. J. Math.*, **2**:287 (1879)] he used only a galvanometer to observe the polarity of the voltage and did not measure its value. It is also of interest to note that in order to get the small value of w needed for a detectable Hall voltage, he used gold foil, since that was one of the few techniques then available for obtaining very thin metallic layers.

measurements. To minimize this time, the shielded-lead effective capacitance can be reduced by means of unity-gain high-input-impedance amplifiers driving the shields.

When dc equipment is used over an extended temperature range, data acquisition becomes a very time-consuming process. As a consequence, a variety of automatic methods have been devised.[16,58-60] The temperature may be allowed to drift slowly to room temperature, with data being recorded as it drifts, or low-inertia holders can be combined with a temperature controller. Control of the sequencing can be by timer, local minicomputer, or off-site time-shared computer.

There are many variations of ac, dc field measurements dating back to at least 1912.[61] The earliest ones were made in order to study the dependence of the Hall effect on frequency. Many of the later ones, however, have used alternating current in order to simplify measurements. Alternating current will eliminate or attenuate most of the spurious voltages. Amplification is always easier with alternating than with direct current, and the noise figure of amplifiers generally improves as the frequency increases from zero to a few hundred hertz. Circuit variations have differed primarily in the method of bucking out the unbalance due to probe misplacement, and in the method of providing a narrow-band detector.[62-67] The sign of the Hall voltage is not directly available for those systems using simple voltmeters for detection, but an oscilloscope may be used to compare the phase of V_H with the sample current, or the balance network can be deliberately unbalanced. Depending on the sign of V_H, the sum of that voltage and V_H will be either greater or less than the voltage at balance (V_H alone). If a phase-sensitive detector rather than a voltmeter is used, it will give the phase directly.

Probably the most popular ac system is that of Dauphinee and Mooser.[68] It uses a dc field and low-frequency alternating current for the sample. It depends on a series of synchronized choppers for simultaneously converting the sample current to alternating current and rectifying the Hall voltage. There may be problems with excessive stray capacitance or rectifying contacts, since either of these will change the balance point.[69] When the choppers open and close, additional noise is generated, but if a low-noise amplifier is used and then gated so that it is inoperative during the time switching occurs, the overall signal-to-noise ratio can be improved.[70]

The use of an ac field affords the same advantage of yielding an ac signal as does alternating current. In addition, the probe-misplacement voltage will have two terms, dc and 2ω, which can easily be separated from the Hall frequency ω. The major difficulty is in providing the alternating magnetic field. One way of accomplishing it is to rotate the sample within a fixed field.[71] This necessitates slip rings and their attendant noise, but a low-noise preamplifier mounted to rotate with the sample can minimize the noise.[72] Another, which obviates slip rings, is to rotate a permanent magnet about the sample.[71]

Should both alternating current of frequency ω_1 and alternating field frequency ω_2 be used, all the spurious voltages of Eq. (5.12) can be eliminated. Therefore, for those cases where the Hall voltage is small and where noise is a problem, the double-frequency method may be advantageous.[28,48,73-79] The frequencies used are generally quite low. The magnetic field, for example, may range from a fraction of a hertz to 60 Hz, and the current from 10 to 100 Hz. If the same frequency for both field and sample current is used, many of the advantages disappear, but the Hall voltage will have a dc term. When V_H as a function of frequency is to be studied and no good ac voltmeters are at hand, the dc term is extraordinarily useful; this is probably why the alternating-field–alternating-current method was used as early as 1901.[80]

The various alternating-current and -field methods are unfortunately not a complete panacea, since such things as vibration and amplifier distortion can still produce voltages of the same frequency as the Hall voltage. As an example, using double alternating current, if the field is driven by 60 Hz and the sample-current power amplifier is powered by 60 Hz, any amplifier 60-Hz intermodulation will produce sum and difference sample currents which will in turn cause a V_M contribution of the same frequency as V_H.[81] If the sample and heads are not very carefully anchored, a variety of voltages can be generated which will be indistinguishable from V_H in cases where dc field or fractional-hertz square-wave fields are used.[82] The user of the ac variations should therefore examine his particular circuitry very carefully before assuming that all competing voltages have been eliminated.

Special Equipment. For specialized application, still more variations have been devised. For example, if there is difficulty in making ohmic contacts, the sample can be cut in a ring shape and rotated in the field.[71] This will induce current directly into the sample so that only voltage contacts are required. When slip rings are undesirable, an ac field can be used to induce the sample current. Irregularly shaped samples can sometimes be more advantageously evaluated by using the magnetoresistance of a spreading resistance probe.[100]

For some materials, e.g., liquid metals and semiconductors, and amorphous semiconductors, it may be more advantageous to use the Corbino effect to measure mobility. No voltage contacts are necessary. The values are independent of sample thickness, and no independent measurement of resistivity is required. The current is fed into a disk sample coaxially, and the voltage is obtained by inductive coupling.[83,84] In the low-frequency region where sample thickness is much less than skin depth the voltage induced in a single turn is proportional to the product of the Hall mobility, the frequency, the sample current, the field, and the sample geometry. However, because of difficulties in analytically evaluating the complete expression, a standard sample of the same geometry and known mobility can be used for calibration.[83] For details concerning the application of conventional Hall measurements to melts, see Ref. 85 and others therein.

Photoconductors are generally characterized by high resistivity and nonohmic contacts. Nonohmic contacts in turn can cause space-charge effects large enough to prevent meaningful Hall measurement. To minimize these problems, various combinations of alternating or steady magnetic field, [87,88] electric field, and incident-light source have been used.[65]

Hall voltage may be measured for fields orders of magnitude greater than the volts per centimeter recommended for normal usage. Under such circumstances, pulsed current is required to prevent excessive heating. For very short pulses, a sampling scope may be used as the Hall-voltage detector.[88] The frequencies of measurement can also be extended far beyond the few hertz ordinarily used, and have in fact been made well into the microwave region.[89]

5.3 CONDUCTIVITY TYPE

Unlike the other measurements in this chapter, conductivity type is routinely determined and can be done with quite simple equipment. There are several basic methods. The choice will largely depend on the specific material and the resistivity range of interest. In some cases, two modes of operation can be combined into one instrument in order to extend the range.

Rectification. If a dc microammeter is placed in series with an ac source (60-Hz transformer, for example) and contacts are made to the semiconductor as shown in Fig. 5.6*a*,[90] the direction of current flow will indicate type. One contact must be ohmic, and the other rectifying. The rectifying contact is a metal point. The ohmic contact may be more difficult to arrange but is often just a large-area clamp.* An oversize battery clip can be used for Si and Ge, although more elegant methods are preferable.

The difficulty of the ohmic contact can be eliminated by the three-probe configuration of Fig. 5.6*b*.[91,92] To check the performance of either of these variations, it may be necessary to observe the waveform on an oscilloscope. If the trace is symmetrical, some other typing system must be used. Such symmetry may occur because the resistivity is very low or, in the case of Fig. 5.6*a*, because both contacts are rectifying equally as well. A hot probe will be less rectifying than a cold one; so an arrangement similar to Fig. 5.6*a* can be used with a hot probe replacing the ohmic contact as in Fig. 5.6*c* (see the following discussion for hot-probe details).

Thermal EMF. A hot probe touching an n-type semiconductor becomes positive with respect to an ambient-temperature contact placed on the same material.[90] For p-type it will be negative. A small heating coil can be placed around one of the probes, or a miniature soldering iron can itself be used. A simple millivoltmeter may suffice for a measuring instrument, or more sensitive electronic instrumentation may be used if required. A collinear four-point-probe system can also be used in such a fashion that current flow between an end probe and one adjacent generates a thermal gradient in the semiconductor. The other two probes will then be at different temperatures and can be used for typing.[91] These possibilities are sketched

*Perfect ohmicity is not required. As long as the point contact is a better rectifier than the other one, the method will work. However, the less the difference, the poorer the sensitivity.

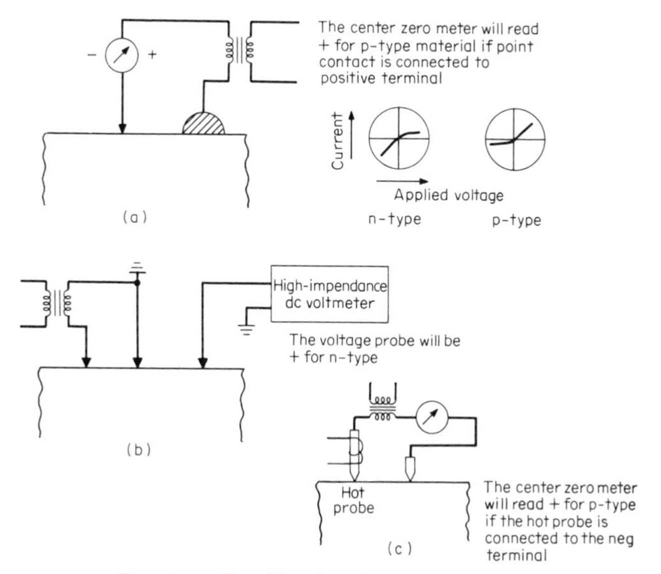

Fig. 5.6. Rectification for type checking.

in Fig. 5.7. The thermal emf system is generally restricted to low-resistivity material, and indeed, if the resistivity becomes high enough, the hot probe may make it intrinsic. Then, any material which has a higher electron than hole mobility will always read n-type. To prevent this occurrence, a cold probe, e.g., one thermoelectrically cooled, may be used instead of a hot one.[93]

Combinations. If four probes are used, for example, a four-point resistivity probe head, they may be connected as in Fig. 5.6*b* for rectification or as in Fig. 5.7*c* for thermal emf.[91] For Si, thermal emf is applicable over the resistivity range of from 10^{-3} to 10^2 or 10^3 Ω-cm while rectification is applicable from 10^{-2} to 10^3 Ω-cm or higher.[91]

The thermal voltage as in Fig. 5.7 and the rectification of Fig. 5.6*a* are additive, so that in fact Fig. 5.6*c* really represents a combination of Figs. 5.6*a* and 5.7*a*. It is suggested for wide bandgap semiconductors and has been used for checking 10^8 Ω-cm GaAs.[94]

Hall Effect. As discussed above, the sign of the Hall voltage is a direct indication of conductivity type. However, since the equipment involved is more complex than the methods just described, it is recommended only should they fail.

Staining. Chapter 7 describes various chemical stains that may selectively decorate one type.

Photovoltaic Effect. A photocurrent will flow between an illuminated rectifying-point contact and an ohmic contact. Its direction will depend on the material type and can in principle be used for type checking.[91] Practical difficulties, mostly associated with surface preparation, have thus far prevented any widespread usage.

MOS Capacitor. If an MOS capacitor such as that shown in Fig. 5.8*a* is available, the nature of its capacitance-voltage dependence is determined by the conductivity type of the semiconductor. Figure 5.8*b* shows details of the *C–V* apparatus, and Fig. 5.8*c* shows the profiles to be expected for n- or p-type material. The requirement for a capacitor is not necessarily restrictive, since they are often either already available or can be easily added. For example, if oxidized slices are to be checked, a mercury probe can be used instead of a metal dot and a process step saved. When very thin inversion layers are to be typed, there are some difficulties because the

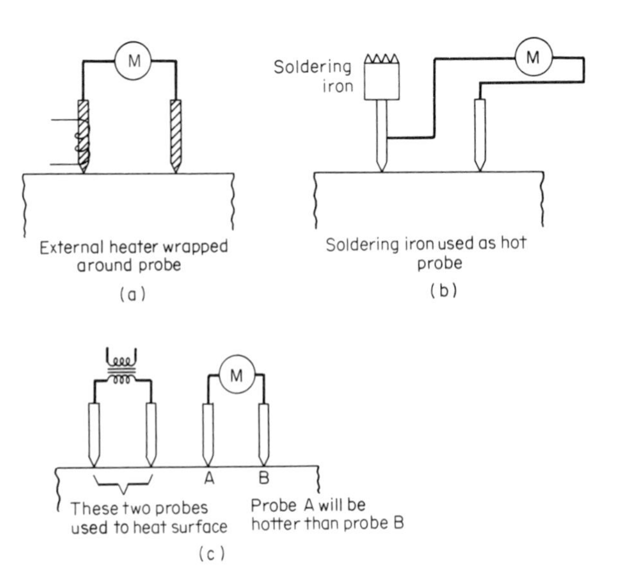

Fig. 5.7. Thermal-probe-type checkers. Hot probe will be + for n-type material.

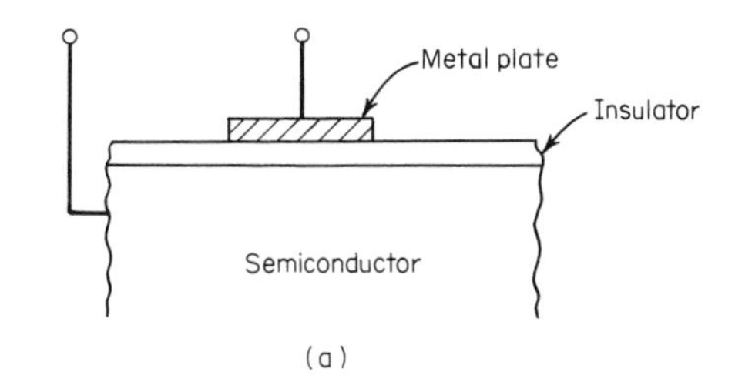

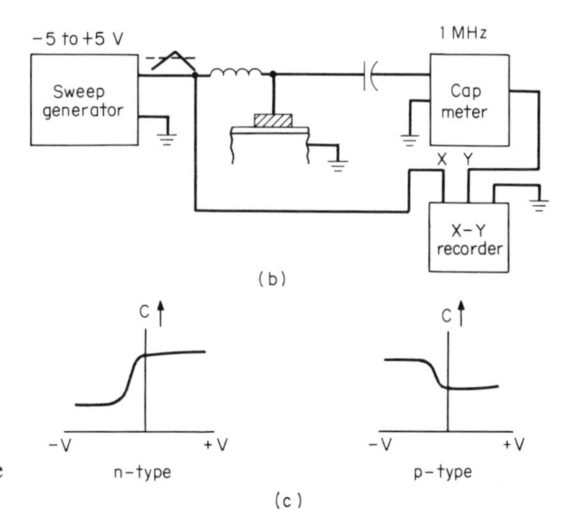

Fig. 5.8. Use of *C-V* plot for type checking.

space-charge region may move through it too quickly. However, they can usually be detected by examination of the detailed shape of the *C–V* curve[95] or the shift of the curve as a function of incident light intensity.[96]

ASTM Recommendations. Three alternates are recommended.

1. Hot probe for n- and p-type Ge less than 50 Ω-cm and n and p Si less than 1,000 Ω-cm.
2. Cold probes for n- and p-type Ge less than 20 Ω-cm and n or p Si less than 1,000 Ω-cm.
3. Rectification for n- and p-type Si between 1 and 1,000 Ω-cm. Not recommended for Ge.

Probe temperature should be held in the 40 to 80°C range. All methods are subject to misreading if probe pressure is too small. The hot-probe material is preferably stainless steel or Ni, and should be terminated with a 60° cone. Shielded leads should be used for resistivities greater than 1 Ω-cm. The center-zero meter should have a sensitivity of at least 200 μA full scale.

REFERENCES

1. R. A. Smith, "Semiconductors," Cambridge University Press, London, 1959.
2. D. Colman, R. T. Bate, and J. P. Mize, Mobility Anisotropy and Piezoresistance in Silicon p-type Inversion Layers, *J. Appl. Phys.*, **39**:1923–1931 (1968).
3. J. R. Haynes and W. Shockley, The Mobility and Life of Injected Holes and Electrons in Germanium, *Phys. Rev.*, **81**:835–843 (1951).

4. J. P. McKelvey, Diffusion Effects in Drift Mobility Measurements in Semiconductors, *J. Appl. Phys.,* **27**:341–343 (1956).
5. A. Many, Measurement of Minority Carrier Lifetime and Contact Injection Ratio on Transistor Materials, *Proc. Phys. Soc. London,* **B67**:9–17 (1954).
6. R. Lawrance and A. F. Gibson, The Measurement of Drift Mobility in Semiconductors, *Proc. Phys. Soc. London,* **B65**:994–995 (1952).
7. M. B. Prince, Drift Mobilities in Semiconductors, I Germanium, *Phys, Rev.,* **92**:681–687 (1953).
8. W. Shockley, "Electrons and Holes in Semiconductors," D. Van Nostrand Company, Inc., Englewood Cliffs, N.J., 1950.
9. H. Y. Fan, Effect of Traps on Carrier Injection in Semiconductors, *Phys. Rev.,* **92**:1424–1428 (1953).
10. R. Lawrance, The Temperature Dependence of Drift Mobility in Germanium, *Phys. Rev.,* **89**:1295 (1953).
11. J. R. Haynes and W. C. Westphal, The Drift Mobility of Electrons in Silicon, *Phys. Rev.,* **85**:680 (1952).
12. C. B. Norris, Jr., and J. F. Gibbons, Measurement of High-Field Carrier Drift Velocities in Silicon by a Time-of-Flight Technique, *IEEE Trans. Electron Devices,* **ED-14**:38–43 (1967); A. G. R. Evans, and P. N. Robson, Drift Mobility Measurements in Thin Epitaxial Semi-conductor Layers Using Time-of-Flight Techniques, *Solid-State Electron.,* **17**:805–812 (1974).
13. C. Canali, G. Ottaviani, and A. Alberigi Quaranta, Drift Velocity of Electrons and Holes and Associated Anistropic Effects in Silicon, *J. Phys. Chem. Solids,* **32**:1707–1720 (1971); and references contained therein.
14. P. A. Tove, G. Andersson, G. Ericsson, and R. Lidholt, Measurement of Drift Velocity of Electrons in Silicon by Exciting a Diode Structure with Short Superradiant Laser Pulses, *IEEE Trans. Electron Devices,* **ED-17**:407–412 (1970).
15. W. E. Spear and J. Mort, Electron and Hole Transport in CdS Crystals. *Proc. Phys. Soc. London,* **6**:130–140 (1963).
16. E. H. Putley, "The Hall Effect and Semiconductor Physics," Butterworth & Co. (Publishers), Ltd., London, 1960; Dover Publications, Inc., New York, 1968.
17. Albert C. Beer, "Galvanomagnetic Effects in Semiconductors," Academic Press, Inc., New York, 1963.
18. L. P. Hunter, Graphical Representation of the Semiconductor Hall Effect, *Phys. Rev.,* **94**:1157–1160 (1954).
19. W. C. Dunlap, Jr., Some Properties of High Resistivity p-Type Germanium, *Phys. Rev.,* **79**:286–292 (1950).
20. Rolf Landauer and John Swanson, Diffusion Currents in the Semiconductor Hall Effect, *Phys. Rev.,* **91**:555–560 (1953).
21. L. Hunter, E. Hulbregtse, and R. Anderson, Current Carrier Lifetimes Deduced from Hall Coefficient and Resistivity Measurements, *Phys. Rev.,* **91**:1315–1320 (1953).
22. P. C. Banbury, H. K. Kenisch, and A. Many, On the Theory of the Isothermal Hall Effect in Semiconductors, *Proc. Phys. Soc. London,* **A66**:753–758 (1953).
23. G. L. Pearson and J. Bardeen, Electrical Properties of Pure Silicon and Silicon Alloys Containing Boron and Phosphorus, *Phys. Rev.,* **75**:865–883 (1949).
24. P. A. Lee, Determination of the Impurity Concentrations in a Semiconductor from Hall Coefficient Measurements, *Brit. J. Appl. Phys.,* **8**:340–343 (1957).
25. Claude A. Klein and W. Deter Straub, The Hall Effect as an Analytical Tool in Ultrapure Silicon and Germanium, in Marvin S. Brooks and John K. Kennedy (eds.), "Ultrapurification of Semiconductor Materials," The Macmillan Company, New York, 1962.
26. Krzysztof Pigon', A Graphical Method for Determination of Mobility Ratio in the Semiconductors from Hall Effect Measurements Only, *J. Appl. Phys.,* **32**:2369–2371 (1961).

27. Olaf Lindberg, Hall Effect, *Proc. IRE,* **40:**1414–1419 (1952).
28. B. Lundberg and G. Bäckström, Hall Voltage and Magnetoresistance of Bi Measured by a Sum Frequency Method in a Belt Apparatus, *Rev. Sci. Instr.,* **43:**872–875 (1972).
29. Richard L. Petritz, Theory of an Experiment for Measuring the Mobility and Density of Carriers in the Space-Charge Region of a Semiconductor Surface, *Phys. Rev.,* **110:**1254–1262 (1958).
30. O. N. Tufte, The Average Conductivity and Hall Effect of Diffused Layers on Silicon, *J. Electrochem. Soc.,* **109:**235–238 (1962).
31. I. Hlásnik, Influence of Carrier Concentration Gradients and Mobility Gradients on Galvanomagnetic Effects in Semiconductors, *Solid State Electron.,* **8:**461–466 (1965).
32. Williams Johnson, Numerical Corrections for Hall Effect Measurements in Silicon Containing Gaussian Dopant Distributions, *Solid State Electron.,* **13:**951–956 (1970).
33. V. K. Subashchiev, and S. A. Poltinnikov, Determination of Carrier Mobility and Density in the Surface Layer of a Semiconductor, *Soviet Phys. Solid State,* **2:**1059–1066 (1960).
34. H. J. Juretschke, R. Landauer, and J. A. Swanson, Hall Effect and Conductivity in Porous Media, *J. Appl. Phys.,* **27:**838 (1956); Moni G. Mathew and Kenneth S. Mendelson, Hall Effect in the "Composite Sphere" Material, *J. Appl. Phys.,* **45:**4370–4372 (1974).
35. Conyers Herring, Effect of Random Inhomogeneities on Electrical and Galvanomagnetic Measurements, *J. Appl. Phys.,* **31:**1939–1953 (1960).
36. I. Isenberg, B. R. Russell, and R. F. Greene, Improved Method for Measuring Hall Coefficients, *Rev. Sci. Instr.,* **19:**685–688 (1948).
37. J. Volger, Note on the Hall Potential across an Inhomogeneous Conductor, *Phys. Rev.,* **79:**1023–1024 (L) (1950).
38. R. F. Wick, Solution of the Field Problem of the Germanium Gyrator, *J. Appl. Phys.,* **25:**741–756 (1954).
39. V. Frank, On the Geometrical Arrangement in Hall Effect Measurements, *Appl. Sci. Res.,* **B3:**129–140 (1953).
40. L. J. van der Pauw, A Method of Measuring Specific Resistivity and Hall Effect of Discs of Arbitrary Shape, *Philips Res. Repts.,* **13:**1–9 (1958).
41. L. J. van der Pauw, A Method of Measuring the Resistivity and Hall Coefficient on Lamellae of Arbitrary Shape, *Philips Tech. Rev.,* **20:**220–224 (1959).
42. M. G. Buehler and G. L. Pearson, Magnetoconductive Correction Factors for an Isotropic Hall Plate with Point Sources, *Solid State Electron.,* **7:**395–407 (1966).
43. M. G. Buehler, A Hall Four-Point Probe on Thin Plates, *Solid State Electron.,* **10:**801–812 (1967).
44. M. A. Green and M. W. Gunn, Four-Point Probe Hall Effect and Resistivity Measurements upon Semiconductors, *Solid State Electron.,* **15:**577–585 (1972).
45. Julius Lange, Method for Hall Mobility and Resistivity Measurements on Thin Layers, *J. Appl. Phys.,* **35:**2659–2664 (1964).
46. H. C. Montgomery, Electrical Noise in Semiconductors, *Bell System Tech. J.,* **31:**950–975 (1952).
47. Francis L. Lummis and Richard L. Petritz, Noise, Time Constant, and Hall Studies on Lead Sulfide Photoconductive Films, *Phys. Rev.,* **105:**502–508 (1957).
48. G. L. Guthrie, Sensitive AC Hall Effect Circuit, *Rev. Sci. Instr.,* **36:**1177–1179 (1965).
49. C. C. Allen and E. G. Bylander, Evaluation Techniques for and Electrical Properties of Silicon Epitaxial Films, in John B. Schroeder (ed.), "Metallurgy of Semiconductor Materials," Interscience Publishers, Inc., New York, 1962.
50. W. J. Patrick, Measurement of Resistivity and Mobility in Silicon Epitaxial Layers on a Control Wafer, *Solid State Electron.,* **9:**203–211 (1966).
51. A. B. M. Elliot and J. C. Anderson, An Investigation of Carrier Transport in Thin Silicon-on-Sapphire Films Using MIS Deep Depletion Hall Effect Structures, *Solid State Electron.,* **15:**531–545 (1972).

52. D. Colman and Don L. Kendall, Effect of Surface Treatments on Silicon Hall Measurements, *J. Appl. Phys.,* **40**:4462–4463 (1969).
53. W. F. Flanagan, P. A. Flinn, and B. L. Averbach, Shorting and Field Corrections in Hall Measurements, *Rev. Sci. Instr.,* **25**:593–595 (1954).
54. J. Dresner, The Photo-Hall Effect in Vitreous Selenium, *J. Phys. Chem. Solids,* **25**:505–511 (1964).
55. Sol E. Harrison, George H. Heilman, and George Warfield, Measurement of the Hall Effect in Metal-Free Phthalocyanine Crystals, *Phys. Rev. Lett.,* **8**:309–311 (1962).
56. Derek Colman, High Resistivity Hall Effect Measurements, *Rev. Sci. Instr.,* **39**:1946–1948 (1968).
57. G. Fischer, D. Greig, and E. Mooser, Apparatus for the Measurement of Galvanomagnetic Effects in High Resistance Semiconductors, *Rev. Sci. Instr.,* **32**:842–846 (1961).
58. R. C. Eden and W. H. Zakrzewski, Semiautomatic Hall Effect Measurement System, *Rev. Sci. Instr.,* **41**:1030–1033 (1970).
59. J. Shewchun, K. M. Ghanekar, R. Yager, H. D. Barber, and D. Thompson, A Computer Controlled Automatic System for Measuring the Conductivity and Hall Effect in Semiconducting Samples, *Rev. Sci. Instr.,* **42**:1797–1807 (1971).
60. W. Bullis, W. R. Thurber, T. N. Pyke, Jr., F. H. Ulmer, and A. L. Koenig, "Use of a Time-shared Computer System to Control a Hall Effect Experiment," National Bureau of Standards, *Tech. Note* 510, 1969.
61. Alpheus W. Smith, The Hall Effect in Bismuth with High Frequency Currents, *Phys. Rev.,* **35**:81–85 (1912).
62. J. J. Donoghue and W. P. Eatherly, A New Method for Precision Measurement of the Hall and Magneto-resistive Coefficients, *Rev. Sci. Instr.,* **22**:513–516 (1951).
63. S. W. Kurnick and R. L. Fitzpatrick, Galvanomagnetic Measurement in Highly Conducting Semiconductors, *Rev. Sci. Instr.,* **32**:452–453 (1961).
64. Eugene E. Olson and John E. Wertz, A High Impedance AC Hall Effect Apparatus, *Rev. Sci. Instr.,* **41**:419–421 (1970).
65. J. Ross Macdonald, and John E. Robinson, AC Hall and Magnetostrictive Effects in Photoconducting Alkali Halides, *Phys. Rev.,* **95**:44–50 (1954).
66. Jerome M. Lavine, Alternate Current Apparatus for Measuring the Ordinary Hall Coefficient of Ferromagnetic Metals and Semiconductors, *Rev. Sci. Instr.,* **29**:970–976 (1958).
67. P. E. Bierstedt and J. E. Hanlon, Apparatus for Measuring the Normal Hall Coefficient in Magnetic Conductors, *Rev. Sci. Instr.,* **42**:1674–1976 (1971).
68. T. M. Dauphinee and E. Mooser, Apparatus for Measuring Resistivity and Hall Coefficient of Semiconductors, *Rev. Sci. Instr.,* **26**:660–664 (1955).
69. L. J. van der Pauw, An Analysis of the Circuit of Dauphinee and Mooser for Measuring Resistivity and Hall Constant, *Rev. Sci. Instr.,* **31**:1189–1192 (1960).
70. H. H. Soonpaa, C. D. Motchenbacher, and H. Dohl, Zero-Point Detector for the Dauphinee-Mooser Circuit, *Rev. Sci. Instr.,* **34**:1341–1344 (1963).
71. F. M. Ryan, Rotating Sample Method for Measuring the Hall Mobility, *Rev. Sci. Instr.,* **33**:76–79 (1962); J. Yahia and G. Perluzzo, A Two Frequency ac Hall Apparatus for Measurements in Metals, *Rev. Sci. Instr.,* **44**:335–337 (1973).
72. A. M. Hermann and J. S. Ham, Apparatus for the Measurement of the Hall Effect in Semiconductors of Low Mobility and High Resistivity, *Rev. Sci. Instr.,* **36**:1553–1555 (1965).
73. Robert G. Pohl, Hall Effect Measurements in Semiconductor Rings, *Rev. Sci. Instr.,* **30**:783–786 (1959).
74. J.-P. Jan, Galvanomagnetic and Thermomagnetic Effects in Metals, in Frederick Seitz and David Turnbull (eds.), "Solid State Physics," vol. 5, Academic Press, Inc., New York, 1957.
75. B. R. Russell and C. Wahlig, A New Method for the Measurement of Hall Coefficients, *Rev. Sci. Instr.,* **21**:1028–1029 (1950).

76. E. M. Pell and R. L. Sproull, Sensitive Recording Alternating Current Hall Effect Apparatus, *Rev. Sci. Instr.,* **23**:548–552 (1952).
77. H. Rzewuski and Z. Werner, New Double Frequency Method for Hall Coefficient Measurements, *Rev. Sci. Instr.,* **36**:235–236 (1965).
78. John L. Levy, Sensitive Hall Measurements on NaCl and on Photoconductive PbTe, *Phys. Rev.,* **92**:215–218 (1953).
79. N. Z. Lupa, N. M. Tallan, and D. S. Tannhauser, Apparatus for Measuring the Hall Effect of Low-Mobility Samples at High Temperatures, *Rev. Sci. Instr.,* **38**:1658–1661 (1967).
80. Des Coudres, *Phys. Z.,* **2**:586 (1901).
81. Rudolf G. Suchannek, Effect of Intermodulation of Measurement of Small Hall Coefficients with Double AC Method, *Rev. Sci. Instr.,* **37**:58a–59x (1966).
82. H. L. McKinzie and D. S. Tannhauser, Systematic Errors in Alternating Current Hall Effect Measurement, *J. Appl. Phys.,* **40**:4954–4958 (1969).
83. G. P. Carver, A Corbino Disc Apparatus to Measure Hall Mobilities in Amorphous Semiconductors, *Rev. Sci. Instr.,* **43**:1257–1263 (1972).
84. P. W. Shackle, Measurement of the Hall Coefficient in Liquid Metals by the Corbino Method, *Phil. Mag.,* **21**:987–1002 (1970).
85. N. E. Cusack, J. E. Ederby, P. W. Kendall, and Y. Tièche, The Measurement of the Hall Coefficient of Liquid Conductors, *J. Sci. Instr.,* **42**:256–259 (1965).
86. Harold E. MacDonald, and Richard H. Bube, Apparatus for Measuring the Temperature Dependence of Photo-Hall Effects in High-Resistivity Photoconductors, *Rev. Sci. Instr.,* **33**:721–723 (1962).
87. I. Eisele and L. Kevan, Double Modulation Method for Hall Effect Measurements on Photoconducting Materials, *Rev. Sci. Instr.,* **43**:189–194 (1972). (See also other references contained therein.)
88. E. Müller and D. K. Ferry, Hall Studies of Instabilities Occurring in Impact Ionization of N-type Indium Antimonide, *J. Phys. Chem. Solids,* **31**:2401–2404 (1970).
89. Y. Nishina and G. C. Danielson, Microwave Measurement of Hall Mobility: Experimental Method, *Rev. Sci. Instr.,* **32**:790–793 (1961).
90. G. Knight, Jr., Measurement of Semiconductor Parameters, in L. P. Hunter (ed.), "Handbook of Semiconductor Electronics," McGraw-Hill Book Company, New York, 1956.
91. W. A. Keenan, C. P. Schneider, and C. A. Pillus, Type-All System for Determining Semiconductor Conductivity Type, *Solid State Tech.,* **14**:51–56 (March 1971).
92. Hakan Hakansson, Conductivity Type Determination for Different Semiconductor Materials, *Rev. Sci. Instr.,* **43**:1380–1381 (1972).
93. Standard Method F 42-69, "ASTM Book of Standards," Part 8, American Society for Testing and Materials, Philadelphia, 1971.
94. K. B. Wolfstirn and M. W. Focht, Thermoelectric n-p Tester Using AC Bias for Gallium Arsenide and Gallium Phosphide, *Rev. Sci. Instr.,* **42**:152–154 (1971).
95. F. P. Heiman, K. H. Zaininger, and G. Warfield, Determination of Conductivity Type from MOS-Capacitance Measurements, *Proc. IEEE,* **52**:863–864 (1964).
96. Lowell E. Clark, Determination of Conductivity Type from Capacitance Measurements on MOS Diodes, *IEEE Trans. Electron Devices,* **ED-12**:390–391 (1965).
97. V. M. Cottles and A. M. Hermann, Hall Probe Loop for Nulling Induced Voltages from a.c. Magnetic Fields, *Rev. Sci. Instr.,* **44**:334 (1973).
98. C. M. Wolfe, G. E. Stillman, and J. A. Rossi, High Apparent Mobility in Inhomogeneous Semiconductors, J. Electrochem. Soc., **119**:250–255 (1972).
99. R. D. Westbrook, Effect of Semiconductor Inhomogeneities on Carrier Mobilities by the van der Pauw Method, *J. Electrochem. Soc.,* **121**:1212–1215 (1974).
100. L. Gutai, Determination of Galvanomagnetic Coefficients by a One-Point Method, *Solid-State Electron.,* **16**:395–406(1973).

6

Thickness Measurements

6.1 INTRODUCTION

Thickness measurements can be broken into three general categories involving (1) the total thickness of a slice or block of semiconductor, (2) the thickness of variously doped regions within the semiconductor, (3) the thickness of external layers of foreign material such as dielectric or metal films deposited on the semiconductor surface. Total thickness may be measured by conventional machine-tool-industry equipment if care is taken not to damage the surface excessively. Semiconductor and dielectric layers are most often measured by optical methods. For the plan view of Fig. 6.1*a*, measurements will depend on the optical constants of each layer being different from the others. If it is viewed on edge, as in Fig. 6.1*b*, or if the wedge of Fig. 6.1*c* is of very low angle, the problem becomes one of delineating the boundaries and then making lateral measurements. Various stains or etchants are often used to improve contrast. When steps such as those in Fig. 6.1*d* are available, optical methods are sometimes abandoned in favor of purely mechanical profilometers which may be drawn across the surface to indicate the step. These same methods may be used for metal layers, but for them a variety of instruments which depend on electrical conductivity are also available.

Tables 6.1 and 6.2 summarize various possibilities and can be used as a guide in choosing the proper method. Table 6.3 gives conversion factors between some of the more common units of thickness. They are by no means metric standards but do conform to the useful philosophy of providing units that will allow most dimensions to be expressed in numbers between 1 and 10.

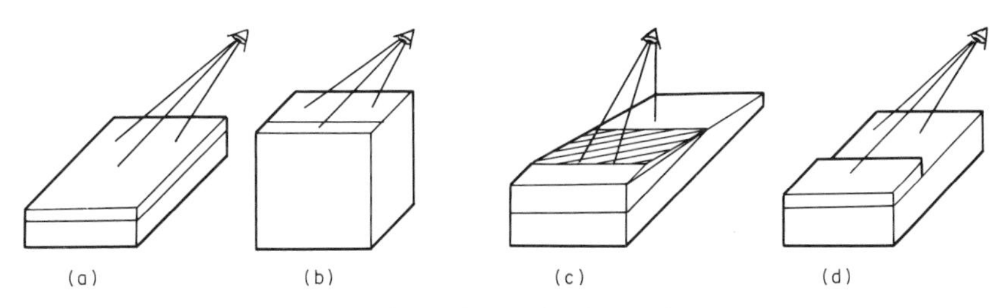

Fig. 6.1. Various views of a layered structure.

Table 6.1. Methods of Measuring Thickness

Method	Applicable to	Advantages
Mechanical micrometer	Slices, dice, Hall bars, etc.	Simplicity
Electrical micrometer	Slices, dice, Hall bars	Less likely to produce damage
Air gage	Slices, dice, Hall bars	Does not produce damage
Absorption	Slices, dice, etc., silicon on sapphire	Nondestructive, not sensitive to particles on surface
Ellipsometry	Epitaxial layers, oxide, nitride, or other dielectric layers on semiconductor surface	Nondestructive, rapid
Interferometry	Epitaxial layers, oxide, nitride, or other dielectric layers on semiconductor surface	Nondestructive, rapid
Interferometry + lap and stain	Epitaxial layers, diffused layers	Convenience, wide application
Interferometry + step	Metal over oxide, oxide over semiconductor	
Light section + step	Metal over oxide over semiconductor	Simplicity
Mechanical profilometer + step	Metal over oxide over semiconductor	Simplicity
High-power microscope + step	Metal over oxide over semiconductor	Availability
High-power microscope + visible boundaries	Transparent layer of known refractive index	Availability
Weight differential	Deposited layers	
Eddy current	Metal layer	
Beta-ray backscattering	Metal layer	
X-ray fluorescence	Thin metal films	Nondestructive, very local
Crystallographic defects	Epitaxial layers	Nondestructive
Device performance	Base and collector widths	Can be used on completed devices
Capacitance bridge	Low-loss materials of known dielectric constants	Noncontacting

6.2 MICROMETERS AND OTHER GAGES

Conventional hand-held micrometers are available which will read directly to the nearest thousandth inch, and with vernier will read to 0.0001 in. Such tools are therefore suitable for measuring slice thickness, although they have several disadvantages. The two major ones are the likelihood of fracturing brittle slices and the fact that the anvils are usually approximately ¼ in in diameter so that dust and/or grit may add to the measured thickness. The addition of ball anvils will minimize the particle problem, but the smaller contact area increases the chance of mechanical damage.

A better choice is a dial-gage indicator mounted over a flat work surface which has a small spherical protrusion directly below the indicator, as shown in Fig. 6.2. Only the spherical tip rises above the work surface a few mils, so that for slices of reasonable size the error because the slice is not perpendicular to the instrument centerline is negligible. Dial indicators use a gear train for magnification and are spring-loaded. However, this loading is usually less than that exerted by a hand

Table 6.2. Guide to Applicable Methods

Problem	Possible approaches
epi on n^+	Infrared spectrophotometer*
epi on p^+	Angle lap and stain, stacking-fault dimensions*
Very thin epi	Infrared ellipsometer; angle lap and stain; overcoat on material of different index of refraction, plus visible spectrophotometer
Very thin Si epi on sapphire or similar substrate	Visible spectrophotometer, visible-light absorption
Oxide layer	Color chart; ellipsometer; visible spectrophotometer; VAMFO; etched step plus profilometer or microscope; interferometer
Nitride layer	Color chart; ellipsometer; visible spectrophotometer; VAMFO; etched step plus profilometer or microscope; interferometer
Nitride over oxide	Ellipsometer, plus program for computing double-layer thickness
Metal layer	Angle lap, etched step plus profilometer or interferometer, beta-ray gage, eddy-current gage
Transistor base width	Angle lap and stain, base transit time
Collector width	Angle lap and stain, breakdown voltage (applicable only over certain combinations of resistivity, base width, and collector width)

*Applies to both epi on n^+ and epi on p^+.

micrometer. Alternatives include electronic, optical, and air gages. Some electronic instruments have a stylus coupled to the movable core of a differential transformer (Fig. 6.3). Very small displacements can be readily measured, and because of the small size of the armature and stylus, quite light loading is possible. Others use the slice to vary the gap between plates and relate the capacitance change to thickness.[1] Optical gages contact the surface with a stem much like a dial indicator, but use an optical lever for magnification. The air gage is shown schematically in

Table 6.3. Thickness Conversion Factors

0.1 mil =	100 μin
	2.54 μm
	25,400 Å
	9.3 Hg lines (5,460 Å)
	8.6 Na lines (5,896, 5,890 Å)
1 μm =	0.04 mil
=	10,000 Å
=	3.66 Hg lines
=	3.39 Na lines
1 Hg line =	0.011 mil
=	0.27 μm
=	2,700 Å
=	0.9 Na lines
1 Na line =	0.012 mil
=	0.29 μm
=	2,900 Å
=	1.06 Hg lines

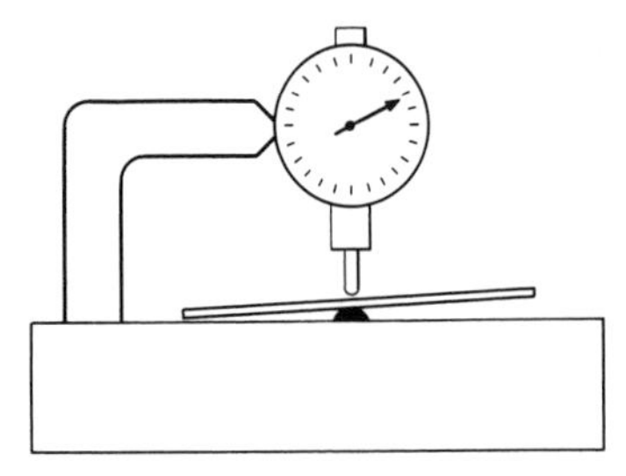

Fig. 6.2. Use of a dial indicator for measuring slice thickness.

Fig. 6.4. The back pressure caused by the variable spacing between a gas jet and a flat surface is sensed. Thus, if samples are inserted between the jet and reference (Fig. 6.4*a*), the distance between the jet and the slice surface can be measured. Should the slice be bowed upward as in Fig. 6.4*b*, a false reading will be obtained; so for slice measurements it is customary to use two opposing jets as shown in Fig. 6.4*c*. However, the air-gage jets should be turned off before thin slices are slipped between them to reduce the likelihood of breakage. When electronic or air gages are used on very sensitive scales, the total range is normally quite small (e.g., 0.3 mil for a 0.005-mil-per-division gage); so reference blocks will be required for zero setting. Because of the great sensitivity of these instruments, they were sometimes used in the early days of silicon epitaxy for layer-thickness measurements.[1] This involved keeping track of individual slices and making measurements before and after growth, as well as minimizing growth on the back of the slice. Such applications have now been superseded by more elegant and more practical methods such as infrared interferometry.

Should slices be slightly tapered, the location of the measurements must be standardized if comparisons between, for example, the manufacturer and the user are to be made. One procedure suggests a center measurement and four additional ones $\frac{1}{8}$ in from the circumference.[2] If the slice has a flat, it is labeled 6 o'clock, and the readings are taken at the 2, 4, 8, and 10 o'clock positions.

6.3 ABSORPTION METHODS

The reduction in intensity of a plane wave passing through a material is given by

$$I = I_0 e^{-\alpha t} \tag{6.1}$$

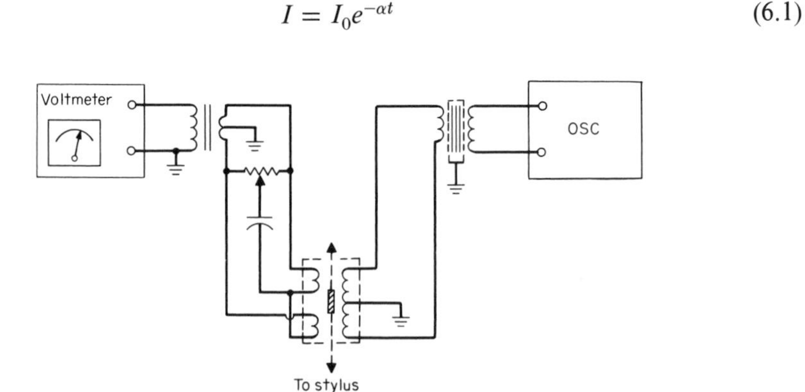

Fig. 6.3. Use of a differential transformer to measure mechanical displacement. Such instruments can be designed to measure step heights of only a few hundred angstroms.

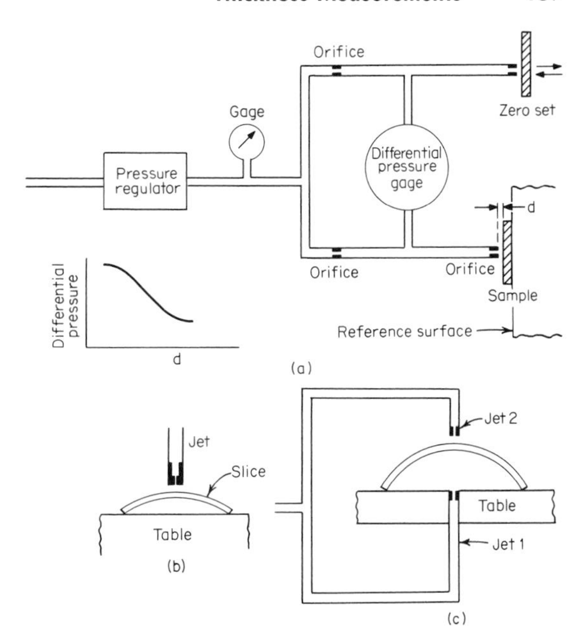

Fig. 6.4. Air gage for measuring slice thickness.

where I_0 is the original intensity, I the value after passing through a thickness t, and α the absorption coefficient. If the sample has an appreciable reflection coefficient R, the basic equation above must be modified to

$$I = \frac{I_0 e^{-\alpha t}(1 - R)^2}{1 - R^2 e^{-2\alpha t}} \tag{6.2}$$

Therefore, if α and R are known for a given radiation, a measurement of I and I_0 allows the thickness t to be calculated. Depending on whether x-ray, β-ray, infrared, or visible light is used, the absorption coefficient will vary over many orders of magnitude. Some, such as x-rays, and light of wavelength larger than the band-edge wavelength, are appropriate for thicknesses of millimeters to centimeters. X-ray measurements have been used[3] for germanium wafers but for most applications have no advantages and the disadvantage of expensive and specialized equipment. Long wavelengths (greater than 1.1 μm in the case of silicon) have absorption coefficients that depend on the impurity level. Because of this, thickness measurements cannot be independently made. For short wavelengths, α is relatively independent of impurity concentration. Thus, for thin layers either self-supporting or on transparent substrates, e.g., Si on sapphire, short-wavelength light might be used, and even if the absorption coefficient is in the 10^5 to 10^6 per centimeter range, there will be sufficient transmission for satisfactory measurements. Visible-light absorption can be used for estimating the thickness of very thin metal films. Most films (e.g., Pt, Au, and Al) will have transmissions in the 20 to 80 percent range when they are a few hundred angstroms thick.[4]

Should the absorption coefficient be so low that $\exp(-\alpha t)$ at $t = \lambda/4$ is greater than perhaps 0.1, the thickness will be a multivalued function of transmission, and interferometry rather than absorption should be used. An additional complication that must be considered is the fact that some films (e.g., Ag) have such a marked change in optical constants in the 50- to 200-Å-thick region that a peak in the absorption-thickness curve may arise and two thickness values can correspond to a single value of transmission.[4] If an absorption band is present, operating at its wavelength can substantially increase α and thus effectively increase the sensitivity in the very thin layers. Similarly, it can be used to minimize interference effects in thicker layers. The 9- and 12-μm bands in SiO_2 have, for example, been used to measure the thickness of SiO_2 films on silicon.[5]

A spectrophotometer is not necessarily required for either visible or infrared absorption measurements. It is, for example, possible to use a laser light source instead of a monochrometer, a simple converging lens for the optical system, and a photodiode for detection.[6] In some systems, even the monochromatic laser light is dispensed with in favor of a broad-band incandescent source. Simple densitometers can in fact be used for many measurements. An intermediate range of complexity uses the broad-band source and an interference filter, although for some infrared ranges the latter are not readily available. There are β-ray instruments that will work with reasonable success in the millimeter and less range for many semiconductors but have not been widely used.

6.4 ELLIPSOMETRY[7-30]

Ellipsometry is most applicable to the thickness measurement* of thin films of a dielectric on a highly absorbing substrate when viewed as in Fig. 6.1*a*. The general principles have been known since before 1900, although the name itself is relatively new (1944).[13,14] Because of the complexity of interpretation, however, the ellipsometer has until recently been used only for very specialized applications. The present availability of high-speed computers makes the equations much easier to deal with, either by allowing many curves covering a wide variety of circumstances to be readily generated, or by coupling the ellipsometer directly into a computer.

General Theory. After light is reflected from a single surface, it will normally be reduced in amplitude and shifted in phase. If multiple reflecting surfaces are involved, as in Fig. 6.5, the various reflected beams will further interact and, depending on relative amplitudes, path differences between the surfaces, and phase shifts at the surfaces, give maxima and minima of intensity (interference effects) as a function of either wavelength and/or spatial position. Should the incident light have

*Ellipsometry is also useful for measuring the optical constants of either the substrate or the layer.

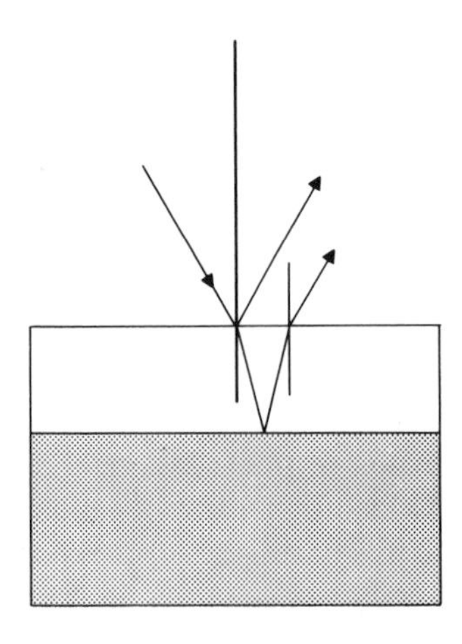

Fig. 6.5. Multiple reflections from a layered structure.

been plane-polarized (which may be resolved into a component p parallel to the plane of incidence and a component s perpendicular to the plane) before impinging on a surface (Fig. 6.6), the two components will usually experience different amounts of phase shift upon reflection and have different reflection coefficients. This phase-shift difference introduces an additional component polarized 90° to the incident beam and thus produces elliptically polarized light as in Fig. 6.6*c*. Projected onto a plane perpendicular to the reflected ray, the resultant E vector of the elliptical light will trace out an ellipse (hence the term elliptical).

Both polarization and interference changes are now widely used to determine various properties of the materials between the reflecting surfaces and give rise to the distinctively different techniques of ellipsometry and interferometry. The former is most widely applied to the study of very thin (<one wavelength) layers on thick substrates. Interference effects, which are the subject of the next section, are used most often for studying layers which are from one-eighth to four or five wavelengths thick.

The angles Δ and Ψ are the most commonly discussed parameters in ellipsometry and are

$$\Delta = \text{differential phase change} = \Delta_p - \Delta_s$$
$$\Psi = \tan^{-1}\left(\frac{R_p}{R_s}\right) \tag{6.3}$$

where R_p is the reflection coefficient of the p component, R_s is the reflection coefficient of the s component, and $\Delta_{p,s}$ are the respective phase shifts introduced during reflection. If there were no phase shift, only a difference in reflection coefficients, the light would remain plane-polarized after reflection, but the plane could be rotated. Thus the change of analyzer angle required to produce extinction after the sample was introduced would be a measure of (R_p/R_s) or Ψ.* Similarly, the amount of phase shift introduced by an additional optical component necessary to compensate for $\Delta_p - \Delta_s$ and again produce linear polarized light is a direct measure of Δ.

*Several conventions have been used in the literature (see, for example, Ref. 15). The reader should take this into account when attempting to compare various references.

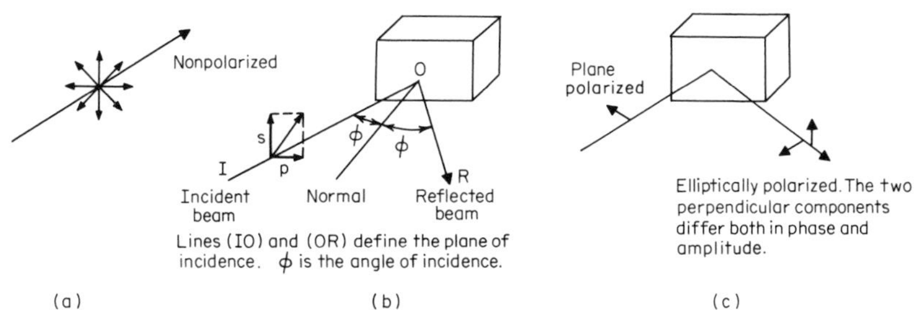

Fig. 6.6. Details of polarized light reflection. (*a*) For plane-wave propagation the E vector must lie perpendicular to the direction of travel but may have any orientation in that plane. Usually there will be E vectors with many different directions. (*b*) and (*c*) For purposes of analysis, polarized light is considered in terms of s and p projections which are in phase but usually differ in amplitude. Component p lies in the plane of incidence, and component s is perpendicular to p.

Having obtained Ψ and Δ experimentally (as will be described later), the problem of equating them to film thickness still remains. This can be accomplished by calculating the reflection coefficients at each interface from Fresnel's equations, the phase shift δ which the light experiences in traversing the film or films, and combining them to give an overall reflection coefficient and phase shift (Ψ and Δ). The film thickness enters the equations only through δ, given by

$$\delta = 2\pi \frac{t}{\lambda}(n^2 - \sin^2 \phi)^{1/2} \tag{6.4}$$

where t is the film thickness and n its index of refraction. Ψ and Δ repeat for every π change in δ, which is why the requirement arises for an independent evaluation of t to within one order. That is, t must be known to within $\lambda/(n^2 - \sin^2 \phi)^{1/2}$.

In principle, simple structures as well as multiple films can be analyzed if all but two of the parameters are known. One may, for example, calculate

1. n, k of a bare substrate (the only case for which an explicit solution is available[11])
2. Thickness and n of a layer on top of a substrate of known n, k
3. Thickness and n of either of two layers if the other layer thickness and refractive index are known, along with n, k of the substrate[7]

The first case is useful because, if no published data are available, it allows the substrate to be easily characterized. Number two is the most common measurement, but the third one can arise if silicon nitride, for example, is being deposited over an SiO_2 layer.

The usual procedure for interpreting the data is to presume that the optical constants of the substrate are known, and then by computer to generate a family of curves such as those shown in Fig. 6.7 which give the film index of refraction and thickness as a function of Ψ and Δ. These curves can then be used for routine thickness measurements.* Over most of the range, given values of Ψ and Δ uniquely

*Since the instrument angle of incidence and the wavelength of light (typically but not necessarily 70° and 5,461 Å) both enter into the calculation of the curves, be sure to use a set which matches the instrument.

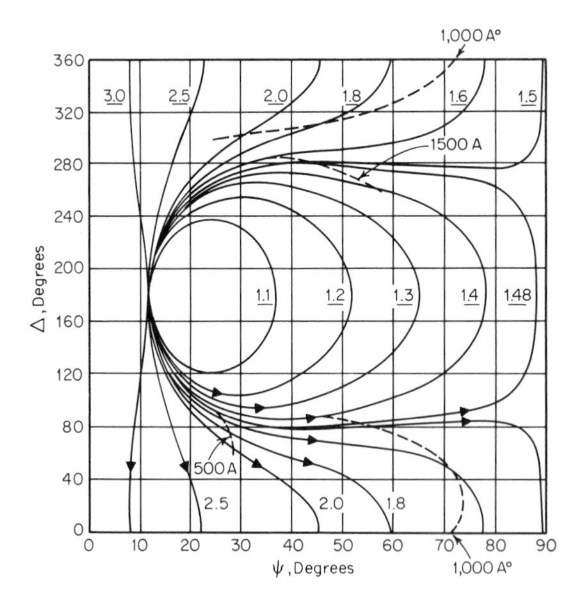

Fig. 6.7. Ψ, Δ for lossless films on a silicon substrate. The solid lines are iso-dielectric-constant curves. The dotted ones are for constant thickness.

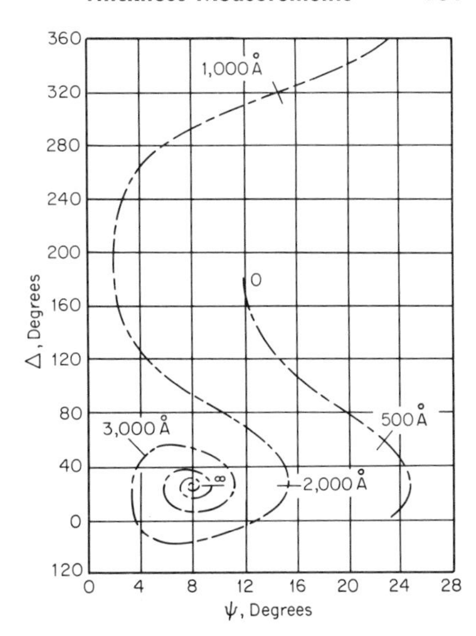

Fig. 6.8. Ψ, Δ as a function of thickness for an absorbing film on silicon. This particular curve assumes that the film index of refraction is 2.2 − *i*0.22. (*After Archer.*[10])

determine n and t. However, very high index curves (e.g,, $n > 9$ on a Si substrate) may overlap the curves for n very close to 1. Again, some background information regarding the film must be available to prevent misinterpretation, although the likelihood of having such index films is rather remote.

If the film is absorbing, Ψ and Δ depend on k as well as n and t, and the Ψ, Δ curve takes on a much different character, as shown in Fig. 6.8. It now converges to the values appropriate for the "film" behaving as a substrate (case 1) as the optical thickness increases, rather than being cyclic. Also, a unique set of properties are not defined by a set of Ψ, Δ readings, since there are now the three unknowns n, k, and t. This problem can be resolved by taking readings at two different values of ϕ,[11] but of course the difficulty of interpretation is considerably increased.

Case 3 and others of similar complexity are best handled by feeding the ellipsometer angles and other pertinent data into a computer and directly performing the calculations.

The optical elements required of an ellipsometer to measure the necessary angles are shown in Fig. 6.9 and include a monochromatic light, polarizer,* quarter-wave compensator, analyzer, and detector. The two operations of compensation and determining the amount of rotation are not performed separately, since "complete"

*Polarizers are often referred to as "Nicols." The Nicol polarizer was invented in 1828 and for many decades was one of the most popular. It has now, however, been largely superseded by other types such as the Glan-Thomson.

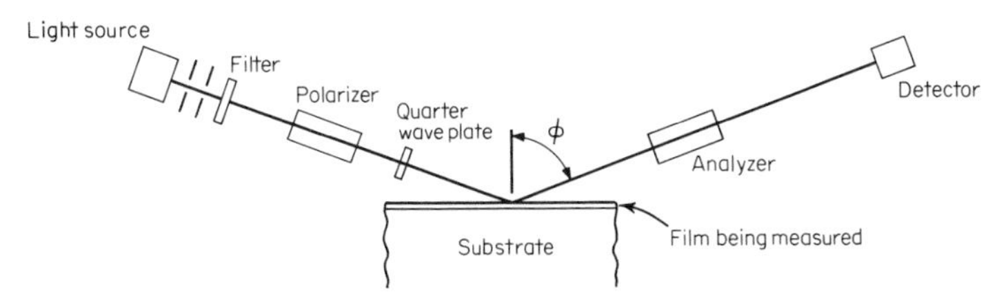

Fig. 6.9. Basic ellipsometer optics.

extinction is the only condition that can be conveniently measured,* and it will occur only when both the necessary compensation has been added and the analyzer properly set. Thus one of the three polarizing elements (polarizer, quarter-wave plate, and analyzer) is fixed, and the other two are simultaneously varied until a null (extinction) is produced. The two most common combinations are:

1. The polarizer fixed at 45° and the compensator and analyzer adjusted for extinction
2. The fast axis of the quarter-wave plate fixed at 45° and the polarizer and analyzer adjusted for extinction

In either case, the quarter-wave plate can be put before or after the sample. With the second combination, if at extinction the polarizer angle is P and the analyzer angle is A, a_R, a_S, and p are defined through Table 6.4.

The angular convention usually used is that all azimuthal angles are measured as positive counterclockwise from the plane of incidence (plane IOR of Fig. 6.6) when looking into the light beam and that the polarizer angle will be adjusted to read zero when the plane of transmission is in the plane of incidence. There are actually several sets of readings that should be equivalent. Sixteen of these arise when the compensator is set at plus and minus 45° and are customarily grouped into the four zones of Table 6.4.[25] Sixteen more readings are obtained if the compensator is also set at ±135° in addition to ±45°.

When the compensator has exactly 90° retardation,

$$\begin{aligned} \Delta &= 90° + 2p \\ \Psi &= a_p = a_s \end{aligned} \tag{6.5}$$

* By the use of an additional optical element, the operations can be sequentially performed. The two halves of a biplate placed between the reflecting surface and the analyzer will appear equally bright when the beam is plane-polarized.

Table 6.4. Relation between A, P, a_s, a_p, and p

Zone	Compensator	p	a_p	a_s	P	A
I	−45°	P	A		p	a_p
		$P - 180°$	A		$p + 180°$	a_p
		P	$A - 180°$		p	$a_p + 180°$
		$P - 180°$	$A - 180°$		$p + 180°$	$a_p + 180°$
III	−45°	$P - 90°$		$180° - A$	$p + 90°$	$180° - a_s$
		$P - 270°$		$180° - A$	$p + 270°$	$180° - a_s$
		$P - 90°$		$360° - A$	$p + 90°$	$360° - a_s$
		$P - 270°$		$360° - A$	$p + 270°$	$360° - a_s$
II	+45°	$90° - P$		A	$90° - p$	a_s
		$270° - P$		A	$270° - p$	a_s
		$90° - P$		$A - 180°$	$90° - p$	$a_s + 180°$
		$270° - P$		$A - 180°$	$270° - p$	$a_s + 180°$
IV	+45°	$180° - P$	$180° - A$		$180° - p$	$180° - a_s$
		$360° - P$	$180° - A$		$360° - p$	$180° - a_s$
		$180° - P$	$360° - A$		$180° - p$	$360° - a_p$
		$360° - P$	$360° - A$		$360° - p$	$360° - a_p$

$\Delta = 90° + 2p$, $\Psi = a_p = a_s$.

Should the retardation of the compensator not be exactly 90°, but rather some value δ,

$$\begin{aligned} \tan \Delta &= \sin \delta \cot (2p) \\ \tan^2 \Psi &= \tan (a_p) \tan (a_s) \end{aligned} \tag{6.6}$$

However, δ is usually close enough to 90° that Eq. (6.6) reduces to Eq. (6.5).

In principle only one value each for p and a is necessary, but in practice the calculated values of p and a will generally be slightly different from zone to zone. It is experimentally observed that averaging values from the appropriate zones will minimize the errors.[26] Thus

$$p(\text{zone I}) + p(\text{zone III}) = 2p \tag{6.7}$$

and

$$p(\text{zone II}) + p(\text{zone IV}) = 2p \tag{6.8}$$

In the case of the a's, $a_p \neq a_s$ from zone to zone unless the compensator is exactly 90°, and in zone 1 a_p is read, and in zone III, a_s. They may be combined, however, through the expression

$$(\tan a_p \tan a_s)^{1/2} = \tan \frac{a_p + a_s}{2} = \tan \Psi \tag{6.9}$$

$$\Psi = \frac{a_p + a_s}{2} = \frac{180 - A_{\text{III}} + A_{\text{I}}}{2} \tag{6.10}$$

For additional details relating to errors, see Refs. 27 and 28.

With the large number of possibilities listed in Table 6.4, there is a very real problem in determining for any given set of null conditions which set of relations should be used.

Since Ψ must always lie between 0 and 90°, the initial search for extinction should be made with A restricted to the first quadrant (zone I). This will ensure that $A = a$, although there will be two values of $P(p$ and $p + \pi)$. Next look for extinction with $90° < A < 180°$, i.e., at $P_1 - 90°$ (zone III). In this case, only $A = 180° - a$ will be found, although again there will be two values of $P(p + 90°$ and $p + 270°)$. For the correct set,

$$P_{\text{I}} + P_{\text{III}} = p + p + 90° \equiv \Delta \tag{6.11}$$

where P_{I} is one of the two P values read in zone I and P_{III} is one of the two P values of zone III. Should they not be properly matched,

$$P_{\text{I}} + P_{\text{III}} = p + 180° + p + 90° = \Delta + 180° \tag{6.12}$$

$$P_{\text{I}} + P_{\text{III}} = p + p + 270° = \Delta + 180° \tag{6.13}$$

In case of either Eq. (6.12) or (6.13), the thickness–refractive index combination determined from Δ and Ψ will be of unreasonable value; so if the instrument is being used for routine checks, it is simple enough to subtract 180° and redetermine the thickness. A multiplicity of thickness values (orders) for each correct set of Δ, Ψ values are given by $t_{\text{actual}} = mt_o + t_m$, where t_m is the measured thickness, m is the order, and t_o is the order thickness given in Table 6.5. Note that color orders and ellipsometer orders do not ordinarily coincide and that the thickness must be approximately known from some independent means.

Instrumentation. The light source for most instruments is the 5,461-Å mercury

Table 6.5. Ellipsometer Order Thickness

Index of refraction	Thickness multiple
1.4	2,627 Å
1.5	2,332 Å
1.6	2,105 Å
1.7	1,925 Å
1.8	1,776 Å
1.9	1,650 Å
2.0	1,544 Å
2.1	1,451 Å

70° angle of incidence. 5,461 Å wavelength (mercury line).

line, but occasionally a sodium line or a laser is used. The latter has the advantage of requiring no filtering to separate other lines and no collimating lenses. If layers other than the common transparent dielectrics are to be studied, other wavelengths may be more appropriate. For example, if a high-resistivity epitaxial silicon layer on a low-resistivity substrate is to be measured, wavelengths in the 30- to 100-μm region are required in order to produce a discernible difference in the optical properties of the substrate and layer.[16]

Quarter-wave compensating plates for use in the visible region are standard optical components and are made of a birefringent material such as mica, quartz, or calcite, which have a pronounced difference in velocity for the two components of polarization. The thickness is adjusted so that the phase difference between the slow and fast directions is 90° for some particular wavelength. If the difference is not exactly 90°, it can still be used and generally will produce negligible errors.

The detector can be the eye, in which case some aids are necessary, or a photodetector such as a photomultiplier. The difficulty with visual detection is that the eye is not very sensitive to the null condition. Accordingly, to compensate for this, a Nakamura biplate may be inserted between the reflecting surface and the analyzer. The two halves of the biplate will appear equally bright when the analyzer is adjusted for extinction, and the eye can judge equal brightness more accurately than it can pinpoint the null. With a photodetector several levels of sophistication can be used. The dc output of the detector, combined with a dark-current-suppression circuit, can be read directly on a microammeter. For limited accuracy like that required for reading layer thickness to a few angstroms, the null can be found directly. For increased accuracy, meter readings can be plotted near and on either side of the null, and the minimum estimated from the curve. The light may be chopped to give an ac signal, or the plane of polarization may be modulated before it enters the analyzer.[17-19] In the latter case, it can be shown that if the angular sweep of the modulation is centered about the null, only direct current and a double frequency will be detected; otherwise odd harmonics will also be present and can be used for balancing. The procedures for the use of the equipment just described are relatively simple but nevertheless slow and cumbersome. In order to speed up the data taking, which may be desirable either because of large numbers of samples or in order to observe layers as they grow, various modifications have developed. They can generally be grouped into three categories. One has used high-speed stepping motors and computer control to perform rapidly the same steps that are required in manual operation.[16,20] A second incorporates optical elements whose plane of polarization can be rotated electrically, and then uses analog procedures to balance the system optically. The currents required to provide the necessary rotation are a measure of the required angles and can be read as desired.[17,21,22] The third depends not on the classical method of interpretation just described but rather on continuously rotating the analyzer and measuring the amplitude and phase shift of the ac signal produced.[23,24]

Should it be necessary to align the instrument, the following procedure can be followed:

1. Without the quarter-wave plate in the instrument set both analyzer and polarizer arms to 90° and align the light source with optics.
2. Set polarizer and analyzer arms to the Brewster angle* of incidence of a reference reflector, e.g., 57° for glass.
3. Remove analyzer or set it so that its plane of transmission is approximately parallel to the plane of incidence. Rotate the polarizer for minimum intensity. This becomes $P \equiv 0$ for the polarizer. Adjust the scale accordingly.
4. Relocate polarizer to 90° incidence angle; adjust analyzer for minimum transmission. Set analyzer scale at 90°, i.e., $A \equiv 90°$.
5. Set polarizer and analyzer arms to the desired angle of incidence, e.g., 70°; raise or lower sample stage until light is centered.
6. Insert clean semiconductor slice.
7. Set P at zero; adjust A for minimum. If reading is not 90°, move A in small increments and readjust P for minimum until the difference of the two readings is 90°. Reset indexes to reflect 0 and 90°.
8. Place quarter-wave compensator in beam. With $P = 0°$, $A = 90°$, adjust compensator for extinction. Set compensator index to zero.
9. If zones I and III (Table 6.4) are to be used, rotate compensator to 315°. Check by taking readings. If the reflective surface is a recently cleaned silicon slice,

$$P_{\mathrm{I}} \simeq 41° \qquad A_{\mathrm{I}} \simeq 12°$$
$$P_{\mathrm{III}} \simeq 131° \qquad A_{\mathrm{III}} \simeq 168°$$

   If P_{I} and P_{III} are approximately 10° higher than these numbers, the compensator was aligned along the slow rather than fast axis.† To correct, rotate it 90° and redo step 8.

6.5 INTERFERENCE EFFECTS

Background. Interference of rays reflected from two different planes can give rise to pronounced maxima and minima in intensity. These in turn can be correlated to the separation of the two planes. Consider first the simplest case of Fig. 6.10*a* in which there is an isolated film surrounded by air. When the path length through

*The Brewster angle ϕ_B is the angle of incidence which gives a zero reflection coefficient for the p component of Fig. 6.6. If the reflecting surface is an air-dielectric interface, $\phi_B = \tan^{-1} n$. By setting the angle at ϕ_B and then adjusting P for a minimum, the polarizer is aligned with its plane of transmission coinciding with the plane of incidence, and the position $P \equiv 0$ is defined.

†The fast and slow axes can also be determined by following the procedure outlined in J. Strong, "Procedures in Experimental Physics," p. 388, Prentice-Hall, Inc., Englewood Cliffs, N.J., 1945.

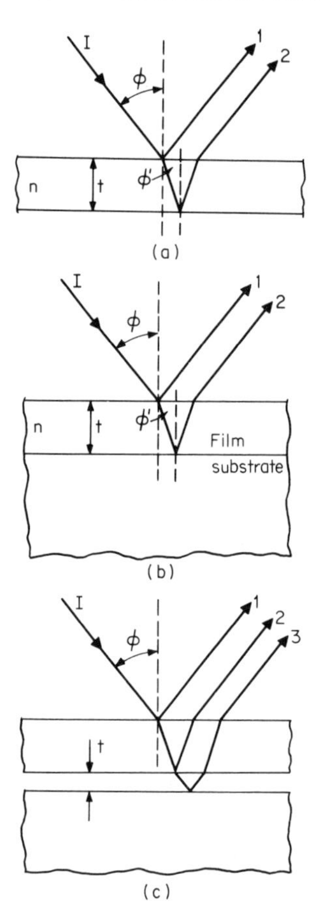

Fig. 6.10. Reflections from various combinations of interfaces.

the sample (neglecting any phase-shift effects at either interface) corresponds to an integral number of wavelengths, constructive interference will occur, and for paths of an odd number of half wavelengths, minima of intensity will occur. Thus the relation between the thickness t and the wavelength λ_0 where a particular maxima or minima occurs is given by

$$t = \frac{m\lambda_0}{2n \cos \phi'} \qquad \text{(maxima)}$$

or

$$t = \frac{(m + \frac{1}{2})\lambda_0}{2n \cos \phi'} \qquad \text{(minima)} \tag{6.14}$$

where m is an integer which becomes progressively smaller as λ increases and ϕ' is the angle of incidence of the beam as it hits the lower reflecting surface.* In principle these effects can be observed for very thick samples, but because of imperfect collimation of the incident beam, the minima become less pronounced as the thickness increases, and eventually become indiscernible.

The configuration of Fig. 6.10b is a common one, and could be, for example, a thin dielectric film on a silicon substrate, or a high resistivity silicon epitaxial layer

*It is also the angle of refraction and is given by $\sin \phi = n \sin \phi'$.

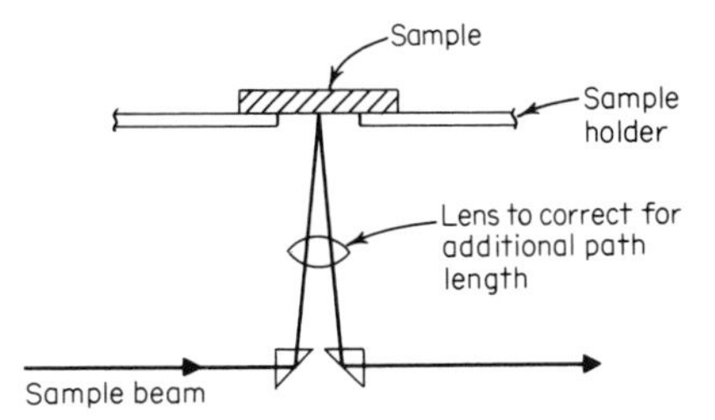

Fig. 6.11. Specular reflection attachment.

on a low-resistivity silicon substrate. The equations for interference maxima and minima are the same as the previous one. However, phase-shift differences between the two interfaces and a change of reflection coefficient with wavelength may need to be considered. The configuration of Fig. 6.10*c* is also likely to arise, and if the top plate is thick enough, interference between rays 1 and 2 will be negligible, and only the interference between rays 2 and 3 will be observed. In this case, the "thickness" measured is that of the air gap, i.e., the separation of the plate and the surface below.

There are numerous optical arrangements that capitalize on interference effects. Many of them are directly applicable to thickness measurement, but oddly enough most interferometers were devised for other purposes and are not necessarily suitable.

Spectrophotometer. By using a specular reflection attachment, shown schematically in Fig. 6.11, to deflect the spectrophotometer beam up to the reflecting surfaces and then back again into the instrument, the wavelength can be varied. For a fixed separation [t of Eq. (6.14) and Fig. 6.10)], the reflected intensity will vary cyclically as in Fig. 6.12. Equation (6.14) could be applied to one of the peaks or valleys of the trace. However, for some randomly chosen peak, m will not be known, although if the recording starts at a long enough wavelength, the first minimum and the first maximum will be observed. If the first minimum ($m = 0$) is recorded, from Eq. (6.14),

$$t = \frac{\lambda}{4n \cos \phi'} \tag{6.15}$$

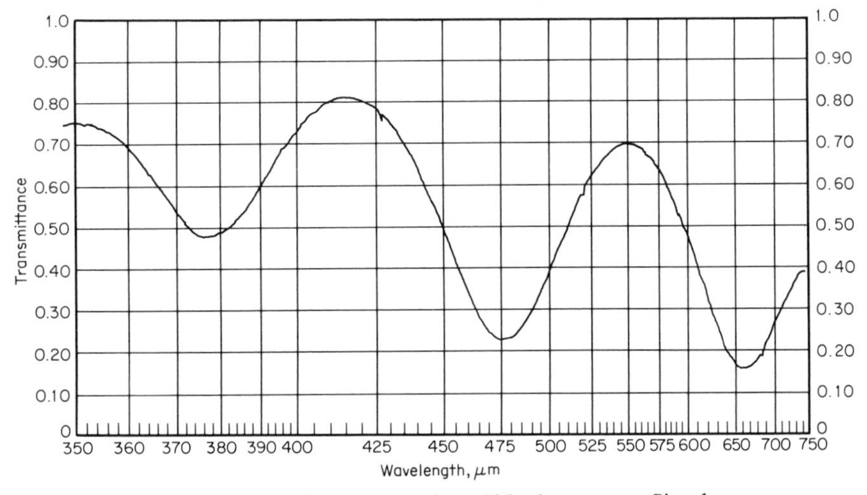

Fig. 6.12. Reflected intensity of an SiO_2 layer on a Si substrate.

For transparent films like oxides, which are usually 2,000 to 10,000 Å thick, the conditions of Eq. (6.15) are relatively easy to fulfill and can be used for measurement.[31] If the film is very thin (e.g., a few hundred angstroms), there will be difficulty in finding an instrument operating at wavelengths short enough to produce even the first minimum. Then the ratio of the reflectivity of the substrate with no film on it to the reflectivity with a film at some fixed wavelength can be used. It is described in more detail in the next section.

By extending Eq. (6.15), the expression for additional consecutive peaks is given as in Eq. (6.16) and has more general application,

$$\begin{aligned} t &= \frac{m\lambda_0}{2n\cos\phi'} \\ t &= \frac{(m+1)\lambda_1}{2n\cos\phi'} \\ t &= \frac{(m+2)\lambda_2}{2n\cos\phi'} \\ &\vdots \\ t &= \frac{(m+i)\lambda_i}{2n\cos\phi'} \end{aligned} \tag{6.16}$$

where λ_i is the wavelength at successive maxima and $\lambda_0 > \lambda_1 > \lambda_2 \ldots$. From these expressions, m is given by

$$m = \frac{i\lambda i}{\lambda_0 - \lambda_i} \tag{6.17}$$

where i is the number of complete cycles from λ_0 to λ_i. Substituting this value of m back into Eq. (6.14) gives Eq. (6.18), which can be used directly for calculating thicknesses,

$$t = \frac{i\lambda_0\lambda_i}{2n(\lambda_0 - \lambda_i)\cos\phi'} \tag{6.18}$$

where i is the number of maxima from λ_0 to λ_i. This expression has the additional advantage of compensating for any phase-shift differences between the two surfaces as long as they are independent of wavelength. $2n \cos \phi'$ is a combined constant of the machine being used and the index of refraction of the film. If more convenient, it can also be expressed in terms of the angle of incidence of the beam onto the first surface (i.e., the beam angle built into the attachment) and replaced by

$$2(n^2 - \sin^2\phi)^{1/2}$$

Equations (6.16), (6.17), and (6.18) will work equally well for minima or one maximum and one minimum as for maxima only. The use of adjacent maximum and minimum positions is of particular importance if the available chart is reduced in length because of some instrument problem, or if the thickness-available wavelength combination is such that only one maximum and minimum is recorded.

These kinds of measurements are widely used both because of the availability of equipment and because they are nondestructive. They have been used for very thin silicon and germanium slices[34] (the configuration in Fig. 6.10*a*), for dielectric films on semiconductor substrates[33,37] and for semiconductor epitaxial films.[38-41]

The fact that most epitaxial-layer thicknesses are measured by spectrophotometer has led to extensive studies of the effects of phase change and substrate resistivity. The following paragraphs summarize the various procedures directly applicable to epitaxial-thickness measurements.

Epitaxial-Film Thickness by Spectrophotometer.[38-50] The use of a spectrophotometer for the specific purpose of measuring epitaxial-film thickness poses several additional difficulties. The major one is that there is little difference in the refractive index of the layer and the substrate. This leads to a low-amplitude reflection from the layer-substrate interface. The index difference becomes more pronounced as the wavelength increases, so that an idealized interference pattern will increase in amplitude as shown in Fig. 6.13. The amplitude will also increase as the concentration of the substrate increases, provided that the concentration of the layer remains fixed. Therefore, there are substrate-layer combinations that cannot be measured this way. For example, ASTM, F 95 71 recommends that for Si, the layer have a resistivity greater than 0.1 Ω-cm and the substrate a resistivity of less than 0.02 Ω-cm.

Equation (6.14) was predicated on the same phase shift at the air-layer interface as at the layer-substrate interface, which is not true for epitaxial layers. Equation (6.18) assumed that the phase shift, although different, was independent of wavelength. Unfortunately, this is probably not true either, and based on theoretical calculations, corrections are usually applied.[44] Such corrections are not required when measuring epitaxial layers overgrown on insulating substrates, e.g., Si on sapphire or on polycrystalline layers overgrown on amorphous dielectric substrates (e.g., Si on SiO_2). The revised equation necessary to include the phase shift θ_i is*

$$t = \frac{(m - \frac{1}{2} + \theta_i/2\pi)\lambda_i}{2(n^2 - \sin^2 \phi)^{1/2}} \tag{6.19}$$

If m is allowed to be $\frac{1}{2}$ order, e.g., $2\frac{1}{2}$, both maxima and minima are combined into one equation. The $(-\frac{1}{2})$ arises because of the phase-shift term. If θ is assumed

*Note that these symbols are different from those in some of the literature. In many cases the order is designated by P, the number of maxima or minima by m, the angle of incidence by θ, and the phase shift by ϕ.

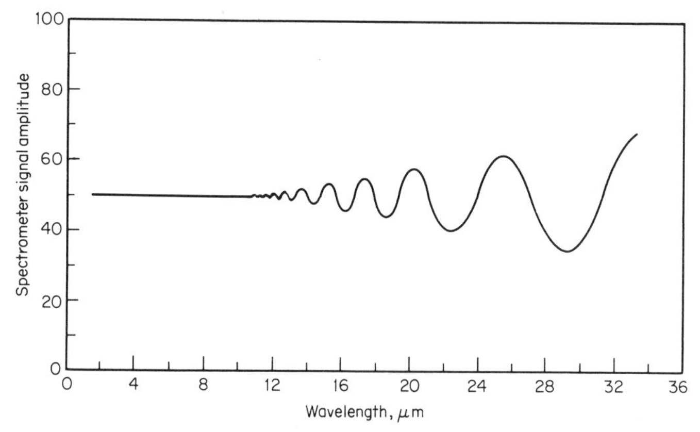

Fig. 6.13. Effect of increasing wavelength on the amplitude of a silicon epitaxial-layer interference pattern.

to be π (i.e., a perfect reflection) then Eq. (6.19) reduces to Eq. (6.14). An equation analogous to Eq. (6.18) but including phase shift can be written which does not require m to be separately determined [Eq. (6.20)]; however, the ASTM procedure recommends the use of Eq. (6.19) and m as determined from Eq. (6.21)

$$t = \frac{i\lambda_0\lambda_i}{2(n^2 - \sin^2\phi)^{1/2}(\lambda_0 - \lambda_i)}\left[\frac{1 - (\theta_0 - \theta_i)}{2i\pi}\right] \tag{6.20}$$

$$m = \frac{i\lambda_0}{\lambda_0 - \lambda_i} + \tfrac{1}{2} - \left[\frac{\theta_0\lambda_0 - \theta_i\lambda_i}{2\pi(\lambda_0 - \lambda_i)}\right] \tag{6.21}$$

The phase shift has been calculated for a range of substrate resistivities and is available in tabular form.[45]

The equations have all been developed by assuming an abrupt interface between the layer and the substrate. If it is not, then the amplitude of reflection will be reduced. In fact the reduction can be used as an indication of interface grading. Figure 6.14 shows reflections from epitaxial slices deposited in a horizontal reactor at different temperatures and illustrates the effect of grading at the higher deposition temperature.

Automated Thickness Measurements. The use of spectrophotometer readouts such as the one shown in Fig. 6.12 has some basic disadvantages. One is that for production checking of thousands of slices, the instrument is slow and prone to operator error. To minimize the operator error and calculation time, tables have long been used. However, automatic readouts are certainly more desirable.[50,51] Even that, however, does not solve the wear problem that is inherent in the linkages and drive mechanisms used in conventional spectrophotometers. Most instruments were designed for laboratory use and not for continuous operation and hence tend to have a relatively high repair rate.

Use of a Michelson interferometer as a Fourier-transform spectrophotometer substantially reduces the number of moving parts and the downtime.[52] Figure 6.15*a* shows the basic interferometer. For epitaxial-film measurement it may be coupled to the sample as in Fig. 6.15*b*. The output (interferogram) of the interferometer detector as shown in Fig. 6.15*c* looks nothing like the conventional spectrometer

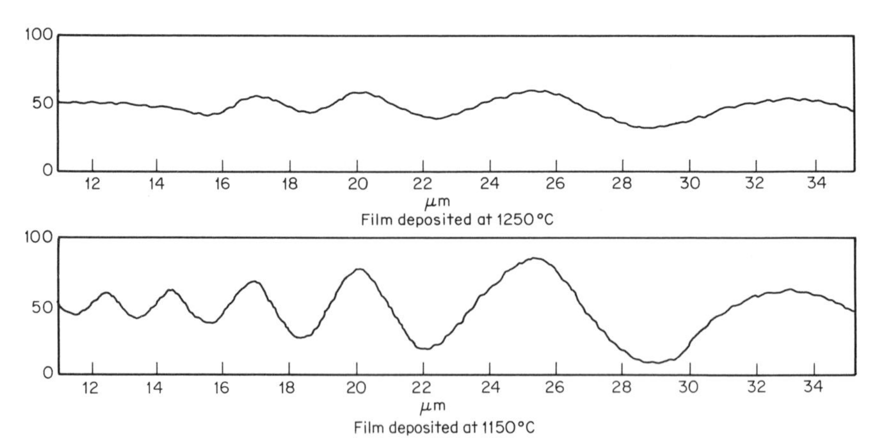

Fig. 6.14. Effect of a diffused interface on a spectrophotometer trace of a silicon epitaxial slice.

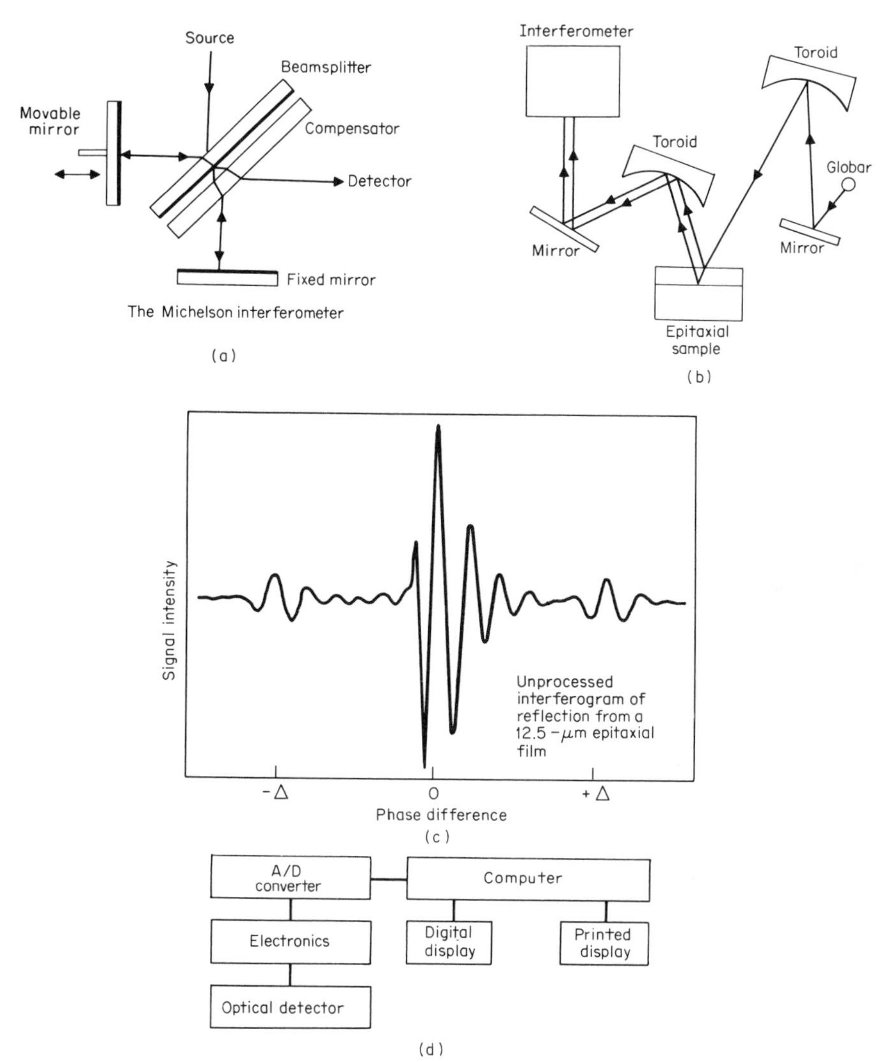

Fig. 6.15. Details of Fourier-transform spectroscopy as used for thickness measurement. (*After Cox and Stalder.*[52])

trace but can be mathematically manipulated to give comparable information.[53] Since there is only one moving part, the mechanical reliability is greatly enhanced. The computational complexity, however, has been considerably increased, and the data must be computer-processed. The data-processing-equipment block diagram is shown in Fig. 6.15*d*. With proper instruction the computer can perform phase shift corrections comparable to those discussed in the previous section.

Specialized Spectrophotometers. In a number of applications a complete spectrophotometer is either not warranted or undesirable. For example, if the wavelength is fixed, and reflection from a layer is observed as a function of layer thickness, the intensity will oscillate as shown in Fig. 6.16. Detailed calculations of reflectivity of SiO_2 and silicon nitride on a silicon surface for several wavelengths are available in the literature.[33,37] By observing this change in amplitude, film-thickness increase can be monitored, and indeed such equipment has been used for many years during

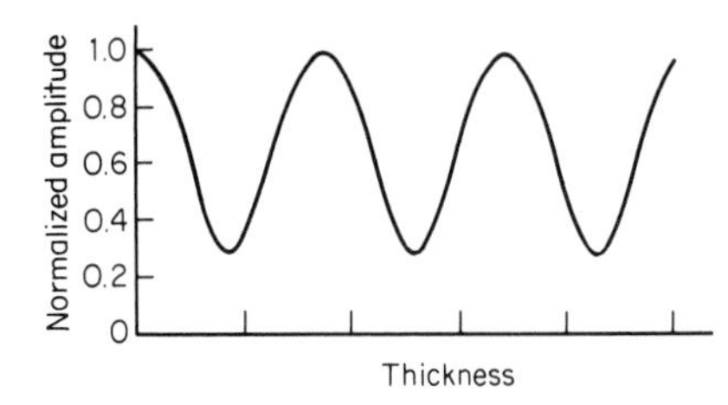

Fig. 6.16. Intensity variation of monochromatic reflected light as the layer thickness increases. The peak-to-valley ratio depends on the relative properties of the film and substrate.

dielectric-film deposition.[54] For the special case of silicon in thin layers, e.g., growing on sapphire, the same approach can be used to measure the silicon as it grows.[55] This is unique in that it is one of the few methods by which epitaxial-layer thickness can be followed during growth.

These techniques can be used with microscope optics so that very small areas can be examined.[33,56] An ordinary spectrophotometer may require anything from 1 mm^2 to 1 cm^2 for observation, while with a high-power metallurgical microscope the thickness of oxide in a single emitter can be measured. For this application, a detector such as a silicon photodiode can be coupled to the microscope where a camera would normally be placed. Either a spot-frequency or a graded-spectrum interference filter can then be inserted in the optical path and the reflectivity measured.

Variable-Angle Measurements.[57] Instead of changing the wavelength to produce the intensity variations of Eq. (6.14), the path length can be changed by choosing different angles of incidence as shown in Fig. 6.17.* In this case Eq. (6.14) still holds but Eq. (6.16) must be changed to

$$\begin{aligned} t &= \frac{m\lambda}{2n \cos \phi_0'} \\ t &= \frac{(m+1)\lambda}{2n \cos \phi_1'} \\ t &= \frac{(m+2)\lambda}{2n \cos \phi_2'} \\ &\vdots \\ t &= \frac{(m+i)\lambda}{2n \cos \phi' i} \end{aligned} \tag{6.22}$$

*See Ref. 4, p. 115.

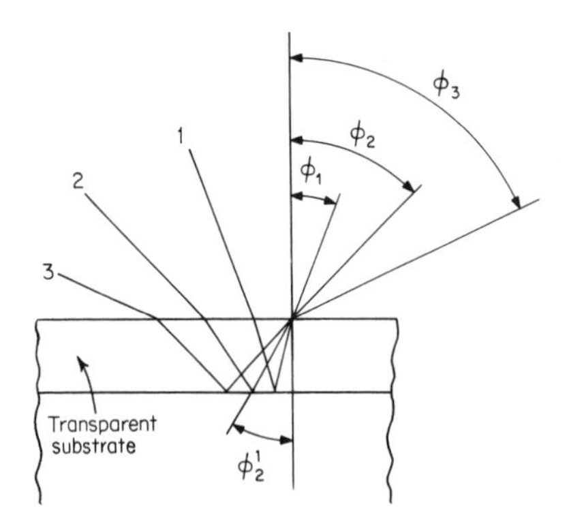

Fig. 6.17. Principle of variable-angle monochrometer. By varying ϕ, the path length in the dielectric is changed until the conditions for interference are fulfilled.

now
$$m = \frac{i \cos \phi_0}{\cos \phi_i' - \cos \phi_0'} \tag{6.23}$$

and the equation analogous to Eq. (6.19) is

$$t = \frac{i\lambda}{2n(\cos \phi_i' - \cos \phi_0')} \tag{6.24}$$

where again i is the number of fringes (bright or dark bands) observed as ϕ' is changed from ϕ_0' to ϕ_i'. Figure 6.18 shows the optical arrangement of one such equipment, which has been referred to as VAMFO (variable-angle monochrometer fringe observation).[57] In use, the angle ϕ is manually varied and the position of maxima and/or minima recorded. The thickness can then be calculated from Eq. (6.24), remembering that ϕ' is required in the equation and ϕ is measured by the instrument. Alternately, after the positions of two or three maxima/minima have been recorded, they can be plotted on a slip of paper and slid along a set of previously calculated curves until the points match. Different curves are required for materials with differing refractive indices. For other simplifying approaches, see Ref. 57.

Visual Determination. When thin transparent layers are viewed either directly by eye or through a microscope, interference effects will give the layer a characteristic color which depends on the film thickness, its index of refraction, and the spectral distribution of the viewing light. The latter is a very important point, because for the same thickness, colors viewed under normal laboratory fluorescent lighting and under an incandescent-lighted metallurgical microscope will be appreciably different.

If calibrated color charts are prepared, these interference effects can be accurate to within 100 to 200 Å. Such charts are widely used to evaluate silicon oxide and nitride thickness. The chart can be descriptive words vs. thickness, as in Tables 6.6 and 6.7 for SiO_2 and Si_3N_4, printed or photographed colors vs. thickness, or a range of actual samples mounted in such a way that the unknown can be placed beside it for comparison. The latter is, of course, by far the most satisfactory, since printing or photography does not give accurate color rendition and words are even worse. When preparing the reference samples, an ellipsometer or spectrophotometer can be used to measure their thickness. One potential source of difficulty is the fact that different orders have substantially the same colors; so if other information does not allow the thickness to be independently estimated to within one order, a gross error could be made. By careful attention to the exact shades it is in principle possible to tell the order directly, but this approach is not recommended for the occasional observer. However, if the unknown and the reference are viewed at angles other than normal, colors will not match unless they are both of the same order. When viewed at an angle of incidence θ,

$$t = \frac{t_0}{\cos \theta}$$

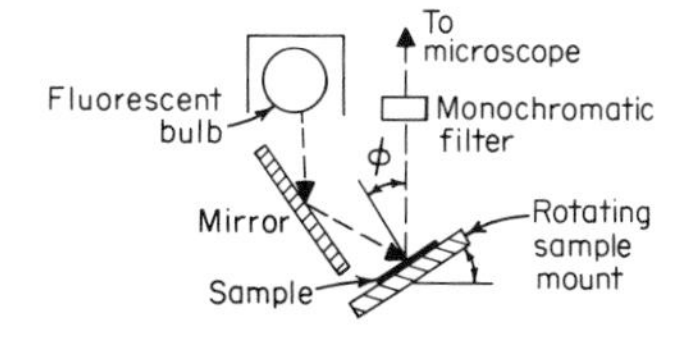

Fig. 6.18. Optical path for VAMFO. Properly situated, the fixed mirror–fluorescent bulb combination will provide lighting independently of the position of the sample. Otherwise it must be adjusted periodically. (*After Pliskin and Conrad.*[57])

Table 6.6. Color Chart for Thermally Grown SiO_2 Films Observed Perpendicularly under Daylight Fluorescent Lighting

Film thickness, μm	Color and comments
0.05	Tan
0.07	Brown
0.10	Dark violet to red-violet
0.12	Royal blue
0.15	Light blue to metallic blue
0.17	Metallic to very light yellow-green
0.20	Light gold or yellow—slightly metallic
0.22	Gold with slight yellow-orange
0.25	Orange to melon
0.27	Red-violet
0.30	Blue to violet-blue
0.31	Blue
0.32	Blue to blue-green
0.34	Light green
0.35	Green to yellow-green
0.36	Yellow-green
0.37	Green-yellow
0.39	Yellow
0.41	Light orange
0.42	Carnation pink
0.44	Violet-red
0.46	Red-violet
0.47	Violet
0.48	Blue-violet
0.49	Blue
0.50	Blue-green
0.52	Green (broad)
0.54	Yellow-green
0.56	Green-yellow
0.57	Yellow to "yellowish" (not yellow but is in the position where yellow is to be expected. At times it appears to be light creamy gray or metallic)
0.58	Light orange or yellow to pink borderline
0.60	Carnation pink
0.63	Violet-red
0.68	"Bluish" (Not blue but borderline between violet and blue-green. It appears more like a mixture between violet-red and blue-green and looks grayish)
0.72	Blue-green to green (quite broad)
0.77	"Yellowish"
0.80	Orange (rather broad for orange)
0.82	Salmon
0.85	Dull, light red-violet
0.86	Violet
0.87	Blue-violet
0.89	Blue
0.92	Blue-green
0.95	Dull yellow-green
0.97	Yellow to "yellowish"
0.99	Orange
1.00	Carnation pink
1.02	Violet-red
1.05	Red-violet
1.06	Violet
1.07	Blue-violet
1.10	Green
1.11	Yellow-green
1.12	Green
1.18	Violet
1.19	Red-violet
1.21	Violet-red
1.24	Carnation pink to salmon
1.25	Orange
1.28	"Yellowish"
1.32	Sky blue to green-blue
1.40	Orange
1.45	Violet
1.46	Blue-violet
1.50	Blue
1.54	Dull yellow-green

After Pliskin and Conrad.[57]

where t_0 is the thickness read from the color chart. The charts may also be used for materials other than the ones for which they were originally intended.[57] In that case,

$$t = \frac{t_0' n_0}{n_f}$$

where n_0 is the index of refraction of the original film and n_f that of the new film.

Table 6.7. Color Chart for Si_3N_4. Tungsten Filament Vertical Illumination and Observation through Low-Power Microscope

Order	Si_3N_4 color	Si_3N_4 thickness range μ
	Silicon	0–0.020
	Brown	0.020–0.040
	Golden brown	0.040–0.055
	Red	0.055–0.073
	Deep blue	0.073–0.077
1st	Blue	0.077–0.093
	Pale blue	0.093–0.10
	Very pale blue	0.10–0.11
	Silicon	0.11–0.12
	Light yellow	0.12–0.13
	Yellow	0.13–0.15
	Orange-red	0.15–0.18
1st	Red	0.18–0.19
	Dark red	0.19–0.21
2d	Blue	0.21–0.23
	Blue-green	0.23–0.25
	Light green	0.25–0.28
	Orange-yellow	0.28–0.30
2d	Red	0.30–0.33

From F. Reizman and W. Van Gelder, *Solid State Electron.*, **10**:625–632 (1967).

6.6 BEVEL MEASUREMENTS BY INTERFEROMETRY AND OTHER METHODS

Interferometry. If the sample has been beveled as shown in Fig. 6.19 to expose some underlying feature, e.g., a stained junction, a partially reflecting reference plane extending out over the incline from the original surface will produce interference fringes between the incline and the reference.[58] The fringe spacing will be very close;

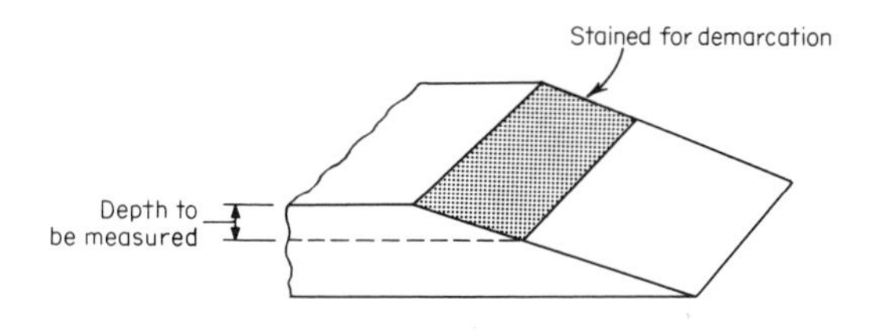

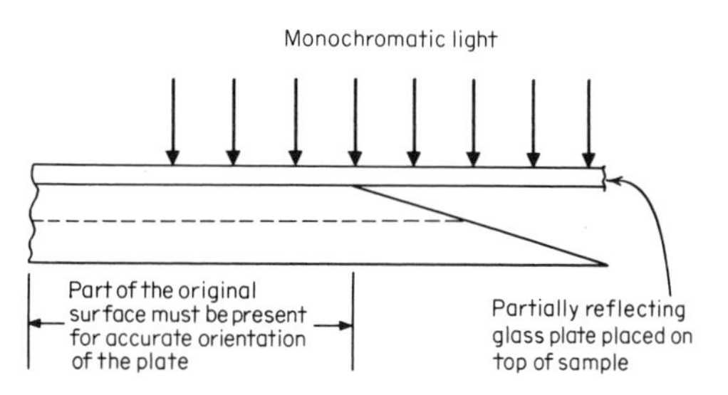

Fig. 6.19. The use of simple interference fringes combined with a bevel to give depth. The mechanics of beveling and staining are covered in Chap. 7.

so a metallurgical microscope with a filter to give reasonably monochromatic light is used for observation. The reference surface can be a small piece of cover glass carefully positioned on the sample, or a reference-plane attachment which is available from some microscope manufacturers. Photographs are usually taken and all necessary measurements and fringe counting done from them. Such bevel interferometry is probably the second most widely made measurement in the semiconductor industry, following only the four-point-probe resistivity determination.

When the reference plane is in perfect coincidence with the original surface and if the incline is not skewed, the fringes (lines) will all be aligned parallel with the line of intersection of the original surface and the incline as shown in Fig. 6.20*a*. By counting the number of lines N from the beginning of the incline out to the point of interest, the vertical depth is determined independently of whether or not the incline is flat or curved, and without knowledge of the incline angle by

$$t = \frac{N\lambda}{2}$$

If the reference is tilted up as shown in Fig. 6.20*b* about a line 3-4 which is parallel to the intersection 1-2, additional fringes will be seen in the region 1-2-3-4, and now the pattern will appear as shown. Since the reference plane is tilted, the number of lines counted must be corrected for the amount the reference rises in the horizontal distance between line 1-2 and the feature of interest. This will be equal to the drop in a similar distance back from line 1-2 toward the pivot. Thus the true depth is given by $N_X - \bar{N}_X$ of Fig. 6.20*c*. If the reference is tilted down, an analogous situation exists except that now the back fringes ($\bar{N}$) must be added. A basic difficulty with

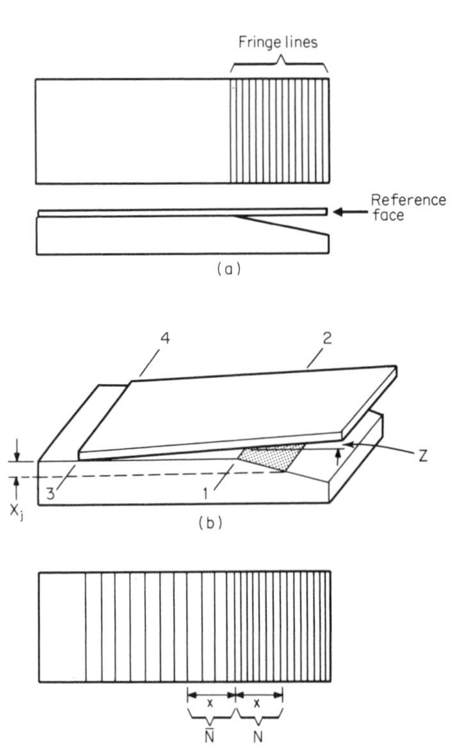

Fig. 6.20. Effect of reference-plane tilt on interference-depth measurements. The depth X_j is given by $(N\text{-}\bar{N})\lambda/2$ where N, $\bar{N}$ is the number of fringes in the distance x.

this approach is that if the lines really are perfectly aligned as shown, there is no way of distinguishing between the two cases. As will be described a little later, if alignment is not perfect, there are procedures for determining the sign of the correction.

The reference plane can also be tilted as shown in Fig. 6.21*a*. In that case the character of the fringes is quite different, as is depicted in Fig. 6.21*b*. This is the configuration normally used when measuring total step height, although it can also be used for measuring partway down (e.g., x_j of Fig. 6.20). The step height t is given by the lateral displacement D of a particular fringe as it goes over the step divided by the normal fringe separation d, all multiplied by $\lambda/2$. That is,

$$t = \frac{D\lambda}{2d}$$

When $\phi \ll \theta$, the fringes over the bevel (step) will be essentially parallel to line 1-2 of Fig. 6.20, and the fringes look like Fig. 6.22*a*. The depth at some point A is determined from the number of fringes between A and B. If there is also tilt up from the bevel (Fig. 6.20*b*), the fringes along the flat region will go off at an obtuse angle as in Fig. 6.22*b* and the correction of Fig. 6.20*b* must be made. If the lines are acute as in Fig. 6.22*c*, the plane tilts down and a positive correction is required. Thus the sign of the correction is determined by the angle the fringes make as they go over the edge of the bevel.[59] An alternate approach is to plot the fringe number, starting with some arbitrary value, as a function of distance along a line drawn perpendicular to the bevel edge.[58] The three shapes corresponding to the three cases of Fig. 6.22 are shown in Fig. 6.23.

No matter which way the fringes are to be interpreted, the reference ideally would be carefully adjusted by gentle, judicious movement to give lines like those of Fig. 6.22*a*. Practically, either of the other positions is acceptable and generally much more quickly obtained. Common procedure is to count the whole number of lines between the two points of interest, but for shallow junctions or narrow bases the error can

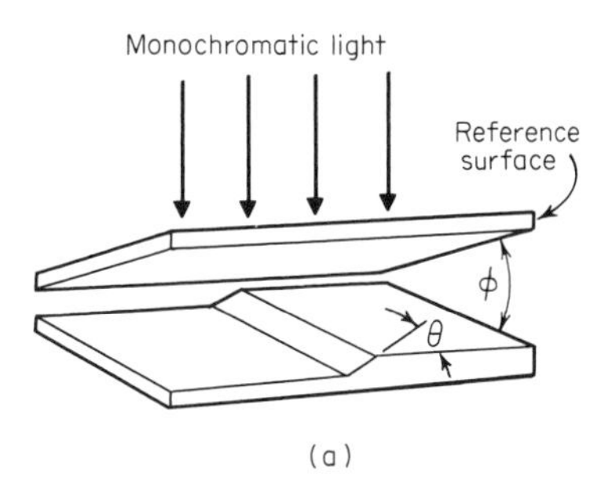

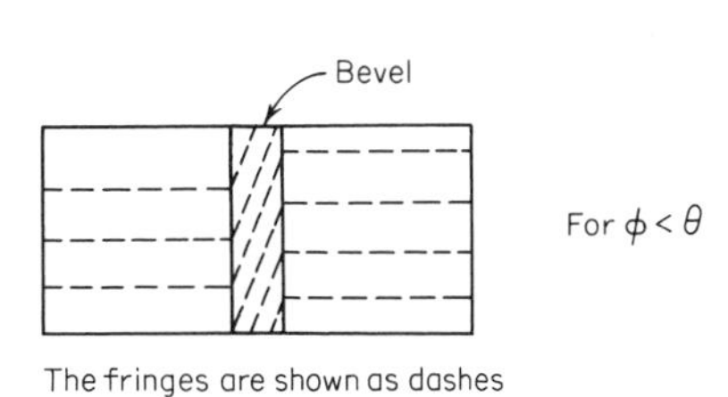

Fig. 6.21. Alternate approach for alignment of the reference plane.

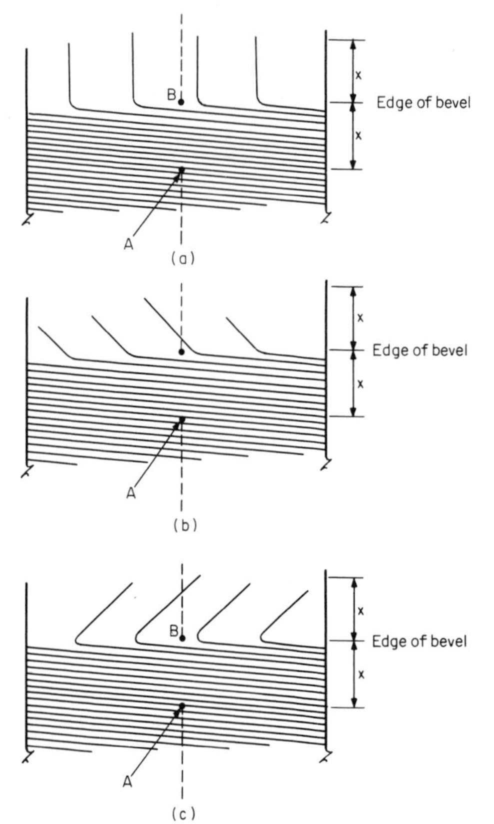

Fig. 6.22. Effect of double tilt on fringe pattern.

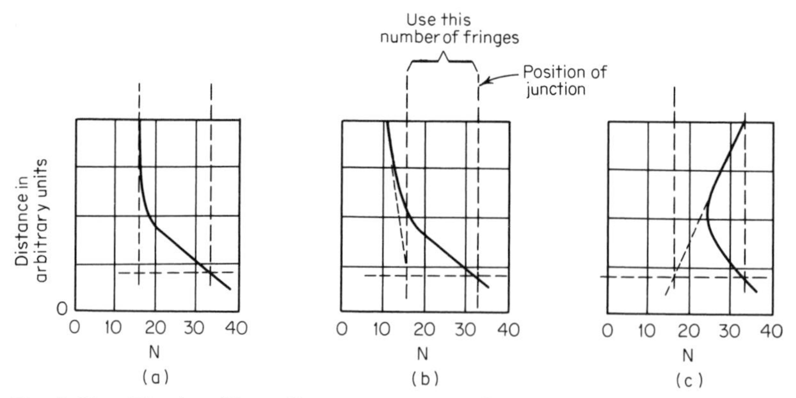

Fig. 6.23. Plotting fringe distance to correct for reference surface tilt. The portion of the curve corresponding to the fringes over the reference surface is extrapolated to the depth to be measured. The double value of X for the same fringe number in (*c*) occurs because the fringes for that case double back (Fig. 6.23*c*) and a line perpendicular to the bevel edge cuts the same fringe twice. (*After Bond and Smits.*[58])

be quite large, particularly on a one-line base. Fractions of lines should be estimated in those cases, and to increase precision, a densitometer can be used to scan the photograph and more accurately define the fringe position.[60]

The fringes have all been depicted on the drawings as very sharp. However, when both surfaces have very low reflectivity, the lines are actually quite diffuse. If a reflecting layer such as aluminum is put on the reference surface and its reflectivity is increased to 90 to 95 percent and the reflectivity of the bottom surface is also increased, the fringe intensity vs. distance changes character, and the fringes really do appear as sharp, well-defined lines. Such a procedure, referred to as *Tolansky multiple-beam interferometry,* can be used to show exceedingly small steps or thickness variations.[61,62]

Interference in the Layer. If the sample is transparent and wedge-shaped, so that t of Eq. (6.18) is changed and λ is held constant, interference fringes will be generated along the wedge. The change in thickness between two consecutive fringes is

$$\Delta t = \frac{\lambda}{2n}$$

so that if the index of refraction n is known, the thickness can be determined by counting the total number of fringes. A metallurgical microscope with a filter can be used for observation. The wedge needs to have a gentle slope so that there is enough separation between fringes for easy resolution. While it is sometimes possible to cleave a silicon slice and have a coincidental fracture of the SiO_2 on top give sufficient taper, the usual procedure for transparent dielectric layers is to etch the step. If a mask that adheres poorly is used deliberately, its lifting will allow the etch to produce the slope necessary to provide adequate fringe separation. For SiO_2 such diverse waxes as Apiezon W in toluene[63] and melted blue china marking pencils[64] have been recommended. The Monsanto evaluation standards[2] specify melted Apiezon W wax followed by a 2-min etch in 49 percent HF (more if the oxide is exceptionally thick). Their procedure for estimating SiO_2 thickness on silicon closer than ± 1 line is to note whether the bare silicon is the same color or lighter than the oxide under the monochromatic light. If it is approximately the same, the thickness is an integral number of fringes thick. If the oxide is darker, the number of dark bands between the bare silicon and the last distinct white band are counted. Then, if an interference filter is used as a monochrometer, it is turned at an angle to the beam in order to change the wavelength. If the oxide appears darker with tilting, the thickness is the number of fringes previously counted plus a half or less. If the oxide gets lighter, it is the count plus a half or more.

Mechanical. If the bevel of Fig. 6.19 is quite flat, the angle θ it makes with the original surface and the lateral dimension can both be measured and the depth x_j calculated by

$$x_j = L \tan \theta$$

The angle may be determined by a gunner's quadrant,[65] a microgoniometer, or a surface profilometer. In the latter case the horizontal and vertical scales of the trace will be different; so the angle must not be measured directly from the trace.

The lateral measurement can be made by moving the microscope stage and thus the image across the field of view by means of a calibrated screw or an attached vernier. In most instruments, toolmaker's microscopes being an exception, the graduations are quite coarse and only gross measurements can be made. A more common approach is to use a filar eyepiece which moves a crosshair across the real-image plane. A calibration must ordinarily be done by counting the number of graduations required to move across an accurately engraved scale on which the objective is focused. Such scales are relatively inexpensive and are a necessity. Instead of a filar eyepiece, the scale can be photographed at the same magnification as are pictures of the object. A new scale can then be cut out of the photograph and used directly as a ruler to measure any desire dimensions on the other photographs. When this approach is used, however, great care must be taken to make sure that the focus is very precise in each case, because the apparent separation of lines will vary with the amount of defocusing. A somewhat more elegant method is to use image shearing and measure the angular displacement necessary to shift one of the images by the width of the object to be measured. By using a television-screen display, an electronic scale can be superimposed and measurements made somewhat easier. However, whichever method is used, the accuracy cannot be better than the limits of resolution of the optics used, no matter how big a photograph or TV display is used.

There are also beveling configurations which do not require a knowledge of the bevel angle. For example, if a cylinder[66] or sphere[67,68] of diameter D is used to grind a depression in the surface as shown in Fig. 6.24, the depth to the demarcation

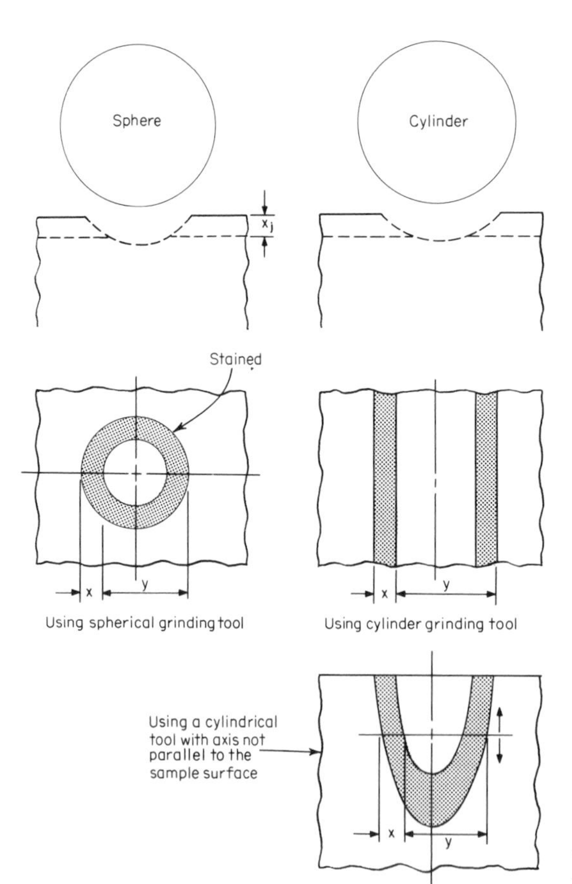

Fig. 6.24. Use of spherical and cylindrical beveling for depth measurements.

line x_j is given approximately by

$$\frac{xy}{D}$$

where x and y are defined in the figure. The lateral measurements can be made by any of the methods just discussed. The mechanics of the grinding operation is not particularly critical, but the abrasive should be very fine, e.g., Linde B. The tool diameter can conveniently be in the $\frac{1}{4}$- to $\frac{1}{2}$-in-diameter range, and for the cylindrical groover, the tool axis may be a few degrees away from parallelism with the surface without producing excessive error.

6.7 STEP-HEIGHT MEASUREMENTS

If the layer can be stripped away in one area so that a step is available (configuration *d* of Fig. 6.1), several mechanical and optical measurements are available.

With Microscope. If the step is large enough, a microscope with a high-power objective and calibrated vertical motion can be focused on each surface and the height measured.

With Light-Section Microscope. Such a microscope is applicable only for steps greater than approximately 1 μm. See Chap. 8 for further details.

With Profilometer. A mechanical stylus drawn across the step affords a rapid and simple way of measuring height of from a few hundred angstroms up to several micrometers. For precise location of the stylus, a stereoscopic microscope is used for positioning the sample under the stylus. Should the step be made by etching a narrow groove or small hole in the layer, consideration should be given to the relative size of the groove and the stylus. With the kinds of microdefinition available in microcircuit technology it is quite possible to produce grooves so narrow that the stylus cannot reveal the bottom. This method is applicable to epitaxial layer thickness measurements since steps can be generated by locally masking the original surface with an oxide[1] or silicon nitride layer[74] and then removing the overgrowth.

6.8 HIGH-POWER MICROSCOPE PLUS VISIBLE BOUNDARIES

If the layers are thicker than a few micrometers and transparent, a microscope can be focused on first one surface and then the other. The thickness is then calculated by multiplying the index of refraction by the measured separation. Either a conventional microscope with calibrated vertical motion or a toolmaker's microscope with dial indicator can be used. In order to reduce the depth of field and thus more accurately define the two surfaces, high-power objectives should be used. For looking at silicon, e.g., Si on SiO_2 on polycrystalline silicon (DI), or GaAs, an infrared microscope can be used. In general it is not possible, however, to see epitaxial- or diffused-layer boundaries by this method, although on occasion the interface may be so poor that some trace will be visible.

6.9 WEIGHT DIFFERENTIAL

By weighing a substrate before and after a layer is deposited, the average thickness can be calculated if the area and the density are known.[1] In the case of epitaxial layers, the density will be known very accurately, but because of the difficulty in preventing all growth from the backside (assuming the slice lies on some sort of heater), the area is seldom known. Conversely, for thin metal deposits by vacuum techniques, the density may not be accurately known, but the areas are generally well defined. By continuously weighing slices during deposition (more applicable to hot-wall systems where the slice can be suspended and simultaneously deposited on both sides), the film thickness vs. time can be measured.[69]

Despite the fact that this system is in principle very simple, it has several rather serious disadvantages such as: (1) It gives only an average value. (2) It requires maintaining the identity of all slices from first weighing through the deposition process to final weighing (very difficult in large-quantity production). (3) It requires control or measurement of the film area. (4) For very thin layers on thick substrates a combination of the sensitivity required and total range may be difficult to achieve in one instrument. (5) When used for epitaxy, it precludes the use of *in situ* substrate etching prior to deposition.

6.10 CRYSTALLOGRAPHIC METHOD[71-73]

Stacking faults originating during epitaxial growth have a predictable geometry. (See Chap. 2.) The size of their outline on the surface is directly proportional to thickness and can be used to calculate the thickness of layers which grew after their inception. Occasionally some difficulties will arise during deposition and initiate additional faults. These will not be as big as the ones which nucleated at the beginning; so before a particular stacking fault is chosen for measurement, the area should be scanned and only the largest ones considered.

The perpendicular distance from the surface to the point of origin of the fault is given by the length of a side of the polygonal fault outline multiplied by the appropriate number from Table 6.8. Except for (111) and (100) surfaces, the sides will not be of equal length. Thus more than one multiplication factor is given in those cases. However, any side multiplied by its own factor will give the same answer as any other. If perchance a layer of orientation other than one of those of Table 6.8 is being measured, Ref. 73 gives details for calculating the appropriate dimensions. There will occasionally be linear faults which will have the same dimensions as a similarly oriented side of a polygon and can also be used* as long as care is taken to understand what their orientation is [no problem on (100) and (111) surfaces, since all sides are of equal length].

Under most growth conditions, stacking faults can be seen with a phase- or interference-contrast microscope without any additional surface treatment. In such cases the method is nondestructive. Should a light etch be required to define them, the details may be found in Chap. 2. If the etch removes a substantial thickness of the film, that amount must be added to the thickness determined from the fault dimensions. To estimate the amount removed, the etch rate is determined in a separate experiment and combined with the sample etch time. The primary disadvantage of using stacking faults is that high-quality depositions have few of them and they may thus be virtually impossible to find. The calculated value differs from almost all other methods in that it measures from surface to the actual epitaxial substrate interface, and not to some other region defined by substrate outdiffusion. There may therefore be correlation problems between this and other methods.

*ASTM Tentative Test Method F 143-71T recommends against using dimensions from incomplete polygons.

Table 6.8. Multiplication Factors to Be Used for Layer Thickness from Stacking-Fault Dimensions

Orientation	Multiplication factor	
	a	*b*
(111)	0.816	
(100)	0.707	
(110)	0.5	0.577

For additional orientations see Ref. 73. Thickness = (length of side) × multiplication factor.

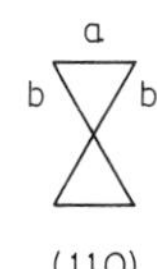

6.11 MISCELLANEOUS METHODS FOR METALLIC FILMS

There are numerous methods which depend on conductivity or magnetic induction which can be used for metal-film-thickness measurement.[70] Most of them are applicable to relatively thick films of considerable lateral extent. Electron backscattering has also been used, as has x-ray fluorescence. The latter does not necessarily depend on the metallic properties but works best on heavy elements and is ordinarily applied to metallic films. The x-ray microprobe allows measurements to be made over exceedingly small areas. Beam size can be as small as a few micrometers in diameter, and sensitivity is such that it can be used to estimate the thickness of Ni/Cr films in the 100-Å range.

REFERENCES

1. C. C. Allen and E. G. Bylander, Evaluation Techniques for and Electrical Properties of Silicon Epitaxial Films, in John B. Schroeder (ed.), "Metallurgy of Semiconductor Materials," Interscience Publishers, Inc., New York, 1962; Robert C. Abbe, Semiconductor Wafer Measurements, *Solid State Tech.*, **17**:47–50 (March 1974).
2. Monsanto Evaluation Standards, Electronic Materials Division, Monsanto, St. Peters, Mo.
3. A. R. Moore, A Method of Accurate Thickness Determination of Germanium Wafers Suitable for Transistor Production, *IRE Trans. Electron Devices,* **ED-4** 309–310 (1957).
4. O. S. Heavens, "Optical Properties of Thin Solid Films," Dover Publications, Inc., New York, 1965.
5. J. E. Dial, R. E. Gong, and J. N. Fordemwalt, Thickness Measurements of Silicon Dioxide Films on Silicon by Infrared Absorption Techniques, *J. Electrochem. Soc.,* **115**:326–327 (1968); C. J. Mogab, Measurement of Film Thickness From Lattice Absorption Bands, *J. Electrochem. Soc.*, **120**:932–937 (1973).
6. S. D. Mittleman and L. Bess, Nondestructive Thickness Measurement in Thin Sections of Ge, *Rev. Sci. Instr.,* **39**:838–840 (1965).
7. H. Yokota, M. Nishibori, and K. Kinosita, Ellipsometric Study of a Thin Transparent Film Overlaid on a Transparent Substrate Having a Surface Layer, *Surface Sci.,* **16**:275–286 (1969); Leif Lundkuist, Ellipsometry of Multilayered Dielectrics on Silicon, Applied to MNOS Structures, *J. Electrochem. Soc.,* **120**:1140–1142 (1973).

8. E. Passaglia, R. R. Stromberg, and J. Kruger (eds.), "Ellipsometry in the Measurement of Surfaces and Thin Films," National Bureau of Standards, *Misc. Pub.* 256, 1964.
9. William A. Shurcliff, "Polarized Light," Harvard University Press, Cambridge, 1962.
10. R. J. Archer, Determination of the Properties of Films on Silicon by the Method of Ellipsometry, *J. Opt. Soc. Am.,* **52**:970–977 (1962).
11. K. H. Zaininger and A. G. Revesz, Ellipsometry—A Valuable Tool in Surface Research, *RCA Rev.,* **25**:85–115 (1964).
12. H. E. Bennett and Jean M. Bennett, Precision Measurements in Thin Film Optics, in Georg Hass and Rudolf E. Thun (eds.), "Physics of Thin Films," vol. 4, Academic Press, Inc., New York, 1967.
13. Paul Drude, "The Theory of Optics," Longmans, Green & Co., Ltd., New York, 1907.
14. Alexandre Rothen, Measurements of the Thickness of Thin Films by Optical Means, from Rayleigh and Drude to Langui, and the Development of the Present Ellipsometer, in E. Passaglia, R. R. Stromberg, and J. Kruger (eds.), "Ellipsometry in the Measurement of Surfaces and Thin Films," National Bureau of Standards, *Misc. Pub.* 256, 1964.
15. Rolf H. Muller, Definitions and Conventions in Ellipsometry, *Surface Sci.,* **16**:14–33 (1969).
16. A. R. Hilton and C. E. Jones, Measurement of Epitaxial Film Thickness Using an Infrared Ellipsometer, *J. Electrochem. Soc.,* **113**:472–478 (1966).
17. A. B. Winterbottom, Increased Scope of Ellipsometric Studies of Surface Film Formation, in E. Passaglia, R. R. Stromberg, and J. Kruger (eds.), "Ellipsometry in the Measurement of Surfaces and Thin Films," National Bureau of Standards, *Misc. Pub.* 256, 1964.
18. Jerome M. Weingart and Alan R. Johnston, Electronic Polarimeter Techniques, in E. Passaglia, R. R. Stromberg, and J. Kruger (eds.), "Ellipsometry in the Measurement of Surfaces and Thin Films," National Bureau of Standards, *Misc. Pub.* 256, 1964.
19. H. G. Jerrard, A High Precision Photoelectric Ellipsometer, *Surface Sci.,* **16**:137–146 (1969).
20. J. L. Ord, An Ellipsometer for Following Film Growth, *Surface Sci.,* **16**:155–165 (1969).
21. Ingo Wilmanns, A Double-Modulation Photoelectric Ellipsometer, *Surface Sci.,* **16**:147–154 (1969).
22. Layer, Howard P., Circuit Design For An Electronic Self-Nulling Ellipsometer, *Surface Sci.,* **16**:177–192 (1969).
23. A. K. N. Reddy and J. O'M. Bockris, Ellipsometry in Electrochemical Studies, in E. Passaglia, R. R. Stromberg, and J. Kruger (eds.), "Ellipsometry in the Measurement of Surfaces and Thin Films," National Bureau of Standards, *Misc. Pub.* 256, 1964.
24. B. D. Cahan and R. F. Spanier, A Highspeed Precision Automatic Ellipsometer, in E. Passaglia, R. R. Stromberg, and J. Kruger (eds.), "Ellipsometry in the Measurement of Surfaces and Thin Films," National Bureau of Standards, *Misc. Pub.* 256, 1964.
25. Frank L. McCrackin, Elio Passaglia, Robert R. Stromberg, and Harold L. Steinberg, Measurement of the Thickness and Refractive Index of Very Thin Films and the Optical Properties of Surfaces by Ellipsometry, *J. Res. Natl. Bur. Std.,* **67A**363–377 (1963).
26. František Lukeš, The Accuracy of the Measurement of the Ellipsometric Parameters Δ and Ψ, *Surface Sci.,* **16**:74–84 (1969).
27. Paul H. Smith, A Theoretical and Experimental Analysis of the Ellipsometer, *Surface Sci.,* **16**:34–66 (1969).
28. H. G. Jerrard, Sources of Error in Ellipsometry, *Surface Sci.,* **16**:67–73 (1969).
29. "Ellipsometer Manual AME-500-1," Applied Materials Technology Inc., Santa Clara, Calif., 1970.
30. R. J. Archer, "Manual on Ellipsometry," Gaertner Scientific Corporation, Chicago, Ill.
31. Lawrence A. Murray and Norman Goldsmith, Nondestructive Determination of Thickness and Perfection of Silica Films, *J. Electrochem. Soc.,* **113**:1297–1300 (1966).
32. N. Goldsmith and L. A. Murray, Determination of Silicon Oxide Thickness, *Solid State Electron.,* **9**:331,332 (1966).

33. Myron J. Rand, Spectrophotometric Thickness Measurements for Very Thin SiO_2 Films on Si, *J. Appl. Phys.,* **41**:787–790 (1970).
34. W. C. Dash and R. Newman, Intrinsic Optical Absorption in Single-Crystal Germanium and Silicon at 77°K and 300°K, *Phys. Rev.,* **99**:1151–1155 (1955).
35. Edwin A. Corl and Hans Wimpfheimer, Thickness Measurement of Silicon Dioxide Layers by Ultraviolet-Visible Interference Method, *Solid State Electron.,* **7**:755–761 (1964).
36. F. Reizman, Optical Thickness Measurement of Thin Transparent Films on Silicon, *J. Appl. Phys.,* **36**:3804–3807 (1965).
37. I. Fränz and W. Langheinrich, A Simple Non-destructive Method of Measuring the Thickness of Transparent Thin Films between 10 and 600 nm, *Solid State Electron.,* **11**:59–64 (1968).
38. W. G. Spitzer and M. Tanenbaum, Interference Method for Measuring the Thickness of Epitaxial Grown Films, *J. Appl. Phys.,* **32**:744–745 (1961).
39. M. P. Albert and J. F. Combs, Thickness Measurement of Epitaxial Films by the Infrared Interference Method, *J. Electrochem. Soc.,* **109**:709–713 (1962).
40. Warren Groves, Measurement of Thickness of Gallium Arsenide, Epitaxial Films by the Infrared Interference Method (Nondestructive), *Semicond. Prod.,* **5**:25–28 (1962).
41. Robert J. Walsh, Measurement of Layer Thickness of Silicon Epitaxial Wafers, *SCP and Solid State Tech.,* **7**:23–27 (August 1964).
42. P. J. Severin, On the Infrared Thickness Measurement of Epitaxially Grown Silicon Layers, *Appl. Opt.,* **9**:2381–2387 (1970).
43. P. A. Schumann, Jr., The Infrared Interference Method of Measuring Epitaxial Layer Thickness, *J. Electrochem. Soc.,* **116**:409–413 (1966).
44. P. A. Schumann, Jr., R. P. Phillips, and P. J. Olshefski, Phase Shift Corrections for Infrared Interference Measurement of Epitaxial Layer Thickness, *J. Electrochem. Soc.,* **113**:368–371 (1966); P. J. Severin, On the Infrared Thickness Measurement of Epitaxially Grown Silicon Layers, *Appl. Opt.,* **9**:2381–2387 (1970); P. J. Severin, Interpretation of the Infrared Thickness Measurement of Epitaxially Grown Layers, *Appl. Opt.,* **11**:691–692 (1972).
45. P. A. Schumann, Jr., and R. P. Phillips, Phase Shifts for Epitaxial Layer Thickness Measurements by the Infrared Interference Method, IBM. *Tech. Rept.* 22.182 (1965).
46. P. A. Schumann, Jr., Thickness Measurements of Very Thin Epitaxial Layers by Infrared Reflectance, in Charles P. Marsden (ed.), "Silicon Device Processing," National Bureau of Standards, *Spec. Pub.* 337, 1970.
47. Toshio Abe and Taketoshi Kato, Infrared Reflectivity of N on N^+ Si Wafers, *Japan. J. Appl. Phys.,* **4**:742–751 (1965).
48. K. Sato, Y. Ishikawa, and K. Sugawara, Infrared Interference Spectra Observed in Silicon Epitaxial Wafers, *Solid State Electron.,* **9**:771–781 (1966).
49. P. A. Schumann, Jr., Current Problems in the Electrical Characterization of Semiconducting Materials, in Rolf R. Haberecht and Edward L. Kern (eds.), "Semiconductor Silicon," Electrochemical Society, New York, 1969.
50. Thomas E. Reichard, Through Thick and Thin with Infrared Beams, *Electronics,* **41**:101–105 (Mar. 18, 1968).
51. Allison Roddan and Vitali Vizir, An Instrument for Automatic Measurement of Epitaxial Layer Thickness, in Charles B. Marsden (ed.), "Silicon Device Processing," National Bureau of Standards, *Spec. Pub.* 337, 1970.
52. Paul F. Cox and Arnold F. Stalder, Fourier Transform Method for Measurement of Epitaxial Layer Thickness, in Huff and Burgess (eds.), "Semiconductor Silicon," Electrochem Society, Princeton, 1973; P. J. Severin, The Influence of the Phase Shift on Thickness Measurements of Silicon Epitaxial Layers with a Fourier Transform Spectrometer, *J. Electrochem. Soc.,* **121**:150–158 (1974).
53. Gary Horlick, Introduction to Fourier Transform Spectroscopy, *Appl. Spectroscopy,* **22**:617–626 (1968).

54. R. T. Kampwirth, An Optical Thickness Monitor for Thin Film Vacuum Deposition Control, *Rev. Sci. Instr.,* **43:**740–743 (1972).
55. D. J. Dumin, Measurement of Film Thickness Using Infrared Interference, *Rev. Sci. Instr.,* **38:**1107–1109 (1967).
56. I. Fränz and W. Langheinrich, Nondestructive Thickness Measurement of Thin Films on Microstructures, *Solid State Electron.,* **11:**987–991 (1968).
57. W. A. Pliskin and E. E. Conrad, Nondestructive Determination of Thickness and Refractive Index of Transparent Films, *IBM J. Res. Develop.,* **8:**43–51 (1964); W. A. Pliskin and R. P. Esch, Refractive Index of SiO_2 Films Grown on Silicon, *J. Appl. Phys.,* **36:**2011–2013 (1965).
58. W. L. Bond and F. M. Smits, The Use of an Interference Microscope for Measurement of Extremely Thin Surface Layers, *Bell System Tech. J.,* **35:**1209–1221 (1956).
59. Martin G. Buehler, personal communication, 1967.
60. R. H. Dudley and T. H. Briggs, Rapid Photometric Method for Increasing Precision of Layer Thickness Measurement by Angle Lap Technique, *Rev. Sci. Instr.,* **37:**1041–1044 (1966).
61. S. Tolansky, "Interferometry," John Wiley & Sons, Inc., New York, 1954.
62. Philip S. Flint, Optimization of the Tolansky Technique for Thin Film Thickness Measurements, in Bertram Schwartz and Newton Schwartz (eds.), "Measurement Techniques for Thin Films," The Electrochemical Society, New York, 1967.
63. G. R. Booker and C. E. Benjamin, Measurement of Thickness and Refractive Index of Oxide Films on Silicon, *J. Electrochem. Soc.,* **109:**1206–1212 (1962).
64. Murray Bloom, Formation of Wedges in the Measurement of Oxide Films on Silicon, *Semicond. Prod.,* **6:**26 (1963).
65. R. M. Burger and R. P. Donovan (eds.), "Fundamentals of Silicon Integrated Device Technology," Prentice-Hall, Inc., Englewood Cliffs, N.J., 1967.
66. B. McDonald and A. Goetzberger, Measurement of the Depth of Diffused Layers in Silicon by the Grooving Method, *J. Electrochem. Soc.,* **109:**141–144 (1962).
67. I. Lagnado and S. M. Polcari, Spherical Drilling, A New Method for the Measurement of Junction Depths in Semiconductor Devices, *Solid State Electron.,* **10:**1219–1220 (1967).
68. S. D. Rosenbaum, Junction Depth Measurement by Spherical Contouring, *Solid State Electron.,* **11:**711–712 (1968).
69. Don W. Shaw, Epitaxial GaAs Kinetic Studies {001} Orientation, *J. Electrochem. Soc.,* **117:**683–687 (1970).
70. D. J. Gillespie, A Survey of Thin Film Thickness Measurement Methods, in Bertram Schwartz and Newton Schwartz (eds.), "Measurement Techniques for Thin Films," The Electrochemical Society, New York, 1967.
71. T. B. Light, Imperfections in Germanium and Silicon Epitaxial Films, in John B. Schroeder (ed.), "Metallurgy of Semiconductor Materials," John Wiley & Sons, Inc., New York, 1962.
72. William C. Dash, A Method for Measuring the Thickness of Epitaxial Silicon Films, *J. Appl. Phys.,* **33:**2395–2396 (1962).
73. S. Mendelson, Stacking Fault Nucleation in Epitaxial Silicon on Variously Oriented Silicon Substrates, *J. Appl. Phys.,* **35:**1570–1581 (1964).
74. F. C. Eversteyn and G. J. P. M. van der Heuvel, Method For Determining the Metallurgical Layer Thickness of Expitaxially Deposited Silicon from SiH_4 Down to 0.5 μm, *J. Electrochem. Soc.,* **120:**699–701 (1973).

7

Preparation of Samples for Microscopic Examination*,†

This chapter deals almost exclusively with the preparation of semiconductor materials and devices for microscopic examination. The microscopic and photographic techniques themselves are covered in another chapter. The preparation can in turn be subdivided into two parts. One is concerned with the mechanics of obtaining the right surface for viewing and the other with various etch, stain, and decoration treatments to cause the desired detail to become visible.

7.1 SECTIONING

Sectioning is used in the examination of transistors and integrated circuits to expose the region below the surface so that metallic interfaces can be seen or so that it can be stained or etched to define p-n junctions or regions of differing crystallographic perfection. In principle, a right-angle section as illustrated in Fig. 7.1*a* should be appropriate. Unfortunately many of the things of interest are so thin that additional magnification is desirable (particularly true of base regions and ion-implanted layers). Such magnification can be achieved by using angle sectioning (beveling) as in Fig. 7.1*b* to obtain vertical magnification. The magnification is given by $1/\sin\theta$ as defined in the figure and can be substantial, since angles of less than a degree are quite feasible.‡ Sectioning is not without its pitfalls, however, since interpretation is more difficult, and if the substructure of interest is of limited horizontal extent, the section may not show its full depth. However, there is often a series of adjacent, repetitive substructures, so that a low-angle section will capture different portions of each as shown in Fig. 7.2 and allow the whole structure to be visualized. Sequential (serial) sectioning as commonly practiced in anatomy and pathology is seldom done intentionally, but the effect just described in Fig. 7.2 is certainly closely akin to it, and the techniques developed in those fields can sometimes be used to advantage.[1,2]

This chapter was coauthored by Stacy B. Watelski.

†Contrary to the precedent established in the other chapters, specific manufacturers' products are mentioned in this one. This does not indicate the authors' endorsement, but rather that these products are known to work and may be used as a starting point for further experimentation.

‡Magnification along the incline. When viewed normal to the original surface, it is $1/\tan\theta$.

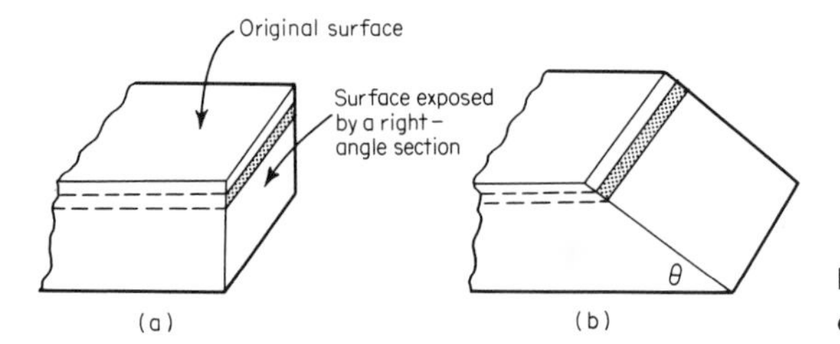

Fig. 7.1. Use of angle sectioning to obtain magnification.

The samples to be sectioned fall into two categories.[3] One includes devices in various stages of assembly. These may range in complexity from a semiconductor die with some monometallic metallization deposited on it (e.g., Al on Si) to a die with metallization, attached to a header with solder, eutectic alloy, or organic adhesive, and connected with various small (often less than 1 mil diameter) wires. The other category is the semiconductor by itself, either before the addition of all its metallic and dielectric accouterments or after they have been stripped away. Clearly the first category will require certain precautions not needed on the second because of the presence of soft metals which smear easily and the likelihood of breaking unsupported wires during sectioning. The second category more often requires low-angle bevel lapping for magnification and subsequent etching and/or staining, and in general is handled differently. In particular, those with metallization are almost always potted in a supporting plastic, and those without are usually not potted. Further, subsequent thickness measurements are often made on the latter, and as is described in Chap. 6, one of the more common methods of doing it is by interferometry. Such a procedure requires that a portion of the original surface be left for a reference as illustrated in Fig. 7.3. Should the details of a metal-metal bond interface be desired, however, the whole sample could be sectioned with no regard for the original surface.

Bevel Polish. There are several variations of the beveling technique, but the fixture shown in Fig. 7.4 will serve as a basic tool. The size of the fixture may be changed to accommodate different sample sizes. The outer ring is made of brass or soft iron. Either of these metals will glide over the glass lapping surface with minimum drag. The circular format further ensures a smooth glide. Of the two metals, brass is

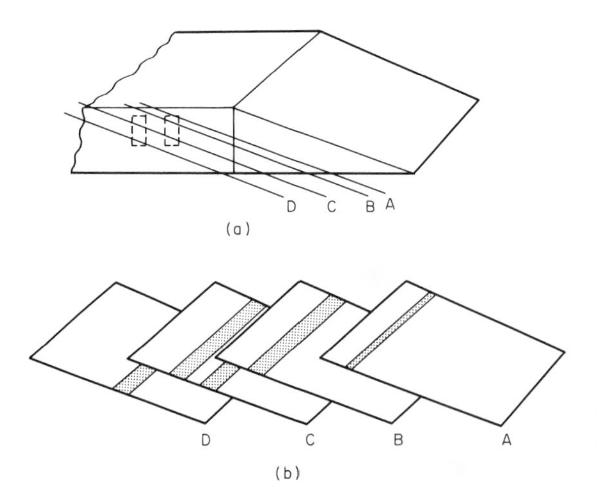

Fig. 7.2. Effect of the position of the section plane on the projected shape of the buried structure.

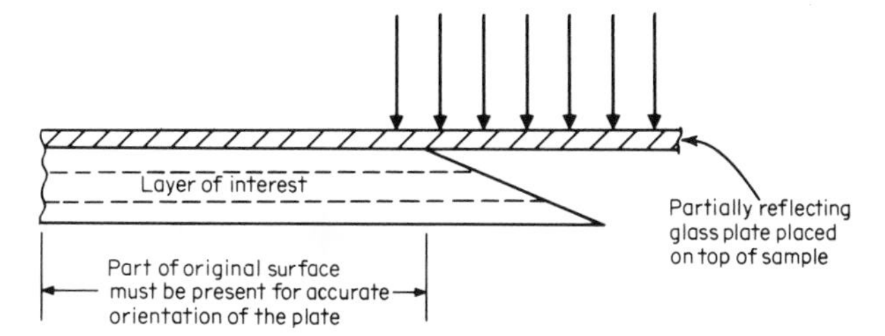

Fig. 7.3. The subsequent use of an angle-lapped specimen for thickness measurements may require some of the original surface to be left.

preferred because of its rust-free nature. If, however, iron is used, it may be left in an aqueous ammoniacal solution to prevent rusting when not in use. The insert is most conveniently made of stainless steel. The sample is affixed to the insert by use of an adhesive. Several acceptable ones are Apiezon W wax, 70C cement,* and Tan wax.† The first is soluble in chlorinated hydrocarbons, the other two in alcohol. The Apiezon wax can be obtained with a range of melting points. For example, W-40 melts at 45°C, W-100 at 55°C, and W at 85°C. The melting points of the other two are 75 and 76°C, respectively.

The choise of adhesive is usually not critical. Each will adequately secure the sample to the insert, although prolonged exposure of the adhesive to etchants (i.e., staining and delineating acidic solutions) will weaken them. This is evidenced by the adhesives lightening in color. The least affected is Apiezon W. Extremely small samples will best be secured by use of Tan wax or 70C cement. The latter is more easily removed from a delicate sample with alcohol; however, the former is less likely to allow the sample to part from the insert under stress. Also, the Apiezon W may be used at a lower temperature than the others; specific samples may require this.

Before mounting the sample, one must decide which bevel should be used. This choice of angle is dependent on the sample size, the depth to be measured or observed, the magnification desired, and somewhat on the sample resistivity. If the sample is very small, a small-angle bevel (i.e., 1°) might utilize all the top surface in order to obtain a sufficient bevel depth. This would be unacceptable if thickness measurements were to be made. The distance lapped back on the top surface and the magnification is approximately 60, 20, and 10 times the depth for 1°, 3°, and 7° bevels, respectively. If the resistivity of the material is greater than 1.0 Ω-cm, a delineation stain may not be very dark in color; in fact it may be a faint gray and extremely difficult to see when magnified sufficiently and superimposed with interference-fringe lines. If a larger angle is used, the critical area expansion is

*Hugh Courtright & Co., Vincennes Ave., Chicago, Ill.
†Fred Lee & Co. Ltd.

(a)

(b)

Fig. 7.4. Bevel-polishing fixture.

Fig. 7.5. Two-piece insert.

sacrificed, but the stain occurs in a narrower region and hence appears darker. If a larger angle cannot be used, the smaller one is used, the faint resulting stain is photographed without the fringe lines, and another photograph is made at the same magnification with the fringe lines. Physical distances from the first photograph are transferred to the second to obtain a proper fringe-line count. The second decision prior to sample mounting is whether to use a one- or two-piece insert. If at all possible, the two-piece insert (Fig. 7.5) is preferred. Because it is very short, it can be easily positioned on a microscope stage. The one-piece insert (Fig. 7.4*a*), because of its height, requires the use of a modified or inverted microscope stage.

If the bevel is to reveal a routine-type delineation in which a large beveled area is not necessary, the sample may be cut and positioned as shown in Fig. 7.6*a*. Only the pointed tip is polished, and because it is a small area, the job is quickly accomplished. Other samples requiring larger areas are cut and positioned such that the bevel will pass through the sample at the desired angle.

After the particular insert to be used has been heated on a hot plate, it is removed and a thin layer of adhesive is melted on its surface. The sample is then placed on the waxed insert end and seated properly by pressing down and at the same time moving the sample in a circular motion. This is easily accomplished by using a pair of tweezers. The tweezer points allow the sample to be properly positioned prior to cooling. The circular sample movement will expel trapped gas and excess wax under the sample. If the wax film is kept very thin and uniform, the subsequent beveling operation will produce a sample with a beveled surface approximately equal to the bevel angle of the insert. The insert is now ready to be cooled. Too rapid cooling will allow some adhesives to craze and the sample will not be tightly secured;

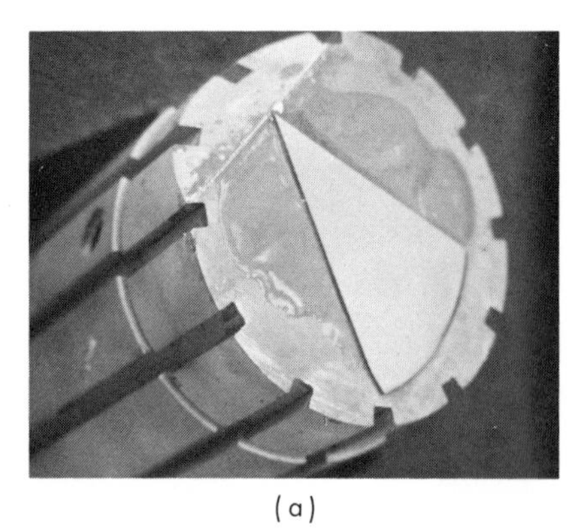

(a)

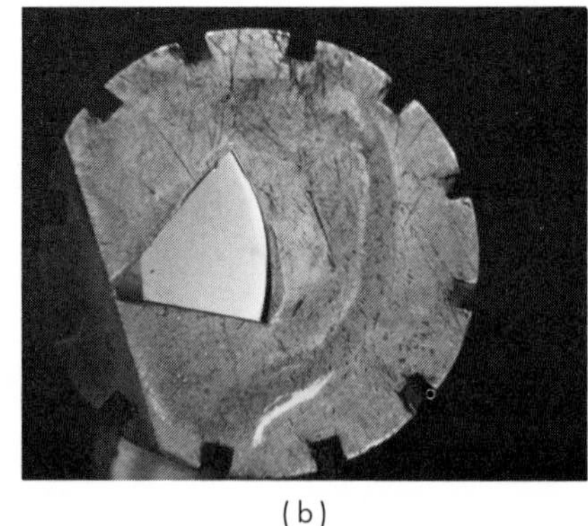

(b)

Fig. 7.6. Mounted sample.

however, cooling may be hastened on an insert by placing the bottom end of it in cool water. After a few seconds the whole insert may be slowly immersed.

The process of sample mounting and positioning usually results in unwanted adhesive around the edges of the sample. This will prevent the polishing of the sample if not removed. A razor blade is used to remove the adhesive from the sides and top of the sample. If the semiconductor material is easily scratched, extreme care must be exercised in mounting and/or cleaning. Silicon samples will not be scratched by scraping their tops with a razor blade to remove excess wax. Germanium, however, will be damaged, and any excess wax should therefore be removed by gently wiping with a cotton swab which has been dipped in a suitable solvent. When most of the excess wax has been removed, the final traces are removed by using a damp swab. Wipe only in one direction, turn the swab top 180°, and wipe again. Discard the swab after this double use. The cleanliness of the surface can be observed by watching the solvent evaporate. The evaporation of the solvent on a wax-free surface will show no residual wax-colored spot as the last traces evaporate.

The polishing plate to be used may be an ordinary piece of plate glass; however, better results are obtained by using a glass plate that has been precision-finished flat to a centerline average (CLA) index of 4.5 μm or better. This surface will allow the bevel fixture to glide with uniform ease when polishing. If an unfinished glass plate has a high or a low spot on its surface, this may cause the polishing fixture to drag, and in turn tilt and bump the sample on the glass surface. Fracturing of the sample then usually occurs. The glass plate should be a minimum size of 9 by 9 in.

The sample is usually polished in an aqueous slurry. An acceptable one is

1 g of Linde A (0.3 μm alumina abrasive)*
2 ml of Joy or other liquid detergent
15 ml of water

This slurry is stirred for rather uniform consistency using the fingers. The insert is then introduced into the fixture and the assembly carefully placed in the polishing slurry. The insert is positioned to touch the glass plate with the sample away from the plate. The fixture is then slowly set upright on the plate. Some of the slurry is added to the top of the fixture and allowed to run down the fluted sides of the insert prior to polishing.

Polishing is accomplished by applying moderate pressure to the insert with one finger while using the remaining fingers to move the fixture in a large figure eight pattern or by using a very gentle mechanical polisher of the vibrating variety. Additional slurry is made at the beginning of each evaluation by adding the separate components to the top of the fixture after it is positioned with the insert on the polishing plate, or by picking up slurry with the fingers and placing it on the top of the insert.

For optimum results:

1. Mix fresh slurry for each sample or two.
2. Clean plate often.
3. Replace a badly scratched plate.
4. Use only one abrasive per plate.

*For the less hard semiconductors such as InSb, softer polishing compounds like magnesium and iron oxide are more appropriate.

5. Keep larger-grit abrasive away from plate. Larger grit could come either from chance contamination or from edges of the sample breaking away.
6. Completely clean fixture often.

Polishing times may range from a few minutes to more than an hour, depending on the size of the beveled face and the depth of bevel required. The bevel polished surface is periodically inspected to see when it is polished deep enough and that it is relatively free of scratches. For demanding subsequent examinations, the polished surface should be observed under dark field to verify that it is quite free of scratches before leaving this operation. This freshly polished surface is ready to be chemically treated and/or microscopically inspected. Procedures for staining the sample will be found in a later section.

To remove the sample from the insert, heat until the adhesive is soft and slide the sample from the insert or cut under the sample with a razor blade. The latter usually destroys the sample but is a quick, effective removal scheme to be used when the sample is no longer needed.

Alternate abrasives which can be used to produce a polished surface are Linde B,* AB Metadi† 3-μm, and AB Metadi† $\frac{1}{4}$-μm diamond polishing compounds. Linde B, a powder, is used exactly the same as Linde A powder. Either of the two diamond polishing compounds is supplied in an oil base. It is advisable to dilute this thick oil with enough light machine oil to allow easy movement of the polishing fixture on the polishing plate. This will also reduce the tendency of the fixture to drag and bump.

Sectioning at Very Small Angles. Occasionally there will be a requirement for even greater magnification. By using lapping blocks with diamond stops, it is possible to mount large-area flat samples, e.g., solar cells or large portions of slices, and lap rather closely to predetermined angles of a few minutes. Polishing must be subsequently done without the stops, but if the block has been filled with dummy slices, the original angle can be reasonably maintained. If any of the original surface is left, the actual angle can be estimated by running a profilometer such as a Rank Talysurf over it and down the bevel. If there is no requirement on the angle other than that it be small, a straightforward polish of the sample and dummy fillers will almost certainly result in a final surface nonparallel with the original one.

Perpendicular Sectioning. Semiconductor material may be perpendicularly sectioned in a manner similar to the bevel polish technique, but using one of the types of modified inserts shown in Fig. 7.7. The modified insert shown on the left can

*Union Carbide Corp. Linde Div., Crystal Products Dept., East Chicago, Ind.
†Buehler, Ltd., Greenwood St., Evanston, Ill.

Fig. 7.7. Modified inserts for perpendicular sectioning.

facilitate a larger sample than the one shown on the right and is often necessary. The time required to polish a large sample to some predetermined position will usually be in excess of 2 h. The sample is mounted so that a portion of it extends beyond the top of the insert. The insert is heated and an adhesive* is melted on only the face of the insert which is parallel to its long axis. The sample is placed on the molten adhesive and pressed with a circular motion to expel the excess gas and adhesive and seat itself against the insert. While still hot, the fixture is inverted such that it rests on a glass microscope cover glass and the sample is positioned to rest its edge on the surface which supports the cover glass.

Thus the sample will overhang the insert top face by an amount equal to the thickness of the cover glass. This procedure also facilitates polishing by fixing the sample edge parallel to the insert edge. Excessive adhesive must be removed from the top edge of the sample and from the top face of the insert so as not to impede the subsequent polishing step. It is preferable to remove the excess adhesive by dissolving it using a cotton swab and a suitable solvent. Because of the large surface area presented to the polishing plate, and the polishing time required, it may be desirous to lap the sample prior to polishing using an 1,800-grit abrasive. The abrasive and water are put on a separate lapping plate whose surface is prepared similar to the polishing plate and mixed to a smooth consistency with the fingers. If too little water is used, the slurry will be thick and the lapping fixture will not move easily through it. If the slurry is too thin, very little material will be removed per unit time. A normal slurry (the consistency of thin syrup) will be colored by the semiconductor being lapped. This darkening of the slurry is normal and gives an indication as to when the slurry should be replaced with a fresh batch. The sample is lapped to expose a region ahead of the desired target region. For silicon, stop 1 to 2 mils ahead; for germanium, stop 3 to 4 mils ahead. For a sample containing a device or fabricated pattern (diffusion, selective epitaxy, alloy, etc.) stop before engaging the pattern. Final polishing should be continued only after the insert and fixture are thoroughly washed free of the large-grit abrasive. Small quantities of large abrasive carried to the polishing plate not only will scratch the plate and the sample but may be responsible for causing chipping of the sample being polished. Approximate polishing times, excluding lapping, are:

1. 10-mil-thick sample 250-mil wide, 5 min
2. 10-mil-thick sample 500-mil wide, 10 min

These are optimum times assuming no difficulties are encountered. Sometimes scratches, chipping, and sample breakage may prolong these times to several hours.

Cleaving. This process allows a slice of semiconductor material to be sectioned by breaking rather than polishing in order to reveal an almost perpendicular section.[50] The cleaving procedure can be varied considerably, but a workable one is as follows:

1. Scribe line across the backside of the slice at the desired cleaving position.
2. Spray both sides of the semiconductor slice with a solution of Apiezon W dissolved in trichloroethylene (0.1 g/ml). Upon drying, this thin wax film will protect both semiconductor surfaces from subsequent scratching.

*Because perpendicular sectioning exerts a greater force on the sample than bevel sectioning, only 70C cement or Tan wax should be used.

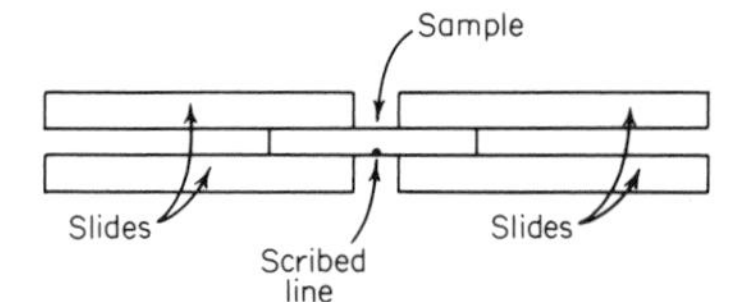

Fig. 7.8. Sample prepared for cleaving.

3. Sandwich the coated slice between pieces of microscope slides as shown in Fig. 7.8.
4. While holding the sample sandwich with both hands, apply a breaking pressure.

Potting in Lucite. Some samples may be extremely fragile, secured to a header, or in some way fixed to a mount such that bevel or perpendicular sectioning as just described would be impossible because of the inability to secure the sample and prevent its breakage during sectioning. Such samples may be potted in Lucite using a specimen-mount press and a metallurgical-grade thermoplastic powder such as AB Transoptic Powder No. 20-3400 (Buehler, Ltd.).

Detailed instruction sheets are given with each press. Following these instructions will assure an acceptable mounted specimen. In addition to mounting irregular-shaped objects, flat specimens, such as a semiconductor slice with one face exposed, may also be mounted within the plastic. An acceptable technique for mounting a slice or slice section and maintaining the slice face parallel to the Lucite face is as follows. Using the manufacturer's instructions, make two blank disks of potting compound each about $\frac{1}{8}$ to $\frac{1}{4}$ in thick. Place them on either side of the slice or slice section in the cylinder on the baseplate. Position the plunger on this assembly, place on press insulating plate, lower heater to position, and start heating; apply approximately 700 lb/in^2 pressure and maintain until the temperature reaches 80°C; then increase pressure to 4,000 lb/in^2, remove heater, position cooling blocks, and maintain pressure of 4,000 lb/in^2 until the temperature reaches 50°C. Then use the recommended procedure to remove the mounted specimen. The advantages of this technique over one employing the slice or slice section surrounded by the plastic powder are (1) positive slice alignment and (2) a very small chance of breaking the slice by excessive pressure applied during the mounting operation.

The mold defects most usually encountered in slice and device mounting and their causes are:

1. Improper wetting of specimen by plastic caused by insufficient temperature.
2. Frosted area around specimen caused by insufficient temperature or time at a proper temperature. Figure 7.9 shows examples of these difficulties.

Precautions to be taken are:

1. Exercise extreme care in orienting sample for ease of future sectioning.
2. Keep specimen clean and dry.
3. Eliminate sharp corners if possible.

Some disadvantages of this form of potting are that the extreme pressure may fracture the sample, and it is difficult to orient the sample and polish it flat. Sectioning details are presented below.

Potting in Teflon. For those samples to be subjected to very strong etchants, potting in Teflon powder has been used. The suggested procedure[4] is to compact Teflon

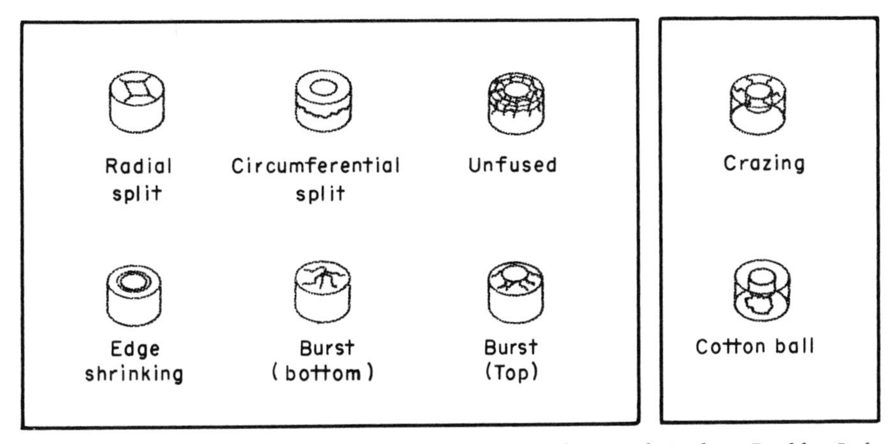

Fig. 7.9. Compression-molded specimens. (*From The Metal Analyst, Buehler Ltd., Evanston, Ill., 1965.*)

No. 8 granular TFE resin at 5,000 lb/in^2, then sinter in nitrogen 2 h at 380°C. The heating and cooling cycle must be very slow ($<$50°C/h) between 280 and 38°C to prevent cracking.

Potting in Casting Resin. Such potting is relatively easily accomplished and may be used for samples either header-mounted or without mount. Some casting resins expand excessively during the curing operation and some tend to crack. There is also a wide range of curing times and temperatures available; so the particular resin used should be carefully evaluated. The sample should be dry and free of any oil or grease. The mold may be either of plastic and expendable, i.e., lapped along with the sample, or of metal coated with a suitable mold release. To simplify lapping to some approximate predetermined angle, a wedge-shaped preform may be positioned in the bottom to support the sample.[5,6] In the subsequent lapping and polishing operation, rounding of the sample is likely to occur unless an outer ring of a harder material such as ceramic is also cast into the resin.[6] This can slow the polishing operation, however, and unless specifically needed, should not be used. The actual potting technique is as follows:

1. Mix 7 parts (by volume) of epoxy* with 1 part (volume) of hardener.†
2. Stir slowly for 1 min to prevent bubbles.
3. Pour over sample in plastic holder.
4. Place in small vacuum chamber and apply three to five short bursts of vacuum such that gas bubbles are evolved.
5. Cure at room temperature for 24 h, or at 70°C for 20 to 30 min—light yellow color when cured.

Sectioning Potted Samples

1. Rough-lap the potted sample on a slow belt surface, having a 180-grit (80-μm) belt. Rinse thoroughly.
2. Medium-lap the sample at slow speed (163 r/min) on a low-speed grinder using (50-μm) 320-grit abrasive. Rinse thoroughly.

*502 ARALDITE Epoxy, CIBA.
†951 Hardener, CIBA, Summit, N.J.

3. Fine-lap the sample at fast speed (1,200 r/min) on a high-speed grinder using (25-μm) 600-grit abrasive. Rinse thoroughly.
4. Final-polish the sample at fast speed (1,200 r/min) on a high-speed grinder using a polishing cloth and 0.03 to 0.05 alumina abrasive. This final polishing step should be kept to as short a time as is necessary to give a polished surface. Prolonged polishing will result in undesired rounding of the sample surface. For very exacting work, a vibrating polisher may be preferred for the final step. For alternate procedures, see Refs. 5, 6, 53, 54.
5. Etch or stain the sample if required (procedures described later).
6. Using a glass slide, some clay, and the prepared sample, mount for microscope viewing using a hand-alignment press to ensure that the polished surface is normal to the microscopic optical axis. If an inverted microscope or an adjustable stage such as is available from various microscope manufacturers is used, the mount operation can be eliminated.

7.2 CHEMICAL TREATMENTS

In addition to or instead of the mechanical sectioning just discussed, a wide variety of chemical treatments are often necessary prior to microscopic examination.

Chemical Safety. When working with the various formulations described, it is advisable to remember that unless properly handled, acids and solvents are dangerous and can be a health hazard. Table 7.1 lists precautions and noticeable results

Table 7.1

Chemical	Precautions	Noticeable results on contact
Hydrofluoric acid	Avoid getting under fingernails*	First pain felt within the hour. Maximum pain occurs in 12 h
Nitric acid	*	Immediate burning sensation. Rapid blistering. Skin is colored brown
Acetic acid	*	Burning sensation. Blistering. White coloration of skin
Hydrochloric acid	*	Burning sensation. Blisters fast
Sulfuric acid	*	Burning sensation. Blisters fast. Slight blackening of skin
Phosphoric acid	*	Burning sensation. Slow to blister
Chromic acid	Avoid breathing dust, will react with moisture in respiratory system and produce severe burns	Discolor and burns
30% hydrogen peroxide	*	Delayed burning. White discoloration of skin
Solvents	*	Mild skin irritation; prolonged breathing may produce after-effects. Chlorinated solvents are particularly hazardous

*Concentrated vapors can cause irritation to eyes, nose, and throat. Use with adequate ventilation. Avoid prolonged or repeated breathing of vapor. Avoid contact with skin, eyes, and clothing.

of contact for most of the chemicals mentioned in this chapter. The most immediate action required in case of accident is flushing with copious quantities of water. Should the burn appear severe, or if HF is involved, medical attention is advisable. Eyeglasses* should be worn during mixing, and rubber gloves can be used to prevent hand burns. The gloves should never be considered as adequate protection for immersing hands in corrosive liquids, but rather only as splatter protection, since a small hole can allow chemicals to get inside the glove unnoticed and produce severe burning.

Acids and organic solvents should be kept separate during disposal, as they may react with one another and cause fires or explosions. Seventy percent perchloric acid (not suggested in any of the formulations in this chapter) should be treated with extreme care in this regard. The mixing of the etchants should proceed carefully, and of course, concentrated acids should be slowly added to water or more diluted solutions in order to prevent a rapid heat buildup and possible explosion. When new formulations are investigated, the choice of chemicals should be carefully reviewed, because it is possible to produce potentially explosive or poisonous mixtures. For example, glycerin is often used with various etchants to adjust the viscosity. However, when nitric acid, which is a prime constituent of most etchants, is mixed with glycerin, nitroglycerin may be formed.[7] In particular 100 percent white fuming nitric acid, or nitric acid plus sulfuric acid, combined with glycerin is to be avoided. Cyanide-metal-plating solutions are reasonably safe when alkaline but if they should be combined with acids may evolve deadly hydrogen cyanide gas. Some of the etch components themselves, particularly H_2O_2, may decompose rather violently.[8] More insidious is the possibility of a delayed reaction such as has been reported for the mixture 5 HNO_3, 2 H_2O, 1 HF, 5 lactic acid. It is unstable because of an autocatalytic reaction between the lactic acid and the HNO_3 which causes a delayed (by up to 12 h) rise in temperature and gas evolution sufficient to rupture storage bottles.[9]

Solution Preparation. When mixing etches, be sure to use the correct method for determining the proportion of each chemical used. The most common method of expressing composition lists the constituents by parts, e.g., 1 part HF, 3 parts HNO_3, and 10 parts H_2O. However, two additional things must be considered: (1) is it parts by volume or parts by weight; (2) is it 1 part pure HF, or 1 part 49 percent HF, 51 percent H_2O, and is it constant-boiling nitric acid, dilute nitric acid, or fuming nitric acid? Considerable confusion can arise concerning the latter point because of the various commercially available mixtures of most acids. Table 7.2 lists some of these and illustrates the range possible.

Other possible ways of expressing composition are in terms of normal solutions, molar solutions, and molal solutions. A one molar solution contains one mole (gram-molecular weight) of solvent per liter of solution, while a one molal solution contains one mole per 1,000 g of solvent. A one normal solution is one gram-equivalent weight (GEW) of solvent per liter of solution. For acids, a GEW is defined as the gram-molecular weight divided by the number of available H^+ ions per molecule, while for bases, it is the gram-molecular weight divided by the number of available (OH) ions per molecule. Thus one mole of HCl is one GEW while one mole of H_2SO_4 is 2.0 GEW, and one mole of $H_3(PO_4)$ is 3.0 GEW.

*The frames of some eyeglasses are themselves a fire hazard, and when combined with organic solvents are particularly dangerous.

Table 7.2

Acetic acid (CH_3COOH)	Commercial Glacial Acetic anhydride	36% acetic acid, remainder water 99.5–100% CH_3COOH $(CH_3CO)_2O$, slowly soluble in water to form acetic acid
Hydrochloric acid (HCl)	Gas Constant-boiling Commercial	100% HCl 20% HCl, remainder H_2O 38% HCl, remainder water
Hydrofluoric acid (HF)	Gas Commercial Commercial	100% HF 49% HF, remainder H_2O 53% HF, 47% H_2O
Hydrogen peroxide (H_2O_2)	Anhydrous USP Hydrogen peroxide 30%	100% H_2O_2 2.5–3.5% by wt. of H_2O_2, remainder H_2O 30% by wt. H_2O_2, remainder H_2O
Nitric acid (HNO_3)	Constant-boiling Dilute Fuming	68–70% HNO_3, remainder H_2O Commonly 10%, but may be different 87–92% HNO_3, with some N_2O_4, remainder H_2O
Phosphoric acid	USP Diluted USP Glacial	85% H_3PO_4, remainder H_2O 10% H_3PO_4, remainder H_2O 100% HPO_3, sometimes called meta phosphoric acid
Sulfuric acid (H_2SO_4)	Commercial Dilute Fuming	93–98% H_2SO_4, remainder H_2O Commonly 10% H_2SO_4, but may be different H_2SO_4, free SO_3. Composition varies, available up to 80% SO_3

HF will attack glass; so any etchants using it must be measured, mixed, and used in containers other than glass. Graduated cylinders, beakers, etc., must be made of materials such as Teflon or polypropylene. During etching, the temperature rise of the solution may be enough to soften some plastics; therefore, either cooling or the use of alternate containers may be necessary. Should it be desirable to maintain a reasonably fixed pH during etching, buffering can sometimes be used. This is accomplished by dissolving a salt of the acid (usually a weak acid) in the same solution so that it can react with small quantities of either acids or bases without changing the pH appreciably. One of the more commonly used buffered etchants is the familiar $HF—H_2O—NH_4F$ solution.

Etchants for Specific Applications. Table 7.3 gives the composition of the various etches referred to by name only in other chapters, and Table 7.4 gives some etches useful for chemically polishing semiconductor surfaces. These are usually required before etching to reveal dislocations. Samples to be used for transmission electron microscopy must be thinned; however, the various etchants used for that purpose are summarized in the electron-microscopy portion of Chap. 9.

Table 7.3. Composition of Etchants

Name	Composition (parts by volume unless otherwise indicated)	Reference
A-B etch	1 ml HF, 2 ml H_2O, 1 g CrO_3, 8 mg $AgNO_3$	10
ASTM etch	Modified copper etch No. 1. See "Silicon Dislocation Density Determination," according to ASTM F 47, for details	
Copper etch No. 1	20 ml HF, 10 ml HNO_3, 20 ml H_2O, 1 g $Cu(NO_3)_2 \cdot 3\,H_2O$. Same as Purdue etch	11, 52
Bell No. 2 (buffered etch)	54% H_2O, 36% NH_4F, 10% HF	
Copper etch No. 2	40 ml HF, 35 ml HNO_3, 10 ml acetic, 25 ml H_2O, 1 g $Cu(NO_3)_2 \cdot 3\,H_2O$	12
CP-4 (chemical polish No. 4)	3 HF, 5 HNO_3, 3 acetic, 0.06 Br_2	13
CP-4a	3 HF, 5 HNO_3, 3 acetic	
Cyanide etch (ferricyanide)	8 g $K_3Fe(CN)_6$, 12 g KOH, 100 ml H_2O	14
Dash etch	1 HF, 3 HNO_3, 10 acetic, sometimes referred to by ratios, e.g., 1-3-10. Other ratios are also used	15
Dow (Secco)	44 g $K_2Cr_2O_7$, 1 l H_2O, mix 1 volume of solution with 2 volumes HF	16
Iodine etch	50 ml HF, 100 ml HNO_3, 110 ml acetic, 0.3 g I_2	17
Mercury etch	3 ml HF, 5 ml HNO_3, 3 ml acetic, 2–3% $Hg(NO_3)_2$ (aged in closed bottle for 6 weeks)	18
p etch	3 ml HF, 20 ml HNO_3, 60 ml H_2O	51
Planar etch	2 ml HF, 15 ml HNO_3, 5 ml acetic	
Purdue etch	See copper etch No. 1	11, 52
Richards-Crockers (modified)	2.4×10^{-3} molar solution of $AgNO_3$, in 2 ml HF, 3 ml HNO_3, 5 ml H_2O	20
Russian etch	10 ml HF, 15 ml HNO_3, 5 ml acetic, 20 ml H_2O, 8 mg I_2, 2 mg KI	21
Sailor's etch	60 ml HF, 30 ml HNO_3, 0.2 ml Br_2, 2.3 g $Cu(NO_3)_2$. Dilute 10:1 with water before using	22
Schell	1 ml HNO_3, 2 ml H_2O	23
Sirtl etch	50 g CrO_3, 100 ml H_2O; mix with 100 ml HF just before using (rate may be varied by adjusting amount of HF)	24
Superoxol	1 ml HF, 1 ml H_2O_2, 4 ml H_2O (proportional variation sometimes indicated as, e.g., 1-1-4 Superoxol)	25
WAg	4 ml HF, 2 ml HNO_3, 4 ml H_2O, 0.2 g $AgNO_3$	26
White etch	1 HF, 4 HNO_3	19
W-R	2 HCl, 1 HNO_3, 2 H_2O	27

All formulas are for 30% H_2O_2, 49% HF, 70% HNO_3, glacial acetic.

Table 7.4. Semiconductor-Polish Etches

Si	CP-4; CP-4a; planar
Ge	CP-4a; CP-4 (will simultaneously show dislocations)
GaAs	1 ml HF, 3 ml HNO_3, 2 ml H_2O; 20 ml H_2SO_4, 75 ml H_2O, 5 ml H_2O_2
InAs	99.6 ml acetic, 0.4 g Br_2
InSb	CP-4a; 1 ml HF, 1 ml HNO_3 for $(\overline{1}\overline{1}\overline{1})$ and (110) only
GaSb	1 ml HF, 9 ml HNO_3

All formulas are for 30% H_2O_2, 49% HF, 70% HNO_3, glacial acetic.

7.3 DELINEATION

This section will treat the art and science of revealing such qualities as p-n junctions; damaged and polycrystalline regions; and insulating and conducting layers in metallographically sectioned material as well as on the surface by staining, etching, and/or plating techniques. The word art is used here in the true sense, for the person who masters delineation is indeed an artist.

Enough detail will be included in describing each operation so that the required artistic qualities of the operator will be minimized. To date there is no substitute for experience; the more samples one investigates, the more proficient he becomes. Soon, with practice, even those occasional samples of considerable difficulty become more or less routine. At last one gains the confidence to tackle any problem and, using a logical sequential approach, usually is rewarded with predictable success or the discovery of a modified approach yielding superior results. Specific information relating to germanium and silicon will be given; however, the technique of delineation will suffice for other materials. Some delineations need only be made good enough to reveal qualitative information when viewed macroscopically; others will be required to be extremely sharp in the 500 to 1,000 × microscopic range so that quantitative data may be obtained from them.

In some instances, many delineates will be known. Although only a few choice ones are considered of value, all will be listed. With the very nature of the delineation and the many sample variables that are possible, a seldom-used delineate may prove invaluable, since the performance of the delineant is strongly dependent on the makeup of the sample and the desired results. It will not be necessary to know the mechanism associated with delineation staining or plating to produce excellent results, that is, as excellent as the process will allow.*

Table 7.5 lists the surface qualities and references the prescribed delineation treatment. Except for faults and dislocations, which are covered in a separate chapter, each quality listed in the table has pertinent notes, hints, and precautions in the text.

*For information concerning probable stain composition and various mechanisms for the formation see, for example, P. J. Archer, *Phys. Chem. Solids,* **14**:104–110 (1960); D. R. Turner, *J. Electrochem. Soc.,* **105**:402–408 (1958); R. Memming and G. Schwandt, *Surface Sci.,* **4**:109–124 (1966).

Table 7.5. Quick Guide to Slice-Surface Delineation for Si or Ge*

Quality	Stain-Etch†
p-n junction or impurity-concentration region	Etch, 1-3-10; stain, 50-6, 1-3-10
Scratches and mechanical damage	Etch; Sirtl, 1-3-6 (see Chap. 2)
Resistivity striations	Etch; Sirtl, HF-Cu, electropolish
Polycrystalline material	Etch; Sirtl, lap (see Chap. 2)
Stacking faults	Etch; Sirtl (see Chap. 2)
Dislocations	Etch; Sirtl, copper etch, 1-3-10 (see Chap. 2)
Incomplete polishing (Si)	Copper-displacement platings, oxidation
Diffusion pipes	Stain; 50-6, 1-3-10, anodic oxidation; etch, Sirtl
Oxide pinholes	Copper decoration

*See text and following tables for more choices.
†Compositions are listed in Tables 7.3 and 7.7.

Junction and/or Localized High-Impurity-Concentration Delineation. Junction delineation can be by noting a difference in etch rate between the two types, by selectively staining one side, or by selective plating metal onto one side. The reactions involved may be either electrolytic or displacement,[28] and the various solutions reported have been in general chosen to favor one or the other of them. The surface preparation is touchy, and some procedures require surface abrasion.[29] Further, it has been demonstrated that the stains are so surface-dependent that selective staining can be made to delineate abraded-unabraded regions of the same type. Thus their performance often appears to depend much more on art and temperament than on good scientific principles. This is primarily because competing reactions are involved in most stains and the results can vary radically with the dominant reaction. Ambient light, temperature, volume of etchant per unit area, residual surface oxides, adsorbed impurities, and surface damage have all been demonstrated to affect the relative behavior of the reactions. Ordinarily, with p^+/p, or n^+/n, the more heavily doped region will stain darkest, and with p-n junctions, the p region will usually be the dark one. Reverse staining is not too common, but does happen. Should it occur, additional information, either physical or electrical, may be necessary to interpret the delineated results accurately. The exact metallurgical position of the line of demarcation is also of concern, since if it does not closely coincide with the metallurgical junction, the results can be misleading. The most common cause of shift is the fact that staining or plating does not occur over the space-charge region. There can be an appreciable distance between the edge of the stain and the metallurgical junction, particularly for the higher-resistivity samples. Also, just because of beveling alone, without any extraneous surface effects, the space-charge position is distorted and in some cases the junction is displaced as shown in Fig. 7.10.[30] The effect will be more pronounced the higher the resistivity. The use of light during staining, which is recommended in almost every case, will forward-bias the junction and reduce the width of the space-charge region. Based on simultaneous capacitance measurements, the intensity of light ordinarily used will usually keep the space-charge width narrow enough to make the errors negligible.[31] Multiple junctions, particularly PIN structures,[55] may pose additional interpretation difficulties; so an independent check of the junction position should be made until the idiosyncrasies of the particular procedure used are well understood.

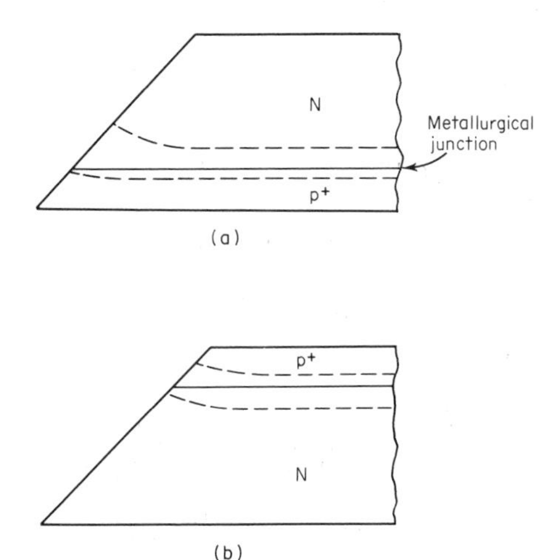

Fig. 7.10. Effect of angle bevel on the zero-voltage space-charge position.

Fig. 7.11. Localized high-surface-impurity concentration region delineated by 50-6 etch. Magnification is 34×.

A typical example to be delineated would be selectively diffused n^+ or p^+ areas into a polished slice of Si or Ge. Oxide or nitride masking as well as any metallization which may have been employed in fabricating this structure should be removed. After water rinsing and drying the sample, it should be masked using Apiezon W wax so that one-fourth of the active area is initially exposed. This allows up to four different delineation methods to be used. The first and most preferred is 50-6. The sample is held within 5 in of a high-intensity tungsten lamp such as a DFA* operated to 90 V ac. A cotton swab is used to paint the unmasked area with 50-6 etch while the surface is observed through a 10- to 15-power microscope. Either silicon or germanium can be stained by this method (Fig. 7.11). When the stain occurs, usually within 15 s, the sample is quickly quenched in water and dried. This stain is very tenacious and can be wiped without rubbing off. Some surfaces are such that the 50-6 etch has no effect and no stain occurs. Should that happen, the area is again painted as before but with a different stain, 50-6 Cu. Stain or mild plating should occur within 5 s (Fig. 7.12). This plated surface is water-rinsed and dried by blowing air over it. It should not be wiped, because the copper will rub off.

If no delineation occurs, the second fourth of the surface is exposed. This necessitates complete removal of the wax on the previously masked three-fourths of the surface. Because surface cleanliness is important, a word of caution is inserted as a reminder to make sure that all traces of wax are removed from the surface prior to rewaxing. Any traces of thin wax films will prevent accurate evaluation of

*DFA lamp 120-V, 150-W General Electric projection lamp.

Fig. 7.12. Localized high-surface-impurity concentration region delineated by 50-6 Cu etch. Magnification is 34×.

subsequent staining. The next preferred etch is 1-3-10. The sample is again positioned within 5 in of the light source, and the 1-3-10 etch is squirted on the surface from a polyethylene squeeze bottle. The surface is microscopically observed. The stain should occur within 1 to 2 s; after 10 s no stain will occur. This is also a mild etch, and etching action can be observed by the evolving gas. Rinse quickly and dry after staining. If the surface remains unstained, another approach is needed.

Another fourth of the surface is exposed and treated as with 1-3-10 but using 1-3-10 Cu. The surface will plate with copper within 5 s, but different concentration areas will plate differently. This plated surface is water-rinsed and dried by blowing air over it. It cannot be wiped, because most of the copper will be removed. Overplating is a common fault and cannot be rectified without repeating the process in a new area. Results may still be unsatisfactory.

The fourth area is exposed and treated with 1-3-6 etch. The sample is immersed in 1-3-6 etch for 3 s, rinsed in water, and dried. If the high-concentration areas are not stained a dark brown, further etching may reveal this. The word "may" is used here because increasing the etch time will either stain, etch away the critical area, or not stain at all. Some surfaces will resist staining; however, the low-resistivity areas in question will be etched more quickly than the surrounding area, forming a step between the two. This step delineation is also important in detecting area boundaries. Should troubles still occur, consult the following paragraphs for additional information.

Delineation of Angle-lapped Specimens. Table 7.6 shows preferred etches for layers that are epitaxial, diffused, alloyed, or combinations thereof. There is included the type of etch application, illumination, sectioned angle, perpendicular section type, visible results, delineation time, usable impurity range, and comments. Table 7.7 gives the composition of these etches, and Table 7.8 is a compendium of additional formulas to be tried if necessary. Included are a few recipes for electroplating. The same general comments made in the previous section are applicable here, and conversely, most of those to follow can be applied to slice staining.

1. If a section is difficult to stain, do not be discouraged—as many as 10 tries per etchant may be necessary.
2. Always use a freshly prepared surface for delineating. A surface that is allowed to remain for several hours prior to staining probably will not stain.
3. Stain all silicon sections dry for best results. If a particular condition requires a slowing down of the etchants action, a wet section may be used. Dry using gas jet. Do not wipe dry.
4. Stain all germanium sections wet.
5. In working with epoxy- and Lucite-mounted samples, application of some etchants may cause the plastic to swell and particles thereof to lie over the sample, obscuring the microscopic view. These may be wiped away using a cotton swab unless the delineation etch contained copper. Swabbing would remove the copper and perhaps the delineation.
6. If the layer in question is thin, do not use a 7° block for sectioning, for the layer will appear too small to be adequately measured.
7. A 7° or larger angle will not be illuminated by a vertical illuminating microscope system along with the unsectioned portion of the sample. If it is necessary to view (microscopically) both areas, the use of a 5° section or less is recommended.

Table 7.6. Solutions for Delineating Layers in Semiconductors

Sample	Delineant	Method of application*	Illumination	Visible results†	Layer delineated	Delineation time, s	Resistivity range
			Silicon				
Diffused							
n^+/p	1-3-10	1, 2, 4	Yes	1	p	1–15	
	50-6						
p^+/n	1-3-10	1, 2, 4	Yes	1	p^+	1–15	
	50-6						$\leq 0.00x / > 0.0x$
n^+/n	1-3-10	1, 2, 4	Yes	1	n^+	1–15	
	50-6						
p^+/p	1-3-10	1, 2, 4	Yes	1	p^+	1–15	
	50-6						
Alloyed							
p^+/n	1-3-6	1, 2, 4	No	3	. . .	10–60	
	Sirtl A	1				1–30	
n^+/p	1-3-6	1, 2, 4	No	3	. . .	10–60	
	Sirtl A	1				1–30	
Epitaxial							
n/n^+	1-3-10	1, 2, 4	Yes	1		1–15	
	1-3-10 Cu	2, 3, 4	Yes	2		1–15	
	1-3-10 Cu-Mo	4	Yes	2	n^+	1–15	$\geq 0.x / \leq 0.0x$
	50-6 Cu	2, 3, 4	Yes	2		1–15	
	1-3-5	1, 2, 4	Yes	1		1–15	
	1-3-6	3	No	1 and/or 3		10–60	
p/p^+	1-3-10	1, 2, 4	Yes	1		1–15	
	1-3-10 Cu	2, 3, 4	Yes	2		1–15	
	1-3-10 Cu-Mo	4	Yes	2	p^+	1–15	$\geq 0.0x / \leq 0.00x$
	50-6 Cu	2, 3, 4	Yes	2		1–15	
	1-3-5	1, 2, 4	Yes	1		1–15	
	1-3-6	3	No	1 and/or 3		10–60	
n/p	1-3-10	1, 2, 4	Yes	1		1–15	
p/n	1-3-10 Cu	2, 3, 4	Yes	2		1–15	
	1-3-10 Cu-Mo	4	Yes	2	p	1–15	All
	50-6 Cu	2, 3, 4	Yes	2		1–15	
	1-3-6	3	No	1 and/or 3		10–60	
	50-6	1, 2, 4	Yes	1		1–15	

Miscellaneous Si-SiO_2	None	. . .	. . .	4			
Si-SiC	None	. . .	. . .	4			
Si-polySi	Sirtl A	3	No	3	. . .	10–30	
Si-ceramic	Sirtl A	1	No	3	. . .	1–30	
			Germanium				
Epitaxial n/n^+	50-6 50-6 Cu 1-3-10 Cu	4	Yes No Yes	2	n^+	<5	
p/p^+	1-3-10 1-3-10 Cu 50-6 Cu 5-1 5-1 Cu	1, 2, 3 4 4 4 5	Yes	4 2 2 1 4	p^+	10–60 <5 <5 <20 <20	$>0 \cdot x / \leq 0.00x$
n/p	50-6 Cu HF-H_2O_2 1-3-10 Cu 1-3-10	4	Yes	2 2 2 3	p n	<5 <30 <60 <60	
p/n	50-6 Cu	4	Yes	2	n	<5	
Ge/GaAs	1-3-10 Cu 1-3-10	4 1, 2, 3	Yes	2, 3 3	Ge Ge	<30 <30	All
Diffused n^+/p p^+/n n^+/n p^+/p	50-6 Cu; 1-3-10 Cu; 5-1 Cu	4; 4; 5, 4	Yes; Yes; Yes	2 and 3; 2 and 3; 2	Plus; Plus; n^+	<30; <30; <20	
Alloyed p^+/n n^+/p	1-3-6 Cu	4	Yes	3	. . .	<30	
Combination (diffused and epitaxial) $n^+/(p/p^+)$	50-6 Cu 1-3-10 Cu 5-1 Cu	4 4 5	Yes	2 3	n^+ n^+ p, p^+	<30	

Table 7.6. Solutions for Delineating Layers in Semiconductors (*Continued*)

Sample	Delineant	Method of application*	Illumination	Visible results†	Layer delineated	Delineation time, s	Resistivity range
$p^+/(n/n^+)$	50-6 1-3-10 Cu 5-1 Cu	4	Yes	2	n^+	<30	
			InAs, GaAs				
Epitaxial InAs/GaAs	1:1 Clorox:water	1	Yes	1	GaAs	1	All
GaAs/GaAs(n/n^+)	200 g KOH 10 g $KAu(CN)_4$ H_2O to make 1 liter	1	Yes	Gold plate	n^+	2	
GaAs/GaAs(n/n^+)	1 ml HNO_3 5 ml HCl Allow to stand 1 h. 3 parts with 2 parts H_2O	1	Yes	1	n^+	900–1,200	
GaAs/GaAs	1 ml HNO_3 9 ml H_2O 0.8 g Fe‡	1	Yes	1	p	5	

Adapted from Charles A. Harper (ed.), "Handbook of Materials and Processes for Electronics," pp. 7-70–7-73, McGraw-Hill Book Company, New York, 1970. Used by permission.

*Methods of application:

1. Swab with cotton swab dipped in delineate.
2. Squirt delineate on sample, using an acid-resistant squirt bottle.
3. Dip in delineate.
4. Drip delineate on sample, using squirt bottle or cotton swab.
5. Dip swab in 5-1 and then in Cu solution, and allow to drip on sample.

†Visible results are:

1. Stain will appear.
2. Stain and/or variable-darkness copper plate.
3. Etched step.
4. Color difference.

‡Sixpenny finishing nail.

Table 7.7. Stain Etchants

Name	Composition	Semiconductor
50-6	50 ml HF, 6 drops HNO_3	Si, Ge
50-6 Cu	10 ml (50-6) + 2 drops Cu*	Si, Ge
5-1	5 ml HF, 1 ml HNO_3	Ge
5-1 Cu	10 ml (5-1) + 2 drops Cu*	Ge
1-3-5	1 ml HF, 3 ml HNO_3, 5 ml acetic	Si
1-3-6	1 ml HF, 3 ml HNO_3, 6 ml acetic	Si
1-3-10	1 ml HF, 3 ml HNO_3, 10 ml acetic	Si, Ge
1-3-10 Cu	10 ml (1-3-10) + 2 drops Cu	Si, Ge
1-3-10 Cu-Mo	10 ml (1-3-10 Cu) + 0.1 g MoO_3	Si
HF-H_2O_2	50 ml HF, 5 drops H_2O_2 (30%)	Ge
Sirtl A	50 g CrO_3, 100 ml H_2O, 75 ml HF	Si

*Cu = 20 g $Cu(NO_3)_2$, 80 ml H_2O, 1 ml HF. All formulas are for 30% H_2O_2, 49% HF, 70% HNO_3.

8. When staining n on p silicon, normally the p layer will stain; however, there are occasions when the reverse is true and the n layer stains. In most cases one can observe the p layer staining first very quickly and dark and then reversing. If the sample is water-quenched during the first darkening, this combination will remain.
9. Sometimes a stain will quickly form and quickly fade away. Do not attempt to restain without forming a new surface or using the copper counterpart of the original delineate.
10. There are cases when the only useful delineate contains copper and plates very quickly. After rinsing in deionized water, the excess copper is swabbed off, usually leaving a semistained area.
11. In application of the etchant containing copper, either layer that stains or copper-plates first, if allowed to remain in contact with the etchant for an excessive period of time, will be obscured by the complete plating of the sample. Careful observation will show the first area that was stained/plated to be the darker one.
12. The application of an etchant containing copper does not necessarily mean that a copper plating will result. Often only a stain will result. Often a stain with a copper overplate will result. Sometimes a copper plate will result with no stain beneath.
13. Copper-stained samples are the most difficult to interpret because they usually contain many bright colors and are extremely sensitive to small surface-variable potentials.
14. In delineating p on p or n on n silicon the use of 1-3-6 is noted as being used without illumination. It may be necessary to remove the sample from the etch and subject it to bright illumination, then return it to the etch. Either the heat of the lamp or the illumination itself seems to enhance the etching step.
15. Delineant containing copper that will not wet the sectioned surface will not stain it. Saturation of the delineant with molybdic acid will allow wetting and enhance staining and/or plating.
16. If an epitaxial n on n^+ or p on p^+ sample has a very slowly changing impurity-concentration gradient, it probably cannot be delineated.

Table 7.8. Additional Junction Delineants

Mat	Composition	Comments	Reference
Si	10 ml HF, 1–2 drops of red fuming HNO_3	p/n, p stains	33
Si	10 g $KAu(CN)_2$, 200 g KOH, deionized H_2O to make 1 liter, temperature 30–70°C, and light. Use lapped surface, 600 or finer	p/n, plating on n	29
Si	10:1 dilution of silver fluoride saturated at 0°C + drop of HF per 5 cm^3 of solution + strong illumination. May plate on p in weak illumination or large volumes of etchant	p/n, plating on n	28
Si	50 ml of diluted copper nitrate, 1–2 drops of concentrated HF, strong illumination	p/n, plating on n	34
Si	Concentrated HF, intense light	p/n, n-type becomes black	34
Si	200 g $CuSO_4 \cdot 5\,H_2O$, 10 cm^3 HF, 1,000 cm^3 H_2O, bright light. May reverse in dark, or if surface is partially polished and partly lapped	p/n, plating on p-type	35
Si	50 ml H_2O, 50 ml HF, 0.1 g $Cu(NO_3)_2 \cdot H_2O$		36
Si	1 ml HF, 20 ml H_2O, 2 g $HIO_4 \cdot 2\,H_2O$. Place drop on surface, watch progress, wash away in 2–3 min	p/n, n bright, p dark; p^+p, p bright; n^+n, n bright	37
Si	40 ml HF, 20 ml HNO_3, 100 ml H_2O, 2 g $AgNO_3$ (store under refrigeration). Put a drop of solution on one surface, await silver plating, rinse and dry, apply drop of diluted HNO_3, rinse, dry	n^+ plates first	38
Ge	Electrolytic etching in 1 g NaOH, 10 ml H_2O. Use approximately 1 A/cm^2. p region etches more rapidly, leaving a step		39
Ge	Place 1 drop of 1 ml HF, 20 g$CuSO_4 \cdot 5\,H_2O$, 80 ml H_2O on surface across junction. Apply voltage across junction in reverse direction. Should plate on p side. Too high voltage plate on n, too low, and plating only near contact	p/n plating on p	40
GaAs	For junction demarcation, 1 drop of 1 ml HF, 1 ml H_2O_2, 10 ml H_2O for 15 s under intense white light or, drop of 0.66 g $HAuCl_4 \cdot 3\,H_2O$, 1,000 ml H_2O on surface for 1 min under intense light followed by 1 above		41
GaP	8 ml HF, 40 mg $AgNO_3$, 5 g CrO_3, 10 ml H_2O, at 75°C, 1 to 5 min gives etch step with n of n/p etching most rapidly, and n^+ of n^+/n		42
GaSb	3 ml HF, 7 ml HNO_3, 10 ml H_2O. p region etches most rapidly, leaving a step at junction		43

Silicon Epitaxial Depositions. For n-type layers deposited on highly doped n^+ substrates, p-type layers deposited on p^+ substrates, n-type layers on p-type substrates, and p-type layers on n-type substrates, delineations should offer no difficulty. Remember that delineated layers less than 1 μm thick may be difficult to see if sectioned at angles greater than 3°. When a p-type layer is delineated on a p-type substrate or an n-type on an n-type substrate using 1-3-6 or 1-3-10 etchants a white line will appear at the interface and a light stain will be seen on both layers.

For n on n^+, after dripping one of the three recommended etchants on the sectioned surface, be prepared to quench quickly with water. The reason for the quick quench is that the n layer plates immediately; prolonged plating will cover the whole surface, thus obscuring the delineated interface. Water quenching will stop the plating action. These plated samples may be water-rinsed and dried in a gas stream; do not swab, blot, or touch the plating, because the copper will be removed. If interferometric techniques are to be used for thickness measurements, exercise caution in positioning the cover glass on the plated area.

The first choice of etchant for p on p^+ is a noncopper one, and the p^+ region will turn frosty in appearance within 60 s after application of the etchant. This is the result of a mild etch on the surface. The p layer will also etch but will not turn frosty. The other two recommended etches contain copper and hence will plate out on the p^+ regions immediately. As with the n-type material, rapid water quenching is essential to prevent overplating the sample with a resultant obscuring of the interface.

The delineation of silicon dioxide, silicon carbide, or silicon nitride requires no etchant. The interface line between the silicon and dioxide, carbide, or nitride can be easily seen, as each layer will have a color different from that of silicon. A polycrystalline silicon layer on an insulating substrate can usually be seen immediately after sectioning without the use of a delineant. However, Sirtl etch will bring out the polycrystalline silicon grain structure (Fig. 7.13). Quite often this grain structure is important from a device point of view, a fine grain being more desirable at times than a coarse grain structure.

Combination of Silicon Diffused and Epitaxial Layer. When delineating a combination diffused and epitaxial layer with 1-3-10 Cu and a cotton swab, constantly swab off the copper layer as it forms so as to produce a darkened stain area.

Germanium Epotaxial Layers. There is no recommendation for metallographically sectioning germanium at a 1° angle or less. The germanium is very easily scratched by the polishing abrasive, and these scratches inhibit any staining. Even when using larger angles, a stain may appear up to a scratch and stop. This is a common effect and should be taken into consideration before the sample is evaluated. In order

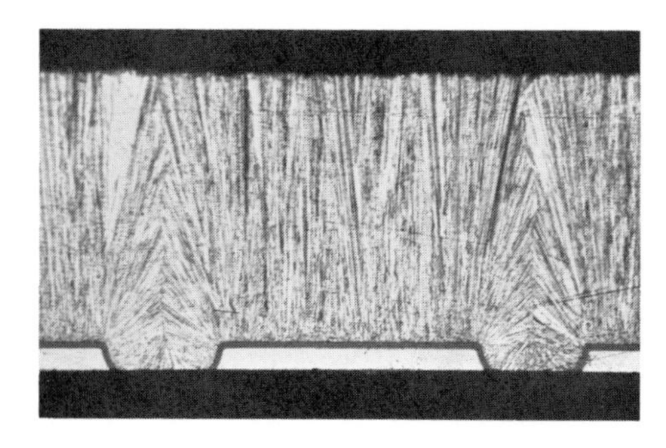

Fig. 7.13. Perpendicular section of polycrystalline silicon deposited on silicon dioxide. Delineation 5 s in Sirtl A etchant. Magnification is 135×.

to prevent this partial staining phenomenon, it is suggested that the polishing strokes be so directed that any scratches formed are perpendicular to the junction or interface line. Then if the junction or interface line is faint, it can still be used. If the scratches were parallel to the interface, it would be difficult to differentiate the scratches from the stain interface.

Damaged Regions (see Chap. 2). Surface damage includes scratches, probe marks, and areas of improper or incomplete mechanical or chemical polishing. Scratches may be microscopically observed without delineating. The use of interference contrast (e.g., the Nomarski interference-contrast attachment used with a Reichert microscope) reveals scratches very easily. They may also be detected using dark-field microscopy. In this case the scratch will appear as an illuminated line on a black background. Although the scratch can be easily seen, not enough detail will be present to judge the scratch qualitatively in aspects other than relative size. Scratches may be destructively delineated by etching. Using the etches shown in Table 7.5, etch time should be kept to a minimum so as not to remove excessive material, thus making the scratch difficult to see. Scratches on a surface prior to an epitaxial deposition will usually be propagated through the layer and result in a surface band which is wider than the original scratch. All the etches shown are acceptable for (111)-oriented material. Etches such as Sirtl may be used on other than (111) Si surfaces but do not work particularly well. Germanium etches may be used on all orientations.

Delineation of thick samples can be accomplished by wax mounting the sample on a glass slide and etching with agitation. Figure 7.14*a* shows scratches delineated in silicon after 6 min in 1-3-6 etchant.

Prior to etching, incompletely polished surfaces appear bright and specular. In reality there are many unpolished regions which have fine semiconductor and abrasive particles packed in these regions and polished. Microscopically, the surfaces look smooth and continuous. A 6-min 1-3-6 etch will remove these packed-in particles and show the unpolished regions (Fig. 7.14*b*). An alternate etch for this purpose is Sirtl; however, the 1-3-6 will delineate without revealing dislocations. For 1-3-10, use an etch time of 1 h, with agitation each 10 min. The same etchants may be used for epitaxial layers by reducing the time in order not to etch off the layer. If the layer is extremely thin (i.e., $<2\ \mu m$), this technique may not be applicable, as no delineation may occur prior to removal of the layer.

To delineate incompletely polished silicon slices without attacking the slice, heat

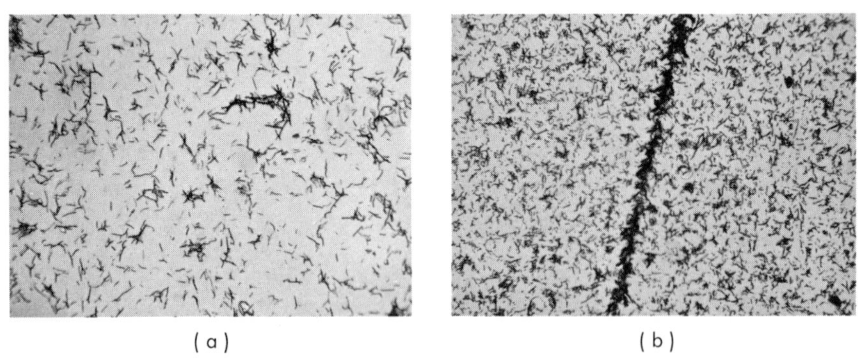

(a) (b)

Fig. 7.14. Delineation of damaged regions by use of 1-3-6 etch. (*a*) Polished surface (magnification is 260×). (*b*) Scratch (magnification is 126×).

the sample in a 10 percent aqueous sodium tetraborate solution at 80°C for 5 min. This will remove the packed-in particles and not etch the sample. At the boiling point, however, the sample will be lightly attacked. Copper-displacement plating for 1 min in a solution of 1,100 ml water, 5 ml HF, and 5.5 g $CuSO_4$-5 H_2O and then removing the plating in nitric acid is also useful in defining polishing compound left embedded in the surface and will remove less than 1,000 Å of silicon. Viewing should be with phase or interference contrast.

When polish damage is not gross but is nevertheless present, silicon can be oxidized (5,000 to 10,000 Å) and then etched. The heat treatment will cause the previous indiscernible defects to become very pronounced. These results should not be confused with the generation of gross slip patterns which occur during high-temperature cycles and uneven heating.

Impurity Striations. Most crystals grown from the melt exhibit impurity striations[44-49] which can be displayed by splitting the crystal lengthwise and treating the exposed face. Such a procedure also has the advantage of showing the shape of the growth interface as a function of crystal length. Alternately, a perpendicular cross section, i.e., a slice, can be examined, in which case spiral or circular patterns are seen. If it is of interest to study the shape of the surface defined by the striations in more detail, various bevel sections at large angles may also be used. Actual delineation may be by etching, staining, or plating, an example of which is shown in Fig. 7.15. Various approaches are summarized in Table 7.9. The striations revealed are generally referred to as "resistivity striations," but "oxygen swirls" and, in epitaxial layers, even "stacking-fault swirls" are occasionally mentioned. The latter may owe their origin to impurities, e.g., carbon, which do not affect etch, plating, or stain rate and hence do not produce readily observable striations directly, or they may be due to some swirling surface treatment prior to epitaxy which caused fault generation. The older literature often considered the resistivity-striation delineation to be due to multiple p-n junctions. Indeed that may have been true, but they can also be seen (in silicon at least) when the average resistivity is less than 0.01 Ω-cm and is almost certainly due to variations less than the amount required for type change.

These techniques by their very nature are subjective and, from the magnitude of resistivity-variation standpoint, are just an indication of a variation. Attempts to correlate them with measurements made by very fine spreading-resistance probes

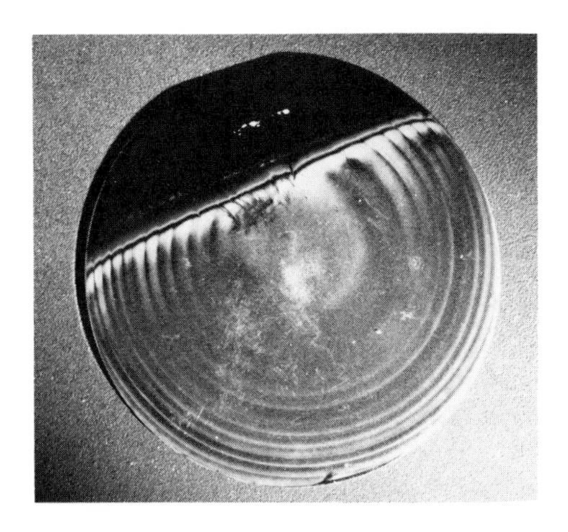

Fig. 7.15. Striations delineated by anodization.

Table 7.9. Resistivity Striation Delineation

Mat	Composition	Reference
Si	3 ml HF, 5 ml HNO_3, 7 ml glacial acetic. Use on polished surface	44
Si	Electrochemical oxidation in 10% KNO_3; 0.1 g molybdic acid for a few minutes. Si is anode, current densities approximately 15 mA/cm^2. Best in 0.00*x* to 0.0*x* range, will work to 4 Ω-cm	
Si	50 ml HF, 15 mg $Cu(NO_3)_2$, remove copper plate with HNO_3, repeat twice more.	
Si	40 ml HF, 1 drop HNO_3 every 30 s for 5–15 min under illumination —use chemically polished surface	63
Si	Wet surface with methanol, place in solution of 40 ml HF, 1 drop of HNO_3. One additional drop after 30 min, and after 1 h under illumination, remove after 2 h. Works on resistivities of from 10 to 100 Ω-cm—use chemically polished surface	45
Si	Kämper etch, very complex, see references for details	64
Ge	Acid solution of copper sulfate. Adjust electrolyte resistivity, pulse length, and repetition rate for best results. Ge is cathode	46
Ge	For n-type. Alkaline plating solution of 16 g NaOH, 1 g $CuSO_4 \cdot 5H_2O$, 19 g tartaric acid, water to make 100 ml. Use Ge as anode, 1,000 V, 10 pulses/s	47
Ge	52 ml H_2SO_4, 210 g $CuSO_4 \cdot 5H_2O$, 948 ml H_2O. Use as an electrolyte with Ge as cathode. Use current densities of from 1.5 to 30 A/cm^2, voltage up to 1,500 V. Use pulse rate of 3 pulses/s	48
InSb	3 HF, 5 HNO_3, 3 acetic, 11 H_2O (used on 211 plane)	49

have met with only limited success. However, they do give excellent spatial resolution of the variations. With the proper choice of etch and multiple-beam interferometric observation, exceedingly closely spaced striations can be observed.

Diffusion Pipes. Pipes are small unplanned diffused regions which provide ohmic paths between transistor collectors and emitters. They are shown schematically in Fig. 7.16*a*. If a cross section of the device can be made which includes such a pipe, then the staining procedures already discussed will show it. Unfortunately, only a few per emitter can cause device failure and yet the chance of blindly stopping the sectioning process at a point which will show one is rather remote. There are, however, two alternate procedures that will often show collector-emitter shorts (besides the electrical tests) and thus, by inference, pipes. One is to etch a slice in Sirtl etch after all metal and oxide have been stripped away. If the collector and emitter are still shorted, enhanced electrochemical etching will cause the pronounced change in the emitter regions shown in Fig. 7.16*b*.

By externally biasing the slice, stain and/or anodic oxide can be grown in the shorted emitters.[56] The advantage of using an external voltage is that low-voltage breakdowns and soft junctions can also be detected. If an integrated-circuit slice is used, the p substrate is made plus. Thus the substrate-collector junction is forward-biased while the collector-base junction is reverse-biased and the emitters will all be floating unless they are shorted to the collector or unless the voltage is high enough to break down the collector-base junction. The back and sides of the slice must be protected by wax from the solution, which may be a 5 percent HF solution for staining or a 10 percent $NaClO_4$ solution for anodic oxidation. When oxidizing, for a fixed current, the voltage will increase with time, or a fixed voltage will result in a pronounced reduction in current. Accordingly some sort of manual or automatic control must be used.

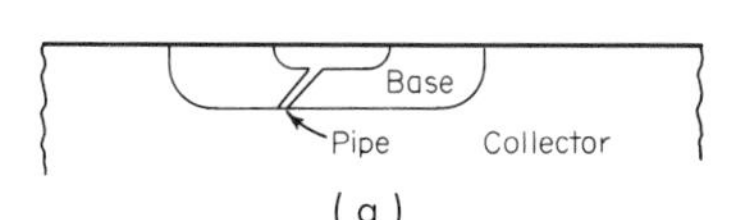

(a)

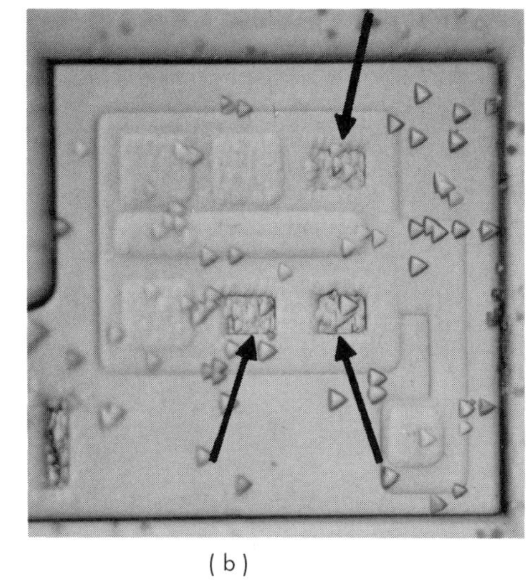

(b)

Fig. 7.16. Diffusion pipes and their effect on Sirtl etching. The arrows indicate the shorted emitters.

p-n-Junction Decoration.[57] The stain and plating methods of displaying junctions depend on delineating either the p or n region and assuming that the junction lies along the boundary. It is also possible to decorate the high-field region at the junction by a line of dielectric particles.[58] In order to sustain the necessary field, the surface must be damage-free (e.g., not a lapped surface), and to obtain the necessary particle mobility the particles must be finely divided and carried in a liquid suspension. The suspending liquid must be nonconductive and have a dielectric constant much less than that of the particles. Three to five-micrometer barium titanate mixed with carbon tetrachloride in the proportions of 1 g of powder per 100 cm^3 of liquid works satisfactorily for Si and Ge. The mixture can be applied by eyedropper onto the surface.

Oxide Pinhole Decoration. Oxides grown or vapor-deposited on silicon can have tiny holes or cracks in them which are too small to be seen without replication electron micrography.* The holes can, however, be decorated and observed with modest magnification. One of the earliest schemes was to put the slice with an oxide layer in a Cl[59] or HCl environment at temperatures above 900°C for 1 to 10 min. Wherever the hot gas contacts the silicon, a hole will be etched that is readily apparent. The main disadvantage of this method is the high-temperature corrosive-gas requirement. The problem is eliminated by the use of either a copper-plating or an electrophoretic solution to deposit readily visible particles about the pinhole. Copper sulfate–water solution will produce copper plating about holes. If dielectric liquids such as acetone, isopropyl alcohol, or methyl alcohol are used and the anode is made of copper, nonconductive copper compounds will be formed and transported to the pinhole site.[60] Ten to two hundred volts for a few minutes is required when the anode is in the form of a wire screen held normal to the slice surface.[61] Electrography can also be used by sandwiching a paper saturated with an aqueous

*For a discussion of the detection of these and other kinds of oxide defects, as well as an extensive bibliography, see Werner Kern, Characterization of Localized Defects in Dielectric Films for Electron Devices, Part I, *Solid State Tech.*, **17**:35–42 (March 1974); Part II, **17**:78–84 (April 1974).

solution of benzidine chloride between the slice and an electrode and applying a voltage across the assembly. Wherever current flows, i.e., at pinholes, the benzidine will be oxidized to a colored state and form a map of the defect locations.[60]

Package Opening. When examining encapsulated units, package opening is sometimes tedious, and without care the device inside can be decimated before there is a chance to examine it. There are many package types, but in general they can be classified as metal, ceramic, or plastic. Metal cans can be opened by using a micromilling machine,[62] sander, file, or for very thin packages, even a knife blade. A vise or chuck to clamp the package should be used, and great care taken to prevent particles from the can or cutter from falling on the semiconductor surface. (A remarkable number of "spurious particles" are found in packages which have an analysis identical to knife blades and tweezers.) Ceramic packages usually have a lid which can be snapped off rather easily using either a knife or diagonal cutters. During such operations it is easy to crack the semiconductor material; so it is well to practice on scrap units before attempting to open the one of interest. Plastic packages (ordinarily epoxy or silicone) can in principle be dissolved, or at least softened so that the plastic can be pulled away. For specific solvents, follow the plastic manufacturer's suggestions. Alternately, virtually all such encapsulants can be removed in hot sulfuric acid. It is important that the acid be dry; otherwise aluminum metallization will be removed in the process. Boiling for several minutes before use will provide a suitably dry acid.

After the package, or at least the package top, has been removed, there may still be a thick layer of deposited SiO_2 over the surface which will make examination difficult. These oxides will almost always dissolve much more rapidly in HF than thermal oxide, and may thus be stripped away with little damage to the underlying device unless considerable overetching is allowed. For oxides deposited over metallization, an eraser, preferably electric, can be used for removal.

REFERENCES

1. Hans Elias, Three-dimensional Structure Identified from Single Sections, *Science,* **174**:993–1000 (1971).
2. W. A. Gaunt, "Microreconstruction," Pitman Publishing Corporation, New York, 1971.
3. J. S. Hanson, Microsectioning: A Metallographic Technique for Semiconductor Devices, *IBM J. Res. Develop.,* **1**:279–288 (1957).
4. Z. Nagy and J. McHardy, Mounting Specimens in Teflon for Electrochemical Studies, *J. Electrochem. Soc.,* **117**:1222–1223 (1970).
5. M. C. Wong, Metallography of Semiconductors, *Metals Rev.,* **38**:(5):5–6 (1965).
6. R. H. Carter and F. P. Gagliano, The Metallographic Preparation of Microminiature Devices, *Microelectron. Reliability,* **7**:301–303 (1968).
7. R. D. Packard, Addendum to Notes on the Chemical Polishing of Gallium Arsenide Surfaces, *J. Electrochem. Soc.,* **112**:871–872 (1965).
8. Cecil V. King, Dangerous Chemicals in the Laboratory, *J. Electrochem. Soc.,* **112**:251C (1965).
9. S. F. Bubar and D. A. Vermilyea, Explosion of a Chemical Polishing Solution, *J. Electrochem. Soc.,* **113**:519 (1966).
10. M. S. Abrahams and C. J. Buiocchi, Etching of Dislocations on the Low-Index Faces of GaAs, *J. Appl. Phys.,* **36**:2855–2863 (1965).
11. D. Navon, R. Bray, and H. Y. Fan, Lifetime of Injected Carriers in Germanium, *Proc. IRE,* **40**:1342–1347 (1952).

12. W. J. Feuerstein, Etch Pit Studies on Silicon, *Trans. AIME,* **212**:210–212 (1958).
13. F. L. Vogel, W. G. Pfann, H. E. Corey, and E. E. Thomas, Observations of Dislocations in Lineage Boundaries in Germanium, *Phys. Rev.,* **90**:489–490 (1953).
14. E. Billig, Some Defects in Crystals Grown from the Melt, I. Defects Caused by Thermal Stresses, *Proc. Roy. Soc. London,* **A235**:37–55 (1956).
15. W. C. Dash, Copper Precipitation on Dislocations in Silicon, *J. Appl. Phys.,* **27**:1193–1195 (1956).
16. F. Secco d'Aragona, Dislocation Etch for (100) Planes in Silicon, *J. Electrochem. Soc.,* **119**:948–951 (1972).
17. P. Wang, Etching of Germanium and Silicon, *Sylvania Tech.,* **11**:50–58 (1958).
18. F. L. Vogel, Jr., and L. Clarice Lovell, Dislocation Etch Pits in Silicon Crystals, *J. Appl. Phys.,* **27**:1413–1415 (1956).
19. Donald H. Lyon, The X-Factor in Germanium, *Western Elec. Engr.,* **7**:3–12 (October 1963).
20. M. S. Abrahams, Dislocation Etch Pits in GaAs, *J. Appl. Phys.,* **35**:3626–3628 (1964).
21. V. N. Vasilevskaya and E. G. Miselyuk, The Problem of Visualization of Dislocations in Germanium by Etching, *Soviet Phys. Solid State,* **3**:313–318 (1961).
22. Allegheny Electric Chemical Co., *Tech. Bull.* 6, June 1958.
23. H. A. Schell, Etch Figures on Gallium Arsenide Single Crystals, *Z. Metallk.,* **48**(4):158–161 (1957).
24. E. Sirtl and A. Adler, Chromic-Hydrofluoric Acid as a Specific System for the Development of Etch Pits on Silicon, *Z. Metallk.,* **52**:529–531 (1961).
25. H. A. Schell, Etch Figures on Germanium Single Crystals, *Z. Metallk.,* **47**:614–620 (1956).
26. R. H. Wynne and C. Goldberg, Preferential Etch For Use in Optical Determination of Germanium Crystal Orientation, *J. Metals,* **5**:436 (1955).
27. J. G. White and W. C. Roth, Polarity of Gallium Arsenide Single Crystals, *J. Appl. Phys.,* **30**:946–947 (1959).
28. P. A. Iles and P. S. Coppen, On the Delineation of p-n Junctions in Silicon, *J. Appl. Phys.,* **29**:1514 (1958).
29. S. J. Silverman and D. R. Benn, Junction Delineation in Silicon by Gold Chemiplating, *J. Electrochem. Soc.,* **105**:170–172 (1958).
30. J. M. Lavine, The Behavior of p- and n-doped Contacts in a Space-Charge Depletion Region, *Solid State Electron.,* **1**:107–122 (1960).
31. D. Eirug Davies, The Implanted Profiles of Boron, Phosphorus and Arsenic in Silicon from Junction Depth Measurements, *Solid State Electron.,* **13**:229–237 (1970).
32. C. S. Fuller and J. A. Ditzenberger, Diffusion of Donor and Acceptor Elements in Silicon, *J. Appl. Phys.,* **27**:544–553 (1956).
33. Harry Robbins, Junction Delineation in Silicon, *J. Electrochem. Soc.,* **109**:63–64 (1962).
34. P. J. Whoriskey, Two Chemical Stains for Marking p-n Junction in Silicon, *J. Appl. Phys.,* **29**:867–868 (1958).
35. D. R. Turner, Junction Delineation on Silicon in Electrochemical Displacement Plating Solutions, *J. Electrochem. Soc.,* **106**:701–705 (1959).
36. John R. Edwards, Evidence of Phosphorus N-Skin on Silicon from Vapor Transport, *J. Electrochem. Soc.,* **116**:866–868 (1969).
37. I. F. Nicolau, Junction Delineation and Dislocation Revealing in Silicon by the HIO_4—HF—H_2O System, *Solid State Electron.,* **12**:446–448 (1969).
38. I. Berman, N^+ N Delineation in Silicon, *J. Electrochem. Soc.,* **109**:1002–1003 (1962).
39. R. W. Jackson, Simple Method of Revealing p-n Junctions in Germanium, *J. Appl. Phys.,* **27**:309–310 (1956).
40. Reinhard Glang, Location of Diffused p-n Junctions on Germanium by Electrodeposition of Copper, *J. Electrochem. Soc.,* **107**:356–357 (1960).
41. J. C. Marinace, Diffused Junctions in GaAs Injection Lasers, *J. Electrochem. Soc.,* **110**:1153–1159 (1963).

42. Robert H. Saul, The Defect Structure of GaP Crystals Grown from Gallium Solutions, Vapor Phase and Liquid Phase Epitaxial Deposition, *J. Electrochem. Soc.,* **115:**1184–1190 (1968).
43. S. J. Silverman, Junction Delineation in GaSb by Differential Chemical Etch Rate, *J. Electrochem. Soc.,* **109:**166–168 (1962).
44. A. F. Witt and H. C. Gatos, Impurity Heterogeneities in Semiconductor Single Crystals, in Rolf R. Haberecht and Edward L. Kern (eds.), "Semiconductor Silicon," The Electrochemical Society, New York, 1969.
45. T. F. Ciszek, Solid-Liquid Interface Morphology of Float Zoned Silicon Crystals, in Rolf R. Haberecht and Edward L. Kern (eds.), "Semiconductor Silicon," The Electrochemical Society, New York, 1969.
46. Paul R. Camp, Resistivity Striations in Germanium Crystals, *J. Appl. Phys.,* **25:**459–463 (1954).
47. J. A. M. Dikhoff, Cross-sectional Resistivity Variations in Germanium Single Crystals, *Solid State Electron.,* **1:**202–210 (1960).
48. Gurion Meltzer, Minute Resistivity Variations in Germanium Crystals and Their Effect on Devices, *J. Electrochem. Soc.,* **109:**947–951 (1962).
49. F. Morizane, A. F. Witt, and H. C. Gatos, Impurity Distributions in Single Crystals, *J. Electrochem. Soc.,* **113:**51–54 (1966).
50. B. Jansen, A Rapid and Accurate Method for Measuring the Thickness of Diffused Layers in Silicon and Germanium, *Solid State Electron.,* **2:**14–17 (1961).
51. W. A. Pliskin and R. P. Gnall, Evidence for Oxidation Growth at the Oxide-Silicon Interface from Controlled Etch Studies, *J. Electrochem. Soc.,* **111:**872–873 (1964).
52. R. V. Jensen and S. M. Christian, Etch Pits and Dislocation Studies in Silicon Crystals, RCA Industry Service, *Lab. Bull.* L13-1023, Mar. 5, 1956.
53. W. A. Hassett and W. R. Hechler, Metallographic Polishing Techniques for Semiconductor Components, *Semicond. Prod.,* **6:**27–30 (October 1963).
54. J. J. Gajda, Evaluation of Semiconductors through Angle Sectioning and Junction Delineation, *SCP and Solid State Tech.,* **7:**17–21 (November 1964).
55. K. Schuster, Determination of the Lifetime from the Stored Carrier Charge in Diffused psn Rectifiers, *Solid State Electron.,* **8:**427–430 (1965).
56. Murlidhar Kulkarni, John C. Hasson, and George A. A. James, Mapping of Electrical Leakage in Transistors by Anodic Oxidation, *IEEE Trans. Electron Devices,* **ED-19:**1098–1102 (1972).
57. P. A. Iles and P. J. Coppen, Location of p-n and l-h junctions in Semiconductors, *Brit. J. Appl. Phys.,* **11:**177–184 (1960).
58. H. E. Bridges, J. H. Scaff, and J. N. Shive (eds.), "Transistor Technology," vol. 1, D. Van Nostrand Company, Inc., Princeton, N.J., 1958.
59. S. W. Ing, Jr., R. E. Morrison, and J. E. Sandor, Gas Permeation Study and Imperfection Detection of Thermally Grown and Deposited Thin Silicon Dioxide Films, *J. Electrochem. Soc.,* **109:**221–226 (1962).
60. P. J. Besser and J. E. Meinhard, Investigation of Methods for the Detection of Structural Defects in Silicon Dioxide Layers, in "Symposium on Manufacturing In-Process Control and Measuring Techniques for Semiconductors," vol. 2, 1966.
61. W. J. Shannon, A Study of Dielectric Defect Detection by Decoration with Copper, *RCA Rev.,* **31:**431–438 (1970).
62. William C. Weger et al., "Reliability Handbook for Silicon Monolithic Microcircuits," prepared on contract NAS 8-20639, Texas Instruments, Inc., 1967.
63. W. D. Edwards, Liquid-Solid Interface Shape Observed in Silicon Crystals Grown by the Czochralski Method, *Can. J. Phys.,* **38:**439–443 (1960).
64. K. R. Mayer, A Simplification of Kämper's Striation Etch for Silicon, *J. Electrochem. Soc.,* **120:**1780–1782 (1973).

8

Microscopy and Photography

One of the major methods of examining semiconductor materials and devices is with optical magnification. This includes hand-held magnifiers, low-power binocular stereoscopic microscopes, and low- and high-power binocular nonstereoscopic microscopes. In addition, various attachments such as interferometers and phase and interference contrast are also useful. Polarization studies, widely used for mineralogical investigations, are normally possible only with infrared-image converters instead of conventional eyepieces, as most semiconductors are not transparent to visible light. For storing the results of the various optical examinations, both low- and high-magnification photography are routinely used, and most laboratory metallurgical microscopes include a companion camera. When greater magnification is required, particularly in combination with increased depth of view, scanning electron microscopes are used. Their cost, complexity, and size preclude their being as widely disseminated as optical microscopes, but their introduction into the semiconductor industry has had a tremendous impact on the understanding of microcircuit interconnection systems.

8.1 BASIC OPTICS[1-4]

The simplest form of magnifier is a single planoconvex lens as in Fig. 8.1*a*, which will, if the object is between the focal point and the lens, present a magnified upright virtual image. Because of spherical and chromatic aberrations, the magnification of such lenses (which is given by $1 + 250/f$, where f is the focal length in millimeters) is usually limited to 10× or less. Aberrations can be minimized by a combination of lenses such as that shown in Fig. 8.1*b*, but still, the practical limit of magnification is 10 to 20×.

When greater power is required, compound microscopes can be used. The simplest embodiment consists of an objective which projects a magnified real image, and an eyepiece which produces a further enlarged virtual image of the real image. The magnification is now given by the product of the magnification of the objective and the magnification of the eyepiece. The objective magnification is approximately given by L (Fig. 8.2*a*) divided by the focal length of the lens. L is the tube length of the microscope and is usually about 200 mm.* This figure depicts both lenses

*The "standard" for biological instruments is 160 mm, but metallurgical microscopes tend to have longer paths.

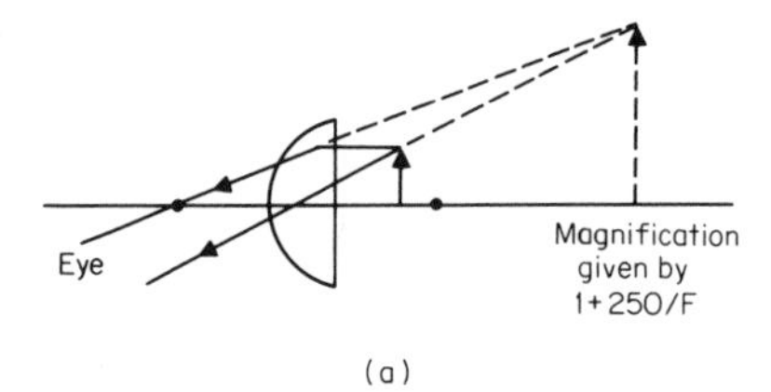

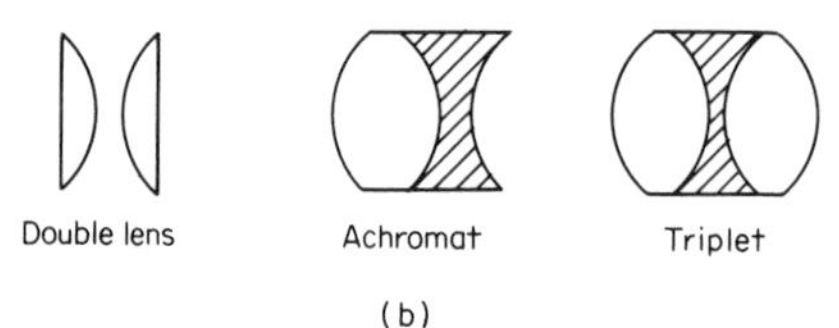

Fig. 8.1. Lens system for a simple microscope. The shaded and clear portions in (*b*) are made of different composition glass. A clear cement is used to join the surfaces.

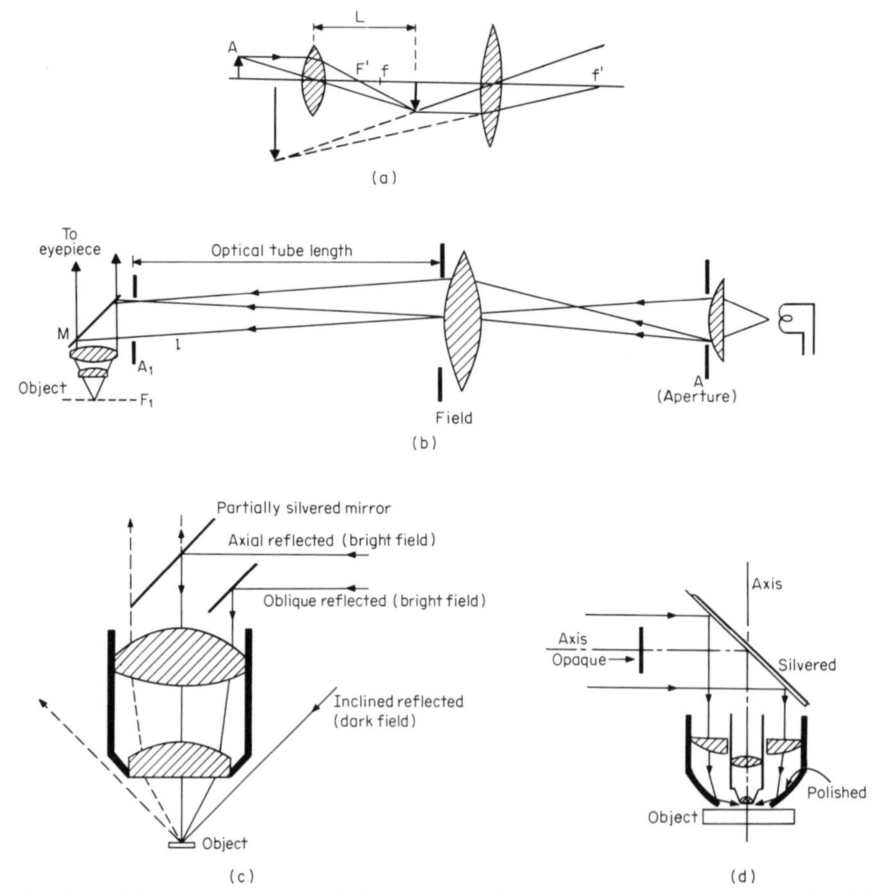

Fig. 8.2. Microscope optics. (*a*) Compound microscopes. (*b*) Arrangement of field and aperture diaphragms and auxiliary condensers for vertical illumination. (*c*) Relationship of various types of illumination. (*d*) Annular dark-field illuminator.

as single elements, but in an actual microscope they will ordinarily be multielement in order to minimize various distortions. Since most metallurgical specimens are opaque, some form of lighting from the top is required. For low-power work, in which the optics are usually quite large and several inches away from the object, an external lamp beamed to the surface is adequate. For high-power work, the vertical illuminator shown in Fig. 8.2*b* is used. A partially silvered mirror *M* is used to direct light down through the objective. Figure 8.2*c* shows more details of the various lighting possibilities.

In particular, it is to be noticed that if the incoming light ray has an angle of incidence greater than can be obtained by projecting through the objective, the reflected ray will also fall outside the objective and will thus not be collected. Therefore, any image formed will be only from light scattered by an irregular surface. Such an arrangement is called *dark field* and is particularly useful in examining an otherwise smooth surface for small holes or protuberances. There are several methods (depending on the manufacturer) of producing the required obliqueness, but Fig. 8.2*d* is typical. Objectives designed for dark field are larger in diameter than those for bright field because of the extra space required for the coaxial lighting. It is possible in some designs to shift the bright-field source from side to side as in Fig. 8.2*c* to introduce shadows which can aid in interpretation. In the case of instruments without this feature, removing the light from its housing and moving it about sometimes helps.

For purposes of both convenience and contrast, a uniform but variable intensity over the sample is required; and for eliminating glare, the light-spot size entering the objective must be controlled. In the most common type of illumination an image of the condensing lens is formed in the objective-image plane (Koehler illumination). The lighting will then appear uniform, regardless of local variations in the source intensity. The openings which control the intensity and area are referred to, respectively, as the aperture and field diaphragms.

To ease eye fatigue and minimize the effects of small particles in the eye fluid drifting across the eye's field of view, binocular microscopes are commonly used. In these, the optical path splits after it leaves the objective and goes to two eyepieces approximately in the manner shown in Fig. 8.3. Separation of the eyepieces to accommodate different interpupil widths may change the optical path length of one eyepiece and require it to be refocused, depending on the exact optical separation used.

True stereoscopic vision requires that slightly different images be presented to the two eyes. This may be accomplished by using a completely separate set of optics

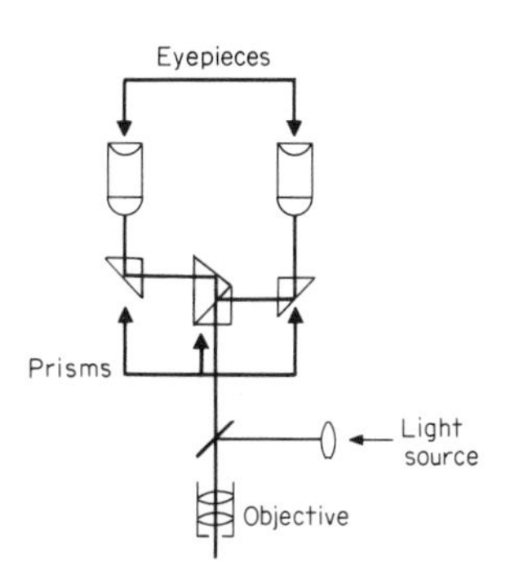

Fig. 8.3. Optical path for binocular microscope.

for each eye and directing them both at the same object, but from slightly different angles. Because the magnification is usually less than 150×, there is considerable distance between the object and the objective and external indirect lighting can be used; however, care must be taken to ensure equal illumination of both eye fields. With the binocular optics just discussed it is possible to get some stereo effect by inserting limiting apertures in the beams, but it is seldom used.

Objectives. Achromatic objectives are corrected spherically for one color, and chromatically for red and blue.* By the use of fluorite positive lenses and barium-flint negative lenses, it is possible to focus red, green, and blue rays simultaneously as well as to correct spherically at two colors and thus considerably improve the color correction. (These are referred to as *apochromats.*) The latter lenses do, however, have to be further compensated by special eyepieces. Fluorites, sometimes called semiapochromats, are intermediate in correction between achromats and apochromats. If a cover glass is being used, the light will be refracted as it goes through it and will introduce spherical aberration in an objective which is fully corrected. Therefore, objectives designed for use with cover glasses should not be used without them (and vice versa). Under normal circumstances metallurgical microscopes do not use cover glasses over specimens; so biological-microscope objectives should not be used with a metallurgical microscope.

Parfocal objectives are mounted so that as the nosepiece is rotated to bring them sequentially into position, little focusing is required from one objective to the next. For a set which is not exactly parfocal, they can often be corrected by putting washers between the individual objectives and the nosepiece. Most manufacturers use standard objective threads and eyepiece-tube diameters, but tube lengths are not standard and care must be exercised in interchanging optics. Objectives are described in terms of the type of correction (achromat, fluorite, etc.), magnification (3.2×, 10×, etc.) or focal length, numerical aperture (which is defined in Sec. 8.2), working distance, whether they are oil- or air-immersion, and whether designed for a specific tube length (L of Fig. 8.2*a*) or are infinity-corrected and require an auxiliary lens built into the tube. Depth of field and image intensity are dependent on numerical aperture (N.A.); they vary as $(1/\text{N.A.})^2$ and $(\text{N.A.})^2$, respectively, and are seldom specified. Typically, for an N.A. of 1.4, the depth of field will be in the quarter-micrometer range.

Eyepieces. Despite the single-lens eyepiece shown in Fig. 8.2, high-quality eyepieces use a field and an eye lens combination, each of which may consist of multiple elements in order to provide for various optical corrections. Three of the more common kinds are the Huygens, the Kellner, and the Ramsden. The first, shown in Fig. 8.4*a*, is characterized by two planoconvex lenses separated by half the sum of their focal lengths. The real image is formed between the two lenses, and while the combination corrects chromatic aberrations of the complete eyepiece, a reticle (which must be placed in the plane of the real image) would be distorted since it is between the lenses and they are not individually corrected. The Ramsden eyepiece removes this difficulty by having the real image fall outside the lens combination as in Fig. 8.4*c*. It also consists of two planoconvex lenses, in this case of equal focal length, separated by two-thirds of a focal length. The Kellner eyepiece (Fig. 8.4*b*)

*Spherical aberration is caused by rays passing through the lens near its center not crossing the optical axis at the same point as rays from farther out. Chromatic aberration is due to a single ray of white light being broken up into a multitude of rays with slightly different angles because of dispersion in the lens.

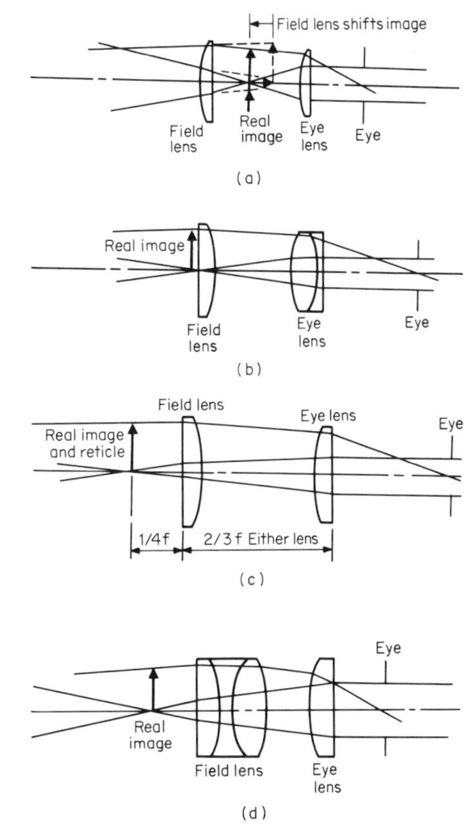

Fig. 8.4. Optics for various eyepieces. (*a*) Huygens. (*b*) Kellner. (*c*) Ramsden. (*d*) Orthoscopic.

has an achromatic doublet for an eye lens, and it and the field lens are separated by an amount which places the real image in the plane surface of the field lens. Therefore, if a reticle is to be used, the Kellner must be slightly misadjusted to place the image just outside the field lens. This may be done and it will still retain most of its advantages of an achromatic correction and an exceptionally wide distortion-free field.

For more complete correction, or to compensate for the objectives, and in some cases to obtain longer eye relief (a particular problem with Huygens as the power is increased much beyond 10×), more complicated configurations are used. Compensating eyepieces are designed for use with apochromatic objectives and are over-color-corrected. For example, periplanes are made by making the eye lens of a Huygens a doublet, and orthoscopes use the structure of Fig. 8.4*d*. These may also be used with fluorite objectives and high-magnification achromats. They do not, however, perform well with low-power achromats. Eyepieces for photographic use are especially designed to produce a flatter field. Some varieties cannot be used for direct viewing by eye, since they will only project a real image.

8.2 OPTICAL IMAGE QUALITY

Objects are made visible because of various combinations of the following:

1. Variation in reflectivity giving rise to reflection images (of most importance in metallurgical examinations)
2. Variation in refractive index giving rise to refraction images (applicable only

if there are transparent particles or layers on top of the opaque reflecting surface)
3. Diffraction patterns
4. Color variation due to absorption within the material
5. Color or intensity contrast due to interference (very useful in semiconductor-device studies)
6. Polarization effects

Successful examination depends on taking the necessary steps to enhance whichever phenomenon is most amenable to interpretation, and to minimize any instrumental weakness. Enhancement by operation on the sample itself, such as etching or staining to produce differences in reflectivity, is covered in detail in Chap. 7. This section is devoted to the instrumentation itself.

Refraction. For transparant objects, refraction at the various surfaces can be obscured if they are illuminated from many directions simultaneously. The use of a sector aperture illuminator or the more conventional adjustable plane-glass reflector (Fig. 8.2*c*) will provide for nonequal lighting.

Diffraction and Resolution. Diffraction effects produce banding which limits resolution and thus determines the maximum magnification that can be obtained.

It can be shown that the wider the beam of light accepted by a lens, the less pronounced the diffraction effects will be. The angular aperture (A.A.), which is the angle between the most divergent rays which can pass through a lens and form an image, is a measure of this beam width. If the object and the front surface of the lens are immersed in a liquid of refractive index n greater than 1, the effective beam width is increased. To account for that possibility, the numerical aperture N.A. defined by

$$\text{N.A.} = n \times \sin \frac{\text{A.A.}}{2} \tag{8.1}$$

is now commonly used instead of angular aperture for describing all lenses, whether or not they are designed for immersion. If they are used in air, $n = 1$, N.A. reduces to sin (A.A./2) and has a maximum value of 1, since 180° is the largest possible collection angle. For the case of $n = 1$, it is most difficult to design quality lenses with an N.A. greater than 0.95. However, by using a high-index-of-refraction oil between the lens and the object, N.A.s of 1.4 are not uncommon.

The minimum distance s of two lines on an object which can just be separated in an image is approximately given by

$$s = \frac{\lambda}{2\ \text{N.A.}} \tag{8.2}$$

where λ is the wavelength of the light being used. Subsequent magnification must then ordinarily be used to allow the eye to resolve the image. The objectives are normally made with as high an N.A. as possible, while the follow-on magnification by eyepiece and/or camera can be with a lower N.A. lens since after initial magnification, lesser resolution is required. The optical system thus behaves very much like a low-noise amplifier in which the first stage is carefully constructed to provide enough gain at the lowest possible noise level to make the signal into the following stages substantially above their noise level.

The resolving limit of the eyes of various individuals varies but is of the order of 250-μm line separation. On this basis, magnification of approximately 1,000 times the numerical aperture is required for the eye to see all possible detail. That is,

$$\frac{s_{\text{eye}}}{s_{\text{microscope}}} = \frac{250}{s} = \frac{250}{\lambda/(2\,\text{N.A.})} \tag{8.3}$$

and for λ of 0.5 μm, which is green light located approximately in the center of the white-light spectrum,

$$\frac{250}{s} = \frac{500(\text{N.A.})}{0.5} = 1{,}000\ \text{N.A.} \tag{8.4}$$

Actually, more magnification is often helpful, both because individual eyes vary, and because it is sometimes more convenient and easier to study somewhat larger images. Table 8.1 gives numerical aperture and resolution for a number of objectives which are representative of various manufacturers. Further maximum eyepiece magnifications based on the foregoing criterion are also included. It is true, however, that one can with care make good useful photographs which have magnifications considerably greater than those listed in the table.

Images of fine points will appear as points surrounded by faint concentric circles. An edge (or other bright-line source) will have dim lines running parallel to it and may sometimes be misinterpreted as a layered structure. Often only the first intensity minimum on either side of the edge or point will be visible and will give rise to a black band surrounding the bright image. Picture framing can also occur because of a sloping region (e.g., an oxide step) reflecting light out of the optical system, from pseudo Becke lines arising from abrupt changes in reflecting surface height,*

*True Becke lines arise when a transparent object with vertical sides is immersed in a medium of different index of refraction and viewed in transmitted light. Some of the light from the object will be refracted in such a manner that when the microscope is defocused, light or dark framing occurs. Pseudo Becke lines are observed in metallurgical microscopes when there is a rather abrupt change in the surface elevation. Because of the defocusing, some light from one elevation is superimposed on the image from the other elevation and thus makes one side appear darker and the other brighter.

Table 8.1. Magnification and N.A. of Typical Metallurgical Microscope Objectives

Magnification	N.A.	Resolution,* μm	Separation, lines/mm	Max usable magnification†	Max usable eyepiece magnification†
3.2	0.06	4.0	250	60	19
5.5	0.15	1.7	590	150	27
6.5	0.2	1.3	770	200	31
8	0.2	1.3	770	200	25
16	0.30	0.8	1,250	300	19
20	0.50	0.5	2,000	500	25
32	0.65	0.4	2,500	650	20
45	0.66	0.4	2,500	660	15
80	0.95	0.26	4,000	950	12
100	1.36	0.18	5,500	1,360, oil immersion	13.6
160	0.95	0.26	4,000	950	6
160	1.40	0.18	5,500	1,400, oil immersion	8.8

*Based on Eq. (8.2).
†Based on Eq. (8.4). Actually, considerably more magnification can often be useful.

from something similar to a true Becke line caused by an abrupt change in the thickness of a high-index transparent layer overlaying the reflecting surface, and from interference effects in oxide and nitride slopes.

Ordinarily such banding only complicates interpretation, particularly when viewing in dark field when the bands are bright, but for low-contrast or extra small objects, diffraction patterns may be required for any visibility. Ultramicroscopy (based on the Tyndall effect) deliberately makes use of diffraction to produce visible images of objects far smaller than can be resolved. Such images have little resemblance to the actual object shape but can be used to estimate numbers. Such microscopes are normally used to study colloids, but various modifications have been used to follow the growth of Si-O chains in silicon.

Inexpensive microscopes often have pronounced diffraction effects, which explains why they may actually show faint height variations better than higher-quality instruments and are sometimes preferred. If lower resolution is desired for additional contrast, the system N.A. can be decreased by stopping down the aperture diaphragm. This same reduction of the aperture diaphragm will also cause a more pronounced pseudo Becke effect. In either case the result is increased contrast. A good place to look for diffraction lines that can be easily identified as such is along the sharp edges or the apex of sharply defined etch pits. When focused on them, the effect of reduced resolution can be demonstrated by closing the aperture diaphragm and noting the increase in the number of lines. Figure 8.5 shows the surface of a low-temperature (100) silicon epitaxial layer taken with a quality metallurgical microscope with the field stop nearly closed, and quite open. Viewing through an inexpensive microscope with fixed stops gives results comparable with Fig. 8.5*b* and demonstrates the earlier thesis that superior optics sometimes give poorer contrast. Such effects do obscure fine detail and under ordinary circumstances should be avoided. If the diaphragm can be varied, for optimum aperture, focus on an object, remove an eyepiece, and adjust until the image of the light source which appears at the back plane of the objective just fills the lens. For any given magnification, an oil-immersion lens will reduce diffraction.

Color. Color filters may sometimes be used to produce better contrast of colored objects. To lighten a particular color, choose a filter of the same color. To darken

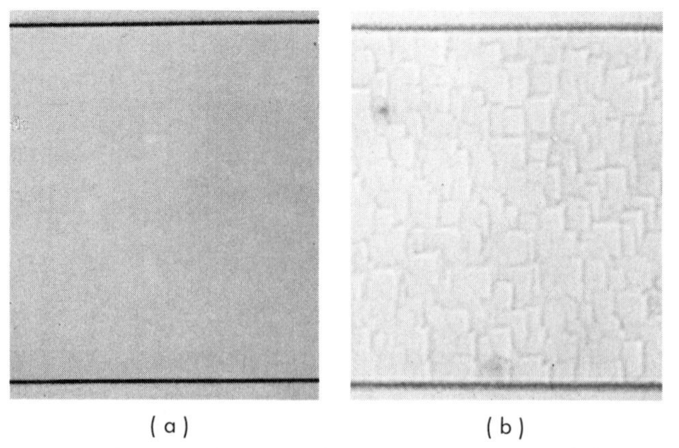

Fig. 8.5. Increased contrast from (*a*) to (*b*) accomplished by stopping aperture diaphragm and slightly defocusing. The subject is a low-temperature silicon epitaxial deposition on a (100) silicon substrate. Magnification is 259×.

Table 8.2. Filters for Improving Contrast

Color of object	Filter
Red	Green
Yellow	Blue
Green	Red
Blue	Yellow or red
Brown	Blue
Purple	Green

it, choose a filter which absorbs light of that color (Table 8.2). They are applicable only if the object coloration arises from absorption and not from interference. It might also be noted that green is useful for reducing eye fatigue. Because a single color will reduce chromatic aberration and give better resolution, narrow-bandpass filters are often used, with green again being recommended. Further, since the limit of resolution is inversely dependent on wavelength, resolution and thus image quality can in principle be increased by using the shortest possible wavelength for which the lens is corrected.

Contrast. Dirty eyepieces or objectives can and often do degrade image quality. In such cases they should be cleaned with a dry lint-free cloth. Lightly attached particulate matter can be removed by blowing with a syringe or pressured, filtered inert gas. For greasy surfaces (eyepieces are particularly susceptible to this), a liquid cleaner may be required. Care should be taken either not to choose one which will dissolve the cement between the lenses or else to be very careful and keep it away from exposed joints. Xylene is usually satisfactory, but if there is any doubt, the specific manufacturer should be consulted. Lens paper which has been touched by fingers will transfer grease back to the lens, but if the paper is first rolled and then torn in two, a fresh, clean surface is exposed. These and various other causes of image degradation are summarized in Table 8.3.

Glare will reduce contrast, but even with minimum glare, if the object is highly reflecting, and if the illumination intensity is too high, detail will be obscured. Glare can arise because of reflections from the first surface of the back lens of the objective (hence one reason for coated optics), from dust on various surfaces, and from miscellaneous reflections caused by metal protuberances in the microscope. It is most serious when looking at surfaces with little contrast, and for any given instrument can be minimized by controlling the field diaphragm.

This diaphragm should never be open more than just enough to illuminate the complete field of the microscope, and for critical cases, it may be reduced much more, and thus illuminate only a small portion of the normal field. Scatter is reduced in proportion to the area reduction, and yet the light intensity of the region covered remains unchanged. Some lenses in a given manufacturer's series may be much worse than others, and in such cases, a different objective is the only correction available. The state of the observer's eyes also contributes to the amount of contrast and detail that can be seen, and to assist them, a darkened room is often helpful.

Occasionally, when image quality suffers, it can be traced to a chipped objective. Some instruments have spring-loaded objectives to minimize this possibility, but the best way to prevent such disasters is to exercise care in focusing, particularly with the higher-power, shorter-focal-length (and much more expensive) objectives. It is good practice to examine the exposed objective surface periodically with a low-power

Table 8.3. Some Causes of Poor Image Definition

Microscope vibration	Particularly noticeable at higher magnifications when operating in multistory buildings
Poor focus	Can arise during photography either from failure to focus on screen properly or because screen and film are not exactly in same plane
Dirty objectives	Scatter of light due to small particulate matter or films on surface
Chipped or etched objectives	The first occurs fairly regularly because many semiconductors are harder than glass and will almost always damage the objective if it is inadvertently lowered onto them. The latter is caused by viewing objects just etched in HF bearing solutions and not properly washed
Cloudy lenses	Cement holding lenses sometimes fails and allows elements to separate
Dust on internal lens, prisms, or mirror surfaces	May cause blurring or dark spots to appear in field of view
Excess glare	See section on contrast
Use of objective designed for cover glass	
Use of oil-immersion lens without oil, or reverse	Most oil-immersion lenses are identified by a black band around the bottom
Use of wrong refractive index oil	
Using eyepieces not properly corrected for objective	It is usually good practice to use objectives and eyepieces from the same manufacturer
Using objectives designed for different tube length	
Stereo-optics misaligned	In addition to blurring, this problem causes severe eyestrain. To check, see if an object in the center of the field remains immobile as first one and then the other eye is closed

magnifier for signs of chipping, dirt, etc. An eyepiece from the same microscope will work quite satisfactorily for this purpose.

Contrast of surface-height variations can be considerably improved by dark field. Figure 8.6 shows the surface of an integrated-circuit bar in normal bright-field and in dark-field illumination. Phase contrast also is very useful for surface-variation contrast, as are microscopes which use diffracted light only to view the object. This can be accomplished by using an annular stop in the incoming beam and blocking

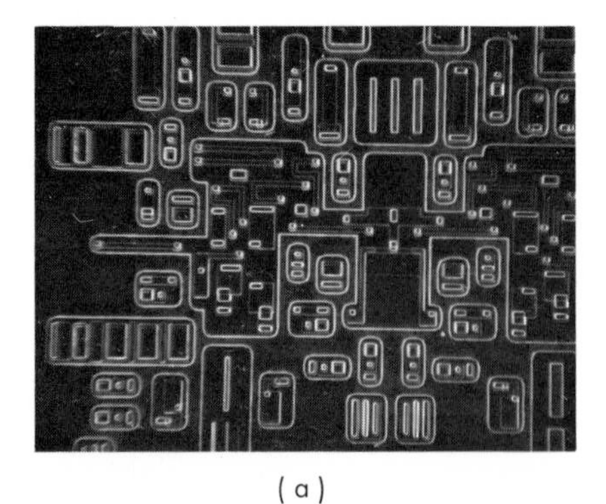

(a)

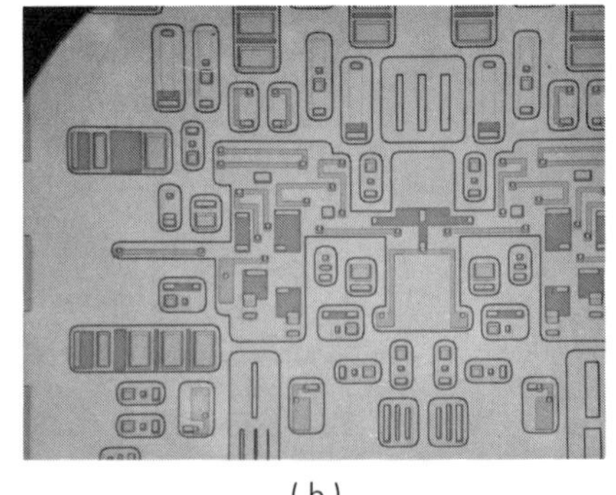

(b)

Fig. 8.6. Comparison of (*a*) dark- and (*b*) light-field images. The subject is a dielectric-isolated integrated circuit.

entry of that direct ring of light after it passes back through the objective after reflection (similar to the phase-contrast system except that opaque material is used instead of a phase shifter). The diffracted beam will be attenuated only slightly and will produce considerably more contrast than when mixed with the direct beam. Probably the best way to obtain maximum height-variation contrast is to use either a multiple-beam interferometer or a polarization interferometer such as the Nomarski. In either case, sensitivities of a small fraction of a wavelength are possible. The polarization varieties show height changes as color differences and give a direct display of the surface, whereas the multiple-beam instruments produce displacements in fringes as they traverse steps. Because of these differences, the polarization interferometers usually give a more interpretable image, but the multiple-beam better allows for quantitative step-height measurements.

Focusing. Focusing with parfocal objectives should always start with the lowest-power objective, since this will give the greatest working distance and will minimize the chance of jamming the objective into the object. When rotating parfocal objectives while in focus, care must be taken if the surface being viewed is in a depression, e.g., an integrated-circuit bar in a delidded package, since high-power objectives may not clear the depression walls. If a fuzzy view of the light-source filament becomes visible, the objective is too close to the object and should be slowly backed away. For lenses that are not parfocal, the objective may be lowered to just above the surface (as observed by placing the eye at object level and watching the spacing) and focused by moving the objective away from the object. For instruments with Koehler illumination, the field-diaphragm leaves (Fig. 8.2*b*) will be in simultaneous focus with the object. It may therefore be stopped down and focusing done on the edge of one of its leaves. This procedure is particularly useful when the object is highly reflecting and difficult to focus on. When all else fails, dust particles may have to be added to the surface to help find it. Small x-y motion of the object while moving the focus control will also sometimes help. When long working distance is required, objectives of low magnification and long focal length must be used. The overall magnification can be maintained by using a higher-power eyepiece, but resolution may be lost (see Table 8.1).

When oil immersion is used, the average slice is so light that oil surface tension will pull it to the lens and prevent focusing unless it has been firmly attached to the stage. Further, the oil itself is better put on the lens than on the edge of a slice, since it will usually run off the latter before the lens can be positioned. Remember also that lenses for oil immersion are especially designed for oil of a given index and should not be used in air or with oils of other indexes. Similarly, dry (air) objectives should not be used with oil, since resolution will suffer.

Binocular and multiple-viewing microscopes have provisions for focusing the remainder of the eyepieces after one eyepiece-objective combination has been focused. This provision is necessary because most eyes are slightly different and individual focusing will minimize eyestrain. The eye will accommodate some degree of misfocusing but will tire easily; so when using the microscope for extended times, the following procedure should be used.

1. Relax eyes.
2. Focus fixed eyepiece-objective combination.
3. Focus the variable eyepiece by moving it all the way out (plus, if it has a scale), then slowly moving it in. When the eyepiece is changed in this direction, the image comes into view from infinity. If the eyepiece is moved

in too far and the image blurs, do not attempt to focus by moving it slowly back out, while viewing, but rather start the procedure all over. Otherwise, "pulling" of the eye may occur.

8.3 SPECIALIZED OPTICAL EQUIPMENT

In addition to the basic microscope, there are a number of specialized instruments and attachments that are quite useful to the semiconductor viewer. By using additional beam splitters, for example, one can add more than one set of eyepieces to a microscope so that two people can simultaneously view and discuss the same object. Projection screens are sometimes used, but with the reflected light used in most metallurgical microscopes, the image is quite dim, and resolution suffers because of screen grain. If a rotating screen composed of lenticular mirrors is used, resolution is considerably improved,* but at the expense of increased complexity. A television pickup and monitor can also be used for multiple viewing, but if it is in black and white, much information is lost, since color is an important identification feature in most microscopic examinations.

For comparing two similar objects, e.g., masks, microscopes with two separate objectives and provision for combining their images in a common binocular head are available. The combined view may be either a split field, i.e., the input from one objective occupies one half of the field and that from the other objective the other half; or the two images can be superimposed. For very high magnification work, metallographs are sometimes used. They differ from other microscopes in being much more rigid as well as more versatile, since they have a wide assortment of attachments which can be precisely located because of the optical bench used for mounting the optics.

Toolmaker's microscopes have x-y specimen stage travel as well as the vertical position controlled by precision micrometers and are monitored by dial indicators so that distances between surface features can be easily measured. When very small crystals are studied, the angles between various exposed planes can be determined by using a microscope (microgoniometer) with both a calibrated rotating stage and calibrated inclinable optical axis.[6] By inclining both the light source and the microscope at angles of 45° to the viewing surface as shown in Fig. 8.7, a light-section microscope is formed which enables step heights to be measured.[7] It does not have as fine a resolution as a microinterferometer but is more applicable to production measurements.

The equipment just described is all derived from modifications of the basic optics of Fig. 8.2. There is, however, other optical equipment dependent on additional phenomena which are also of use in examining semiconductor materials and devices. Examples of these are phase contrast, interference contrast, and polarizers.

Polarization Microscopy. In materials which are not optically isotropic there are three kinds of polarization effects that may occur in transmitted light. The most common is double refraction (the index of refraction is dependent on the direction of the E vector). Some materials are optically active; i.e., the direction is rotated as it traverses the material. Finally, some materials have absorption coefficients which vary with the direction of the E vector (pleochroism).

In order to study these effects, polarizers are used. These are optical elements

*Aerial Image Projection System, Vision Engineering, Ltd.

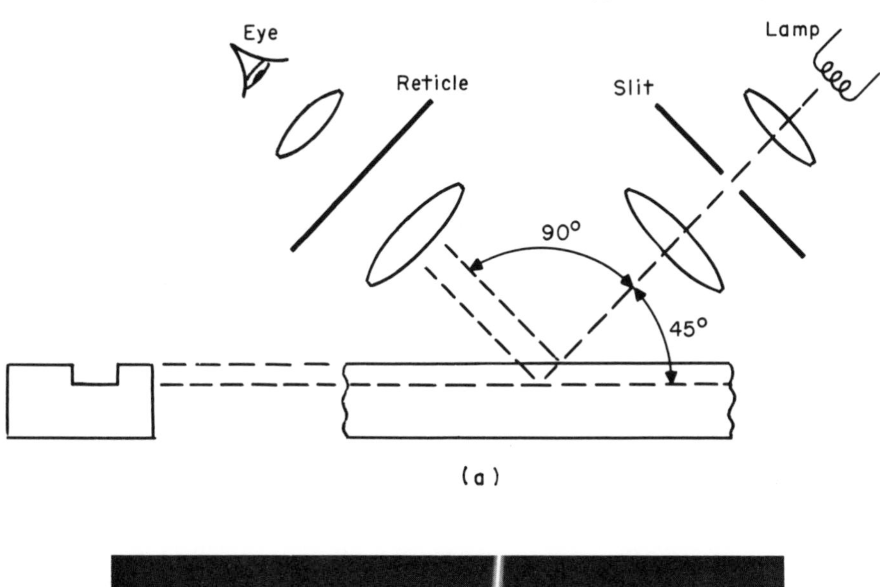

(b)

Fig. 8.7. (*a*) Optical path for a light-section microscope. (*b*) An example of line displacement.

which resolve randomly polarized light into two perpendicular components and then transmit only one of them. The plane of polarization bears a fixed relation to the physical construction of the polarizer and can be changed with respect to a given set of coordinates by rotating the polarizer about its optical axis. If a second polarizer (analyzer) is placed in a linear polarized beam and rotated until the plane of polarization of the light it transmits is 90° to that transmitted by the polarizer, little light will pass, and the combination is referred to as *crossed polarizers.*

Polarization microscopes have two extra optical elements: a rotatable polarizer which is inserted between the light source and objective, and the analyzer, which may be placed anywhere in the optical path between the object and the eye. Usually,

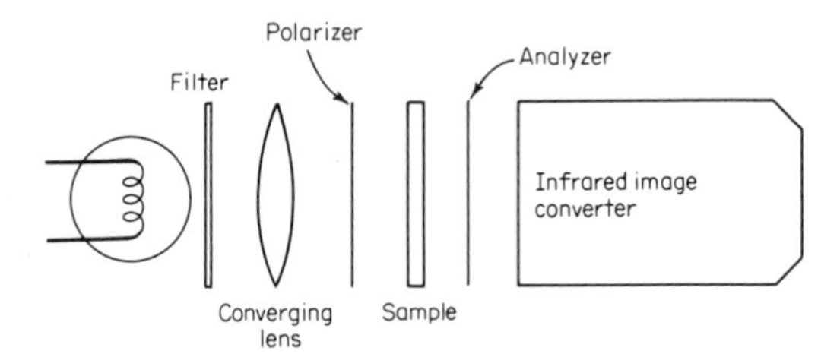

Fig. 8.8. An infrared conoscope for observing strain in silicon. The infrared converter has image-forming optics incorporated into it. For 1.1-μm operation, a silicon filter and Polaroid HR6 polarizing sheet may be used.

however, it is placed between the objective and the eyepiece and is fixed in angular position with respect to the microscope frame. Objectives used with polarizing microscopes must be strain-free and are usually specially manufactured. For that reason, ordinary objectives are not recommended for use in polarization microscopes.

When an isotropic object is viewed between crossed polarizers, the field will remain dark regardless of its orientation. Uni- or biaxial specimens will appear bright in each quadrant as they are rotated unless viewing is along one of the optical axes. To determine quickly whether a sample is isotropic or not, instead of a polarizing microscope, a conoscope may be used. In it the light is converging when it passes through the crystal and gives rise to very distinctive interference patterns. Figure 8.8 is a schematic of a low-power conoscope for large samples, and Fig. 8.9 shows the kind of patterns to be expected. Figure 8.10 shows a conoscopic picture taken from an infrared-image converter of a uniformly strained silicon slice and illustrates the strain conversion of the normally isotropic cubic silicon to uniaxial behavior. If reflected polarized light is used, i.e., a microscope with an opaque illuminator, anisotropic reflection coefficients are observed for all noncubic crystals and can aid in separating various material phases. Conoscopic observations of isotropic materials by reflected light will give cross configurations similar to those observed on uniaxial materials in transmitted light.[8] Polarization microscopy as just described is seldom used in semiconductor studies except to search for strained regions.

Interference Microscopy. Interferometry can be combined with various lens systems to give microinterferometry and is very useful in obtaining the contrast necessary to study small-scale surface steps. In some forms, the conventional displaced fringes associated with interferometers are observed.[9] In others, interference is used to enhance the image contrast.[10,11] Figure 8.11*a* and *b* shows two versions of the first type. The simple arrangement of Fig. 8.11*a* is widely used in conjunction with angle-lap and stain techniques (see Chap. 7). If the reference surface has a low reflectivity and high transmission, two-beam Fizeau interference fringes are observed. The reference may be a thin piece of glass (e.g., a cover glass) or an attachment designed to fit over the end of the microscope objective. The more sophisticated dual-beam instrument of Fig. 8.11*b* has the advantage of not requiring a reference surface in close contact with the surface being observed. It can be used

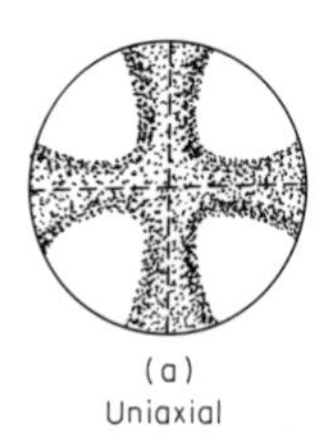

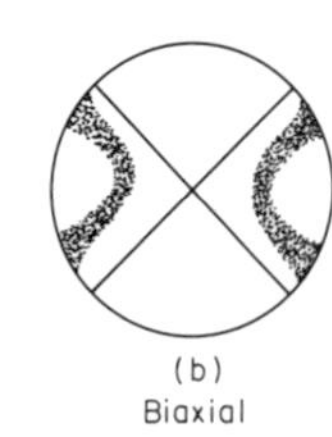

Fig. 8.9. Conoscopic interference patterns observed by transmitted light. (*a*) The optical axis is perpendicular to the surface. (*b*) The optical axes lie in a plane perpendicular to the surface and lying along a line connecting the apexes of the hyperbolas.

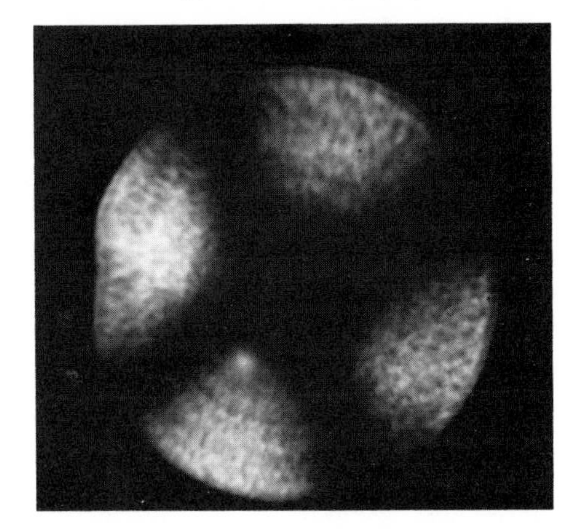

Fig. 8.10. Transmission conoscopic picture of a strained silicon slice.

at relatively low power to study deviation from planarity of slice surfaces.[12,13] Two-beam instrumentation is useful for detecting deviations from planarity down to perhaps one-tenth wavelength. When the reference and the surface to be examined are lightly aluminized to increase reflectivity,[9] multiple reflections occur, the lines are sharpened, and with considerable care the sensitivity can be increased to 5 to 10 Å.

There are also methods of obtaining image contrast by interference. The simplest one is the conventional phase contrast described below. However, more elaborate techniques based on polarization colors are more pleasing to use and usually give a higher-intensity image.

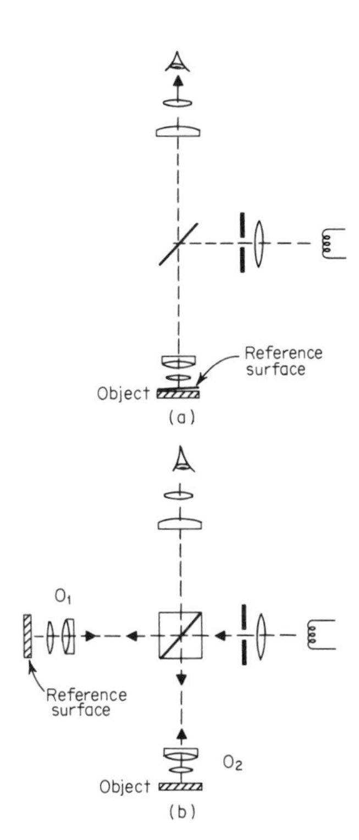

Fig. 8.11. (*a*) A partially reflecting transparent reference surface placed over the object yields interference fringes. (*b*) The reference surface is widely separated from the object through the use of a micro-Michelson interferometer. The optics O_1 and O_2 are matched microscope objectives.

Phase Contrast. One method for increasing contrast is to introduce phase shift in a portion of the beam as shown in Fig. 8.12. The source-light beam has a stop so that only an annular ring of light is transmitted. In the plane between the objective and eyepiece where that ring is imaged, a phase plate ring is placed which shifts the phase of the light going through it by a quarter wavelength. This shift affects all the direct beam, but most of the light reflected or diffracted from the object will not pass through the phase plate and from that point on will be of different phase from the background light. The phase shifter is usually placed on the end of the objective housing, and a phase-contrast objective can be recognized by the plate, as well as by manufacturer's designations on the objective. The annular stop in the incident beam must be changed in size to accommodate each objective; this may be accomplished by rotating a disk with a number of apertures matched to specific objectives. Further contrast can be obtained by simultaneously reducing the intensity of the background by placing a filter in series with the phase ring. In order to provide for a continuous transition from phase to bright field, the source ring can be moved with respect to the objective. If it is far away (Fig. 8.12*b*), the light will pass unimpeded through the center of the phase plate, there will be no phase change, and the illumination is bright-field. Moved closer, it will pass through the phase-shifting region and there will be the phase contrast just described. Closer still, it will be imaged outside the retarding region and a bright field will again be seen. Further movement will begin to shift the light path over to the edge where it will be blocked by the microscope body and give dark field. Phase contrast was originally conceived for improving contrast of colorless specimens in transparent fluids through differences in index of refraction, but when applied some years later to opaque samples, its greatest usefulness proved to be making small variations in height visible. It will also aid in detecting any variations in refractive index of transparent surface films as well as defining regions which produce differing amounts of phase shift when the incident light is reflected from them.

Interference Contrast.[10] Rather than using a flat plate as a reference surface

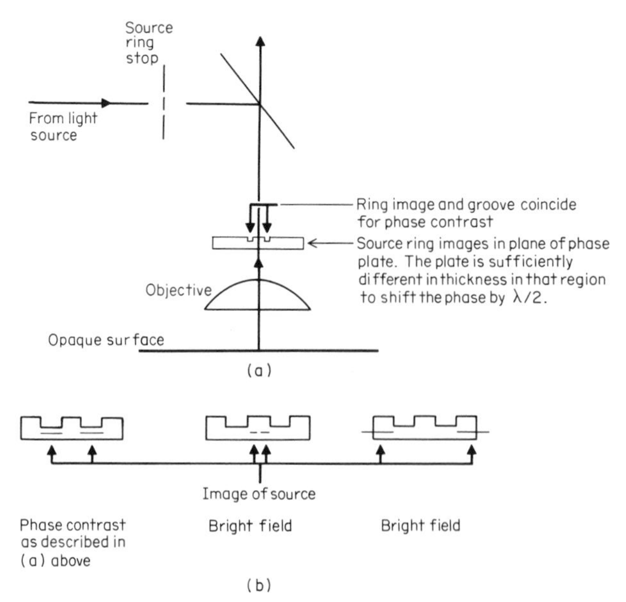

Fig. 8.12. Phase-contrast optics.

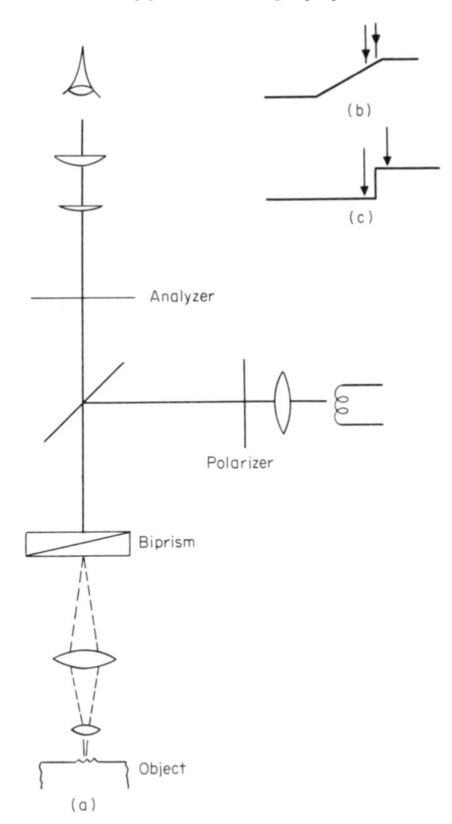

Fig. 8.13. Interference contrast. The light is resolved into two perpendicular components which emerge from the biprism with slight path deviations. The biprism is oriented at a 45° angle with the polarizer. The analyzer and polarizer are crossed. Thus, any phase difference due to a path difference of the two rays will appear as a polarization color.

for obtaining interference as was done in the dual-beam instrument, or shifting the phase of the scattered and diffracted light with respect to the direct beam reflection as in phase contrast, it is possible to separate the incident white light into light with two perpendicular planes of polarization and laterally displace one beam slightly with respect to the other as shown in Fig. 8.13*a*. If the two beams strike a flat surface, they will be reflected in phase and upon recombination will yield a normally colored image. However, if there is an incline as in Fig. 8.13*b*, the two will be out of phase, and upon recombination, will give a polarization color as previously discussed. The color then is a measure of the slope of the incline. Were the step to be abrupt as in Fig. 8.13*c*, the two levels would be of uniform but different color. In order that the beam displacement will not produce a banding effect, it is limited by optical design to less than the limit of resolution of the optics. There are several designs differing in detail, but the one which has been most accepted for semiconductor work is that of Nomarski. It is an exceedingly useful technique for studying slices at various stages of processing. Figure 8.14 is comprised of two photographs of approximately the same area of a silicon slice which had been lightly etched. Figure 8.14*a* is an ordinary photograph taken through a conventional microscope and shows little detail. Figure 8.14*b* was taken using the same microscope with a Nomarski interference attachment and shows much additional detail. It is also possible to provide a form of interference contrast built entirely into an eyepiece, which can then be used with any microscope.[10] Such eyepieces, however, are relatively rare and seldom used.

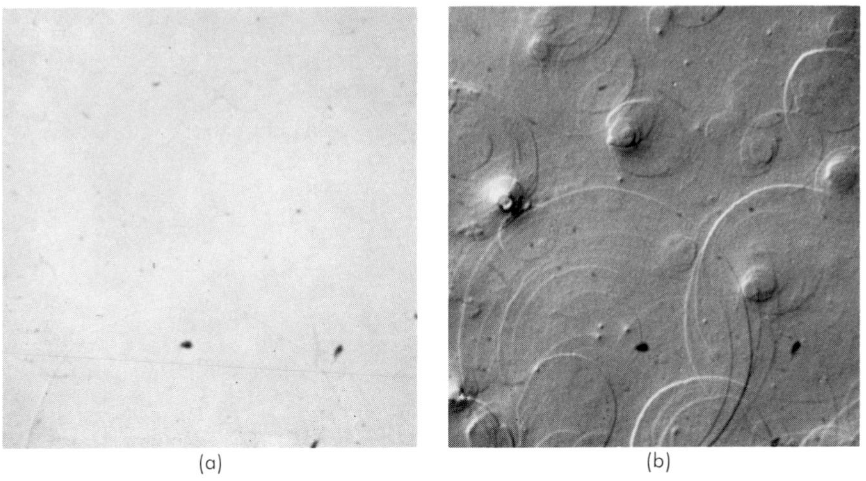

Fig. 8.14. The enhanced contrast from (*a*) to (*b*), available through the use of a Nomarski interference-contrast attachment. Magnification 224× in each case. The subject is a silicon slice etched with Dow etch.

Infrared Microscopy. Infrared imaging is useful for examining semiconductors which are not transparent to visible light. An ordinary microscope can be equipped with a camera and infrared film, e.g., Kodak 1M, or have an infrared-image converter substituted for the eyepiece and used out to the cutoff of the converters, which is at about 1.2 μm. When observing silicon, which will transmit only past 1.1 μm, the light converted is reasonably monochromatic because only wavelengths longer than 1.1 μm will reach the converter, and only light with wavelengths less than 1.2 μm will be converted. Since the light has a rather narrow wavelength spread, lack of chromatic corrections is not serious, and spherical aberration can be corrected by adjusting the spacing between front and rear lenses of the objective in the same manner that cover-glass thickness variations were accounted for around the turn of the century. Reasonable image quality is thus possible, although the converter lens systems have severe barrel distortion.

More sophisticated systems can be constructed using either infrared vidicons and electronic scanning or a single-element detector and mechanical scanning. Considerable additional freedom is afforded with the mechanical scan, since lenses or mirrors and detectors can be chosen for optimum performance over any desired wavelength range.[14-16] Longer wavelengths, for example, are sometimes helpful in delineating impurity-concentration variations, since free-carrier absorption increases with increasing wavelength. Infrared microscopes can also be used in the reflection mode, like a conventional metallurgical microscope, when looking for doping variations since reflectivity, as well as transmissivity, is a function of carrier concentration. When applicable, reflected light is much simpler, since the back of the slice need not be polished. Thin films of opaque semiconductors can be viewed without infrared, and in particular, silicon-on-sapphire from its very nature (thin, on a transparant backing) can be studied directly.[17]

The temperature variations over a transistor or integrated circuit can also be used to give a "picture" by using a suitable scan system and detector and the thermal emission of the object itself as the source. Such pictures or thermal contours are most helpful in locating hot spots and current crowding in new designs.[18,19]

8.4 PHOTOGRAPHY[20-26]

The photography most likely to be encountered is photomicrography, i.e., the taking of high-magnification pictures with the aid of a compound microscope. However, unity magnification to perhaps 50× is very useful for studying slices, small crystals, and packages. Techniques for this magnification range are similar to those of conventional close-up photography and give rise to photomacrography. The latter does not use a compound microscope for image formation but rather relies on a single lens, just as in conventional photography.

Photomicrography is most conveniently combined with a microscope having a separate tube for camera attachment, though conversion kits to slip over an eyepiece are available. There is no requirement for a lens in the camera; however, if it has one, it may be used by focusing the microscope in the normal manner, setting the camera focus to infinity, and placing its lens close to the eyepiece. Some cameras are fixed-length and are designed to be in focus with the microscope, but the most versatile employ a ground glass so that the image can be viewed directly. Photomacrography may use a camera with a specially designed close-up lens or, in some cases, standard lenses with provisions for moving them farther from the film in order to increase magnification. For example, either rings or bellows attachments are available for many 35-mm cameras. Sixteen-millimeter movie-camera lenses are particularly recommended for close-up work.[20] They, as well as conventional lenses, will usually work better used backward because of their basic optical design.

Exposure. The matter of proper exposure is most easily handled by using an exposure meter. They are available commercially or can be made in the case of photomicrography by taping a CdS cell over a spare eyepiece and using it in conjunction with a suitable resistance-measuring circuit. As long as the eyepiece used with the camera is of the same power as the one to which the photocell is attached, meter readings are independent of objective. Calibrations can then be made involving whichever of the variables, time or meter reading, is most convenient. For very low light levels, film speed is often slower than the published index number (failure of the reciprocity law); so appropriate corrections may be required. Table 8.4 indicates the severity of the problem as exposure times exceed 1 s. Sometimes it is tempting to vary the light-source intensity in order to bring the light level at the eyepiece to a standard level to match a fixed film speed and shutter time. However, since this method changes the filament temperature and hence spectral

Table 8.4. Effect of Nonreciprocity on Film Exposure*

Exposure time, s	Normalized time–light intensity product for same density on film
0.001	1
0.01	1
0.1	1
1.0	1.3
10	2
100	4
1,000	16

*Typical values. Manufacturer's recommendations for specific film used should be followed.

distribution of light, it is not recommended for color photography. When a print is relatively dark but detail can be seen, it is usually necessary to quadruple the exposure time or intensity in order to get a usable picture. If the picture is just slightly on the dark side and lacks a little detail by being a slight gray, doubling the time should be adequate. The latitude of various films will vary, and in the interest of film conservation, it is recommended that a chart similar to that of Fig. 8.15 be made for each film type being used. It can then serve as a guide in correcting for improper exposures. Thus, even without an exposure meter or regular use of a given camera-light combination, the proper exposure can be quickly found. Notice that if a completely dark, or a completely washed-out film results, the light or time could be changed by at least a factor of 16 without shifting to the other extreme. When using macrophotographic lenses with f-stop adjustments, remember that each full stop ($f/1.4$, $f/2$, $f/2.8$, $f/4$, $f/5.6$, $f/8$, $f/11$, $f/16$, $f/22$, $f/32$, and $f/45$) changes the light intensity by a factor of 2. The larger the number, the less the intensity. If rings are used between the lens and camera body to increase magnification, the f number is decreased and may be calculated from

$$f_{\text{eff}} = \frac{Vf}{F}$$

where V is the distance from the lens to the film surface and F is the focal length of the lens. The exposure time T will then be

$$T_{\text{new}} = T_{\text{orig}} \frac{V^2}{F^2}$$

Focusing. Accurate focusing can be troublesome, even with a ground-glass screen. The grain of the glass will often be so large that visual resolution is lost. To minimize this effect, a small portion of the screen should be clear. For example, a ½-in-diameter plug of ground glass can be removed by cavitroning or core drilling and a clear one substituted, or a cross can be drawn on the ground surface and a small piece of glass cemented to it. The cement will reduce the effect of the grinding and make the glass in that region appear practically clear. The image is then focused on the clear surface and can even be viewed with a low-power hand magnifier for better definition. This procedure is particularly helpful when the image is very dim,

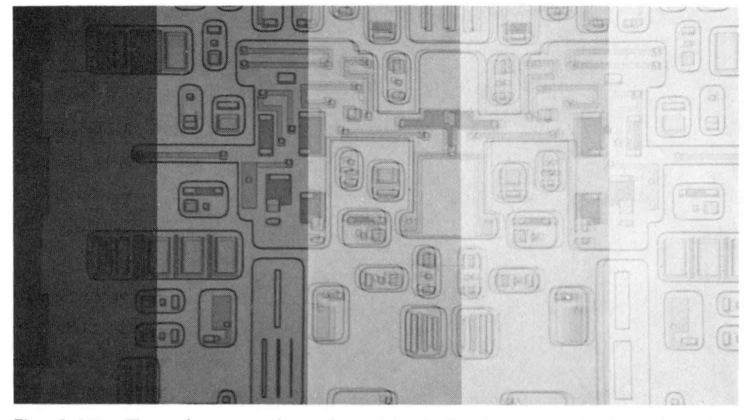

Fig. 8.15. Test photograph to show film latitude and assist in estimating exposure times. Each step doubles the time.

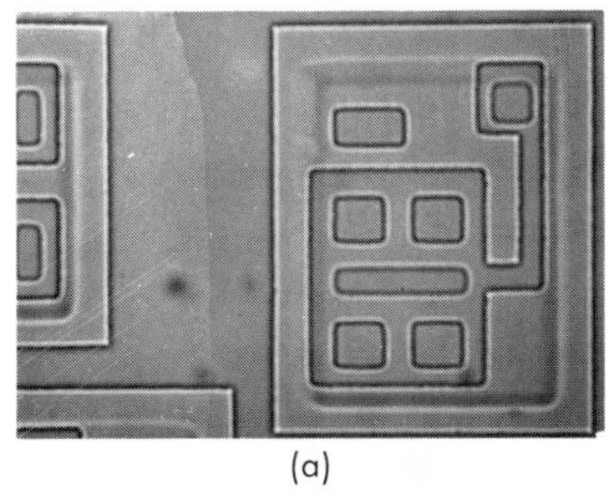
(a)

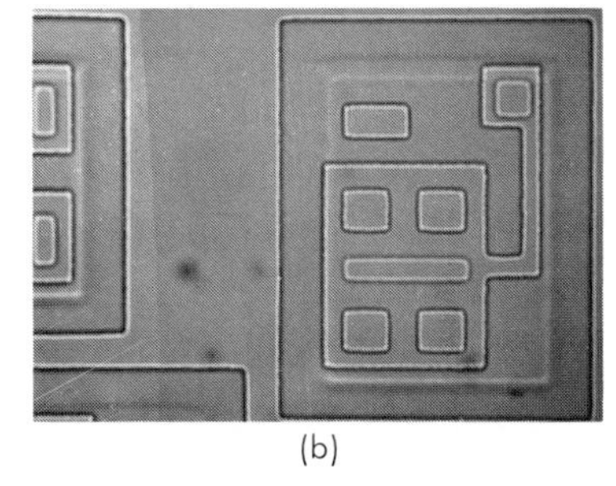
(b)

Fig. 8.16. Picture framing which occurs when the aperture diaphragm is nearly closed and the microscope is slightly defocused. This provides a quick method of determining which side of a microscopic discontinuity is raised since (*a*) if defocusing is by moving the objective toward the surface, the bright line will be over the raised region. (*b*) Moving it away causes the bright line to lie on the low side of the boundary.

but it does require accurate alignment of the eye to the light rays passing through the clear region. Under high magnification it can be extremely difficult on a screen to tell when the system is exactly focused, even with a clear-glass screen. There will appear to be a range of focusing where the image appears no clearer. However, should there be framing of the image by black or white pseudo Becke lines, they can be used as a guide. As the focus is changed, the framing can be seen to change from white to black, to black to white (or vice versa). Proper focus has been reached when the transition occurs and the lines appear neither black nor white. Figure 8.16 shows the two extremes of framing. If the photograph is to be used for subsequent measurements, it is imperative that the proper focus be used; otherwise the framing as illustrated in Fig. 8.16 can cause large errors. When a very exacting focus is required, or if the film plane does not exactly coincide with the glass plane, it may be necessary to take a sequence of pictures with the fine-focus control set at different positions.

At high magnification, the edges of the field may show distortion or fuzziness no matter how careful the focus. This may be because the face under observation is not absolutely perpendicular to the optical axis, and that is therefore the first thing that should be checked. An unmatched eyepiece and objective can cause similar results, as can poor-quality optics. The only choice in such cases is either to accept the quality or to upgrade the optics.

Depth of Field and Detail. In photomicrography, the depth of field is determined by the microscope objective N.A. For instruments without aperture control, the depth is thus controlled exclusively by the objective choice. If the aperture is variable, it can be stopped down to decrease N.A. and increase depth of field. However, resolution may suffer. When high power, maximum resolution, and great depth of field are all required at once, a scanning electron microscope may be the only solution, but sometimes a composite photograph can be made either by multiple exposures on the same film as the point of focus is changed or by taking a series of pictures with different regions focused, cutting the different sections from the photograph, and making a composite. If the height contours are linear, e.g., viewing down a slope, the latter is quite practical. Attempting multiple exposures on the same film will give a washed-out appearance and overexposure if many steps are used.

In photomacrography, the *f* stop of the lens and the magnification can be varied

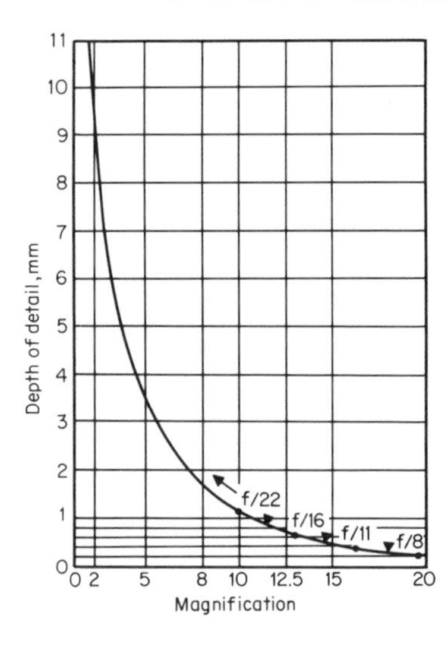

Fig. 8.17. This curve relates the depth of detail attainable at optimum camera aperture as the final magnification at the print is varied. (*Reproduced with permission from copyrighted Kodak Publication N-12B.*)

to change the depth. However, as magnification and *f* number increase, depth of detail may become more important than depth of field. In principle, depth of field should continue to increase as the *f* number increases, but when the point is reached where diffraction effects are predominant, depth of detail suffers. Figure 8.17 shows a curve which relates depth of detail to total photograph magnification, and gives the *f* number required at various magnifications.[20] Total magnification is the product of that of the camera and any subsequent enlargements. However, enlargement magnification should probably not exceed 3×.

Film Resolution. The eye can resolve the order of 5 lines/mm; so for no subsequent enlargement, the film should have at least that capability. Table 8.5 summarizes typical film resolution. However, if the total magnification, objective × eyepiece × enlargement, exceeds the magnification shown in Table 8.1, empty magnification will still result, regardless of the resolution. The choice of magnification might depend on aesthetics or on the desire to provide visible separation of some special features of the subject. In the latter case, remember that the limit of resolution of the eye should be the guide for minimum magnification.

Color Rendition. It is important when photographing objects such as oxide colors or colors on specific devices that color contrast be maintained on the film. In general, when photographed in daylight flash or fluorescent light, greens, reds, and blues

Table 8.5. Resolution of More Common Film Types

Type	Film speed (ASA)	Manufacturer	Resolution, lines/mm*
52	400	Polaroid	22–28
51	125	Polaroid	22–28
57	3,000	Polaroid	22–28
55/P/N	64	Polaroid	14–17 (positive) 150–165 (negative)
Pan X	32	Eastman Kodak	136–225 (negative)
Plus X	120	Eastman Kodak	96–135 (negative)
Tri X	160	Eastman Kodak	69–95 (negative)

*Based on manufacturer's specifications.

appear dark gray and magenta, cyan, and yellow appear light gray or white. Sometimes contrast can be enhanced by using light filters (not applicable to interference colors such as in oxide layers), sometimes by changing the background (applicable to photographing large objects where the background color can be changed), and sometimes by the proper choice of film. Table 8.2 summarized the color of filters or lighting to be used. Filters should also be used as necessary to provide color balance during color photography. It may also be desirable to use neutral filters to reduce microscope light intensity, since decreasing filament temperature will shift the spectral distribution. As an aid in maintaining the proper color temperature, some transformers have the approximate color temperature marked on their controls. Tables and nomographs are available to assist in the proper filter choice.[24,25] For the more common film and light sources, Table 8.6 gives suggested filters.

Lighting. When using macroscopic techniques for photographing objects with highly specular surfaces, for example, a polished silicon slice, oblique lighting will usually result in a photograph giving a dark-colored surface. In order to have the surface appear light in color so that minute markings and fine surface detail can be readily seen, a lighting system similar to that used in the metallurgical microscope can be used. For small objects such as dice and headers, a cover glass glued to some support as shown in Fig. 8.18 will suffice. For larger areas such as 2- or 3-in-diameter slices, commensurately larger reflectors are required. In the latter case a large-area light source will also be necessary and may be made of a close-spaced array of low-wattage bulbs with a diffuser of Mylar sheet. The whole area can then be photographed at one time, and all surface markings will show without any reflections from the camera provided it has been adequately dulled or shielded to eliminate any bright reflective metal from its surface. In either case, image quality will be degraded if the reflector and camera optical axis are not carefully aligned[23] as indicated in Fig. 8.18.

Other choices of lighting are diffused and oblique. Simple oblique lighting from a single source will cause a profusion of highlights from rough reflecting surfaces which may obscure detail. In such cases diffused lighting may be required. As an example, Fig. 8.19 shows the difference between using two diametrically opposed lights (to reduce shadows) and diffused light from a writing-paper cylinder surrounding the object and lighted from the outside by one of the lights. However, for other types of subjects where there is a dearth of specularly reflecting surfaces,

Table 8.6. Filters to Correct Light Sources to Color Film*

		Wratten filter type or equivalent			
Light source	Approximate color temp	Kodak type B	Kodak type A	Kodak daylight type	Polaroid color
6-V tungsten ribbon filament	3000 K	82A	82C	80A + 82A	80B + CC30B
6-V tungsten coil filament	3100 K	82	83B	80A + 82	
Zirconium arc, 300–750-W tungsten coil filament	3200 K	None	82A	80A	80B + CC20B
Carbon arc	3700 K	81C	81A	80C	80B
Xenon arc	6500 K	85B	85	None	None

*From data in Refs. 24 and 25.

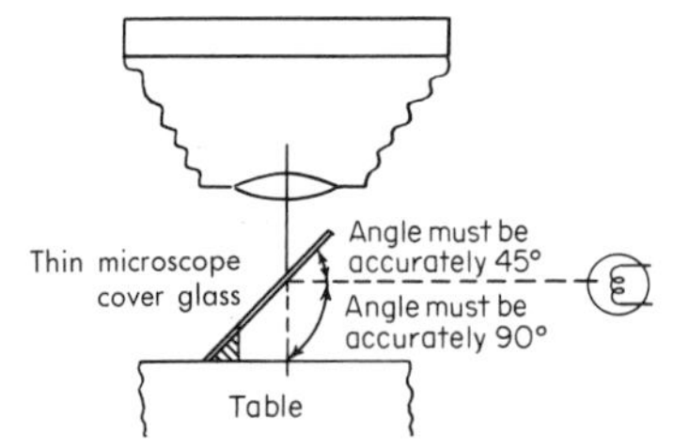

Fig. 8.18. Vertical illuminator for photomacrography.

a single small bare bulb may produce greater detail. For large subjects and troublesome reflections, spray-on films are available to produce a dull surface.

Single lights can be moved from position to position and multiple exposures made, or a combination of diffuse lighting and a single bulb to produce some particular highlight may be helpful. If the total exposure time is long enough and if the camera is in a dark room, the shutter can be placed on "time" and the light moved slowly from one position to another, leaving it on those preferred areas for longer periods of time. With color photography different-colored lights beamed from different directions may help accentuate, and yet keep separate, many different highlights.

Stereophotographs.[26] It may sometimes be desirable to take stereoscopic-pair photographs either with a camera or through the microscope. Dual cameras with the lenses separated a few centimeters are available which have been used for several years, and stereo aerial photography is accomplished by moving the camera over the terrain and sequentially photographing with a single lens. Neither of these approaches is particularly applicable for photomacrography, but by taking a photograph, rotating the subject approximately 8°, and rephotographing, satisfactory stereo pairs can be generated. Either diffused lighting or rotation of the lights with the sample may be required in order not to change the highlighting between the two photographs radically. Ordinarily two conventional photographs and a standard viewer would be used, but for special applications, one photograph might be taken through a red filter and the other through green, and both then viewed with a red filter over one eye and a blue one over the other.

Fig. 8.19. Photographs of a $\frac{3}{8}$-in-diameter rod of polycrystalline silicon showing the effect of lighting. (*a*) Taken with two high-intensity reading lamps arranged to minimize shadows. Notice that various reflections dominate the picture. (*b*) Taken by surrounding the section with a tube of writing paper and shining one of the lights previously used on the outside of the tube. Now considerable detail of the individual spherulites is visible.

REFERENCES

1. Emile Monnin Chamot and Clyde Walter Mason, "Handbook of Chemical Microscopy," vol. 1, John Wiley & Sons, Inc., New York, 1954.
2. L. C. Martin and B. K. Johnson, "Practical Microscopy," Chemical Publishing Company, Inc., New York, 1951.
3. Harold Schaeffer, "Microscopy for Chemistry," D. Van Nostrand Company, Inc., New York, 1953.
4. Bureau of Naval Personnel, "Basic Optics and Optical Instruments," Dover Publications, Inc., New York, 1969.
5. George L. Clark (ed.), "The Encyclopedia of Microscopy," Reinhold Publishing Corporation, New York, 1961.
6. Tadami Taoka, Eiichi Furubayashi, and Shin Takeuchi: Gonio-microscope and Its Metallurgical Applications, *Japan. J. Appl. Phys.,* **4**:120–128 (1965).
7. W. E. Degenhard, Light-Section Microscope Measures Thin-Film Thickness, *Electron. Prod.,* May 1965.
8. Eugene N. Cameron, The Study of Opaque Minerals in Reflected Light, in "Symposium on Microscopy," *Spec. Tech. Pub.* 257, American Society for Testing and Materials, Philadelphia, 1959.
9. S. Tolansky, "Multiple-Beam Interferometry of Surfaces and Films," Dover Publications, Inc., New York, 1970.
10. M. Françon, "Progress in Microscopy," Row, Peterson & Company, New York, 1961.
11. B. W. Mott, Metallurgical Aspects of Microscopy, in A. E. J. Vickers (ed.), "Modern Methods of Microscopy," Butterworths Scientific Publications, London, 1956.
12. Walter Kinder, The Interference Flatness Tester and Interference Instruments for Surface Testing, *Zeiss Inform.,* **58**:136–140 (1965).
13. A. E. Feuersanger, Interference Microscopy of Thin Films and Semiconductor Structures, *Solid State Des.,* **4**:29–32 (October 1963).
14. J. F. Black, B. Sherman, and V. Fowler, Examination of Semiconductor Wafers with a Scanned Laser Infrared Microscope, in Rolf R. Haberecht and Edward L. Kern (eds.), "Semiconductor Silicon," The Electrochemical Society, New York, 1969.
15. D. C. Gupta, B. Sherman, Ed Jungbluth, and J. F. Black, Non-destructor S/C Testing Using Scanned Laser Techniques, *Solid State Tech.,* **14**:44–50 (March 1971).
16. Richard A. Sunshine and Norman Goldsmith, Infrared Transmission Microscopy Utilizing a High-Resolution Video Display, *RCA Rev.,* **33**:383–392 (1972).
17. Richard A. Sunshine, Optical Techniques for Detecting Defects in Silicon-on-Insulator Devices, *RCA Rev.,* **32**:263–278 (1971).
18. David Peterman and Wilton Workman, Infrared Radiometry of Semiconductor Devices, *Microelectron. Reliability,* **6**:307–315 (1967).
19. Anon., IR Scanner Detects Flaws in Microcircuit, *Electronics,* **43**:99–100 (Nov. 23, 1970).
20. "Photomacrography," Kodak Scientific Publication N-12B, 1969.
21. "Basic Scientific Photography," Kodak Scientific Data Book, 1970.
22. "Photomicrography of Metals," Kodak Scientific Publication, P-39, 1971.
23. "Photography through the Microscope," Kodak Scientific Publication P-2, 1970.
24. "Kodak Filters for Scientific and Technical Uses," Kodak Publication B-3, 1972.
25. "How to Use Polaroid Polacolor Land Film for Technical and Industrial Purposes," Polaroid Corp., 1970.
26. William C. Hyzer, Taking Stereo Photos in the Laboratory, *Res./Develop.,* **23**:51–56 (December 1972).

9

The Electron Microscope and Other Analytical Instruments

Unlike other chapters which have dealt extensively with one instrument or type of measurement, this one discusses a variety of instrumentation, but in much less detail. The rationale is that the other measurements will probably be widely used by many individuals, whereas the equipment described here usually requires specialists to operate and interpret results. Thus the intent of this chapter is not to give full details but rather to provide insight into what the various techniques have to offer and to mention problems peculiar to semiconductor measurements.

9.1 ELECTRON MICROSCOPY, GENERAL

Optical microscopy has an upper limit of magnification of about 1,200× and, at that magnification, a very small depth of field. These two limitations, coupled with the fact that most microcircuits have spacings between diffusions, metallization, etc., that are comparable with the wavelength of light used by the microscope, have led to the increased popularity of electron microscopy. These instruments use electrons accelerated by tens of kilovolts to produce the image and resolve much smaller objects. The increased resolution arises from the much shorter wavelength λ_e of electrons, given approximately by

$$\lambda_e(\text{Å}) = \sqrt{\frac{150}{V}}$$

where V is the electron-acceleration voltage. There are two basic types of electron microscopes: transmission (TEM), which is analogous to a light microscope, and scanning (SEM), which sweeps the surface. The TEM is capable of very high resolution (3 to 5 Å), the SEM to only 100 to 200 Å. However, the first can be used only with objects thin enough to allow appreciable transmission of the electron beam. The actual thickness that can be tolerated depends on the material and the electron-accelerating voltage used, but in general is less than 2,000 Å. Thus most of the thin films used in microelectronics are too thick for viewing even if they were not on much thicker substrates. Bulk materials and thicker films must then be thinned or, if only surface texture is of interest, replicated by a film thin enough to be used.

The SEM depends on a different set of phenomena for contrast, and all measuring

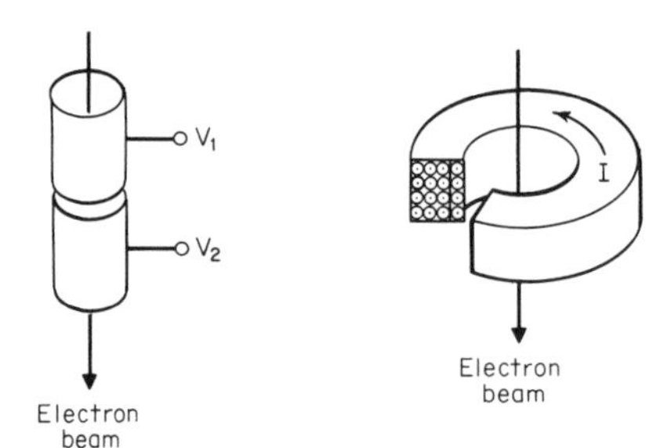

Fig. 9.1. Simple electron-beam lenses.

can be done from one surface. For basic material studies such as looking for precipitates and studying grain structure and dislocations, the TEM is most appropriate. For device process control and failure analysis, the SEM is far more helpful and is used both as an extension of optical microscopy and for detecting defects not readily discernible any other way.

9.2 TRANSMISSION MICROSCOPY[1,2]

The transmission electron microscope is based on the fact that it is possible to focus electrons. That is, it is possible to produce an electrostatic or magnetic field of such shape that electrons leaving one point in somewhat different directions can be brought back together again at another point. The main requirement for such a field (or lens) is that it have axial symmetry. Figure 9.1 shows a simple example of both a magnetic and an electrostatic lens. In practice they will be much more complex, just as the simple optical lenses described in Chap. 8 are seldom used in actual equipment.

With a family of lenses and a source of electrons a microscope can be built which may appear as in Fig. 9.2. Such a microscope has three features not found in optical microscopes. One is the electron source, which is usually a heated tungsten filament (and must be replaced quite frequently) and the high-voltage power supply. Second, the whole beam path must be evacuated, since the mean free path of the relatively low energy electrons is quite short in air. Third, because the eye cannot see the electrons directly, a fluorescent screen must be used for viewing. By substituting film for the screen, the image can be photographed. In order to extend the contrast range, and also to provide for group viewing, the fluorescent screen can be coupled to a TV camera. The brightness and contrast can then be independently varied and electronic signal processing used if desired.[3]

The stage for holding the sample is an insert machined to fit into the objective pole-piece assembly. Actual support is usually by a removable piece of wire mesh (typically 200 per inch) covering the end of the insert. To provide for photo stereo pairs, some stages can be tilted. Others have been designed to operate at high temperatures, at low temperatures, and to provide stress on the sample during observation. These various additions, of course, have to be done without appreciably altering the magnetic field of the lenses and thus degrading performance.

Since most samples are not thin enough in their normal state to be studied directly, an important TEM experimental procedure is sample thinning. If a single thin layer which is part of a thicker composite is to be studied, stripping rather than thinning may be required. For example, if the structure of a 200-Å-thick Ni/Cr resistor of an integrated circuit is to be studied, the silicon and oxide can be sequentially removed from the back in one small area and the protective coating stripped from the front. The resistor film will then be supported by the unremoved Si, and the whole assembly may be mounted for examination. Surface replication should be used only if very high resolution of surface features is required. Otherwise the SEM is more convenient.

Transmission Sample Thinning. Numerous techniques have been developed for metal thinning.[4] Many of these can be adapted to semiconductors if account is taken of their brittle nature and if the appropriate etchant is chosen. Unlike metals, if the thinning is done near room temperature, stress relief during thinning should not cause migration or loss of dislocations. Usually the sample is etched completely away in some places and is too thick in others, but the proper thickness can usually be found in the transition region. Table 9.1 summarizes thinning procedures reported for various semiconductors. One pitfall which can cause misinterpretation is the tendency for some etches to give rough surfaces or to leave debris on the surface, either of which may be interpreted as a precipitate.[9] Also, native oxides can grow quite rapidly in air after the thinning operations; so interpretation should consider the possibility of features associated with the oxide.

Replication.[4] The simplest form of replication is shown in Fig. 9.3. The surface

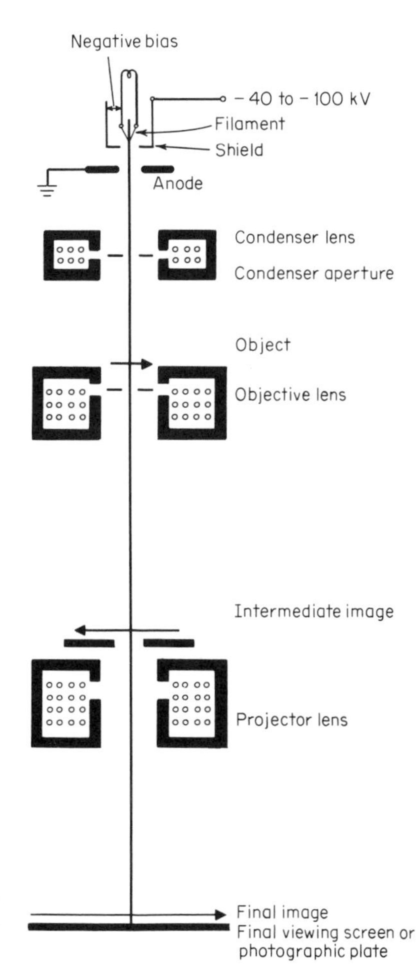

Fig. 9.2. Diagram of an electron microscope. The magnetic lenses are more complex than the one shown in Fig. 9.1. The gap forms two cylindrical magnetic poles. Their exact shape is carefully calculated and precision-machined.

Table 9.1. Etchants for Sample Thinning

Material	Etchant*	Mode	Reference
Si	95 HNO_3, 5 HF, with sample attached to Teflon holder and rotated	Chemical	5,6
	9 HNO_3, 1 HF	Jet	7
	4 g NaOH, 96 ml H_2O, plus sodium hypochlorite to suppress hydrogen evolution	Float specimen on etch	8
Ge	9 HNO_3, 1 HF	Jet	7
	Warm dilute sodium hypochlorite	Float specimen on etch	8
GaAs	40 HCl, 4 H_2O_2, 1 H_2O, at 20°C	Jet	9
	25% perchloric acid, 75% glacial acetic acid, gently flowing from an orifice above sample at 42 V	Electrolytic	10
GaP	Cl in methanol	Jet	11
PbSe	50 ml 45% KOH, 50 ml ethylene glycol, 10 ml H_2O_2. H_2O_2 may be added during etching to maintain rate. Surface stains may be removed with 10 ml acetic acid, 10 ml water	Electrolytic	12
PbTe	20 g KOH, 45 ml H_2O, 35 ml glycerol, 20 ml ethenol 4–6 V, 0.2 A/cm^2	Electrolytic	13

*See Sec. 7.2 for safety precautions.

to be examined is coated with a film of a material such as collodion or carbon thin enough for the electron beam to penetrate easily. Then, in order to provide contrast, shadowing is introduced by evaporating some heavy metal onto the film from a low angle. Without it, the uniform thickness of the film would appear nearly featureless. The replica used in the microscope may be made in a one-step process (as in Fig. 9.3) or in two or more steps by making an intermediate replica and then using it to generate the one actually examined. The first is preferred, but there are a limited number of materials to be used in the microscope, and if they and the object are not compatible, intermediate steps become necessary. The final film is almost always Formvar, collodion, or evaporated SiO, SiO_2, or carbon. The shadowing material is chosen for its electron-stopping power, ability to be deposited with very small grains, and resistance to growing larger grains during the electron-beam heating which occurs during observation. Carbon with platinum shadowing has been used to study initial nucleation of Si when being grown epitaxially onto various substrates such as Si[14] and sapphire. Replication is also widely used in grain-size studies of the various metallization system such as Al, Au, and Ni/Cr which are found in semiconductor devices.

9.3 THE SCANNING MICROSCOPE[15]

Figure 9.4 is a diagram of an SEM. The purpose of the series of lenses is not to provide magnification as was done on the TEM but rather to reduce the diameter

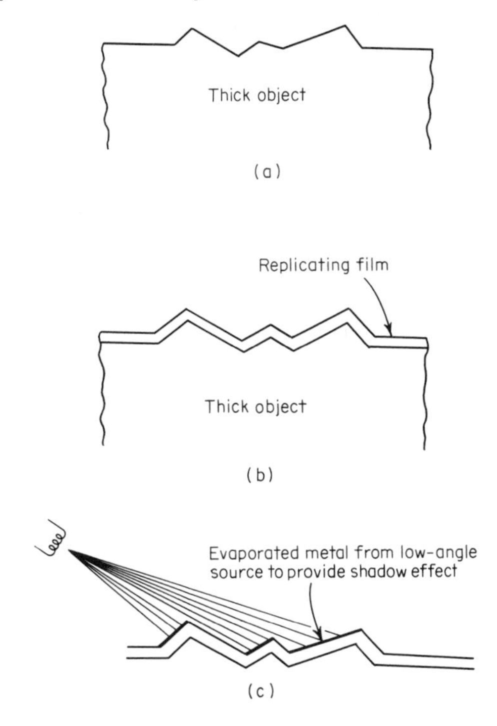

Fig. 9.3. Surface replication and shadowing for TEM. After shadowing, the film is stripped from the object and viewed directly.

of the electron beam so that it is only 200 or 300 Å in diameter when it hits the sample. Further, there must be a means of deflecting the beam over the surface in a raster fashion. This can be done either with electrostatic plates or by magnetic coils. When the beam strikes the sample, there is the possibility of extracting several different kinds of signals as shown in Fig. 9.5. Some of the incident electrons will be backscattered with no appreciable energy loss and may be detected by a sur-

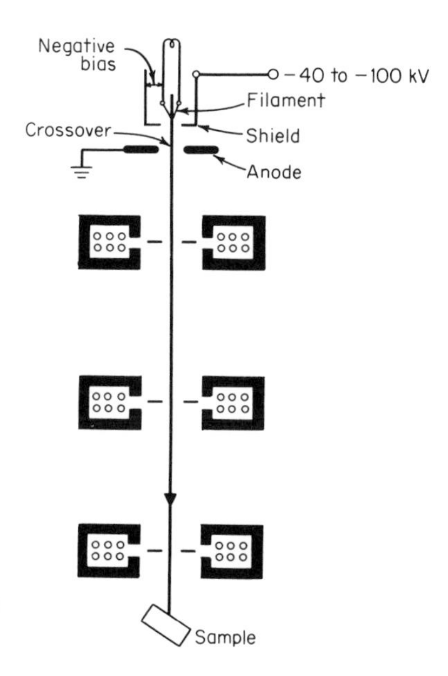

Fig. 9.4. Diagram of an SEM. For detector details see Fig. 9.5.

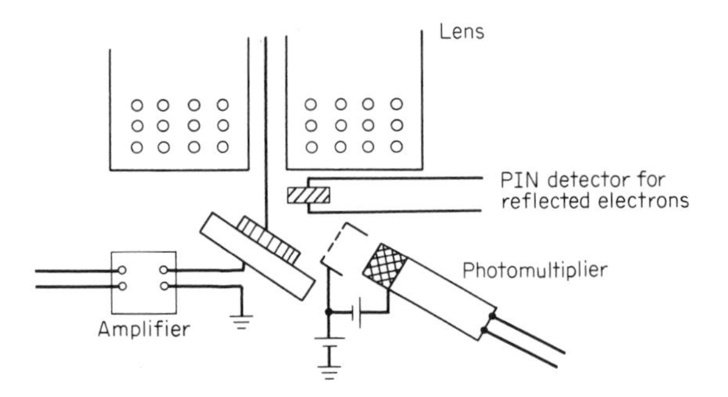

Fig. 9.5. SEM signals. The cathodoluminescence detector is not shown. It must be optically coupled by lens or light pipe to the point of electron-beam impact.

face-barrier PIN diode. Low-energy secondary electrons will be generated and can be collected by biasing a wire mesh a few hundred volts positive. They are then accelerated by several thousand volts before striking a scintillator crystal. The light emitted as they strike is coupled to a photomultiplier tube whose output is proportional to the number of secondary electrons collected. Should the material being studied exhibit cathodoluminescence, the emitted light can be coupled to a detector via a light pipe or a lens system. Finally, the currents and voltages generated in the sample owing to the incident electron beam can be measured. This may be by inserting a low-impedance current amplifier in the sample-ground connection or by monitoring various leads of finished devices.

The chosen signal, which actually may be a composite from two or more of the phenomena just described, can then be used for several sorts of display. Generally a cathode-ray tube (CRT) will be used and one beam scan across its face will be synchronized with one sweep across the sample surface. For two-dimensional coverage of the surface a succession of sweeps like a TV raster is used. Magnification is determined by the ratio of the beam movement on the surface to the spot movement across the face of the CRT. If Z-axis modulation is used, the conventional TV-type picture results. Y-axis modulation of the CRT beam makes the display take on a three-dimensional-like appearance.[16-18] It produces considerable distortion, however, since the coordinates on the screen are a function not only of the position of the electron beam on the sample surface but also of the signal amplitude.

Contrast Mechanisms. Work function and atomic number affect the yield of both secondary electrons and the backscattered beam, but the most important feature from the standpoint of displaying surface topography is the variation of yield with the angle of incidence θ of the impinging electron beam.[19] Over an appreciable range of θ, the number of secondary electrons is proportional to sec θ. For most instruments this transforms into a sensitivity for angular change of from 1 to 4°. For the case of reflected electrons, behavior is similar. However, since the detector can receive only line-of-sight electrons, whereas the secondary-electron collector does not have that limitation, reflected electron displays tend to show greater shadowing.

The potential of the object surface can have considerable effect on the number of secondary electrons collected but will depend somewhat on the actual placement and geometry of the electron collector.[19] Such voltage contrast can be used to delineate p-n junctions and to locate breaks in metallization.[20] A major disadvantage is that charge buildup on insulators will cause loss of detail. Because of this, devices with insulating-layer overcoatings for leads protection usually have it removed before SEM analysis; although if the insulating material is of low atomic number, e.g., SiO_2,

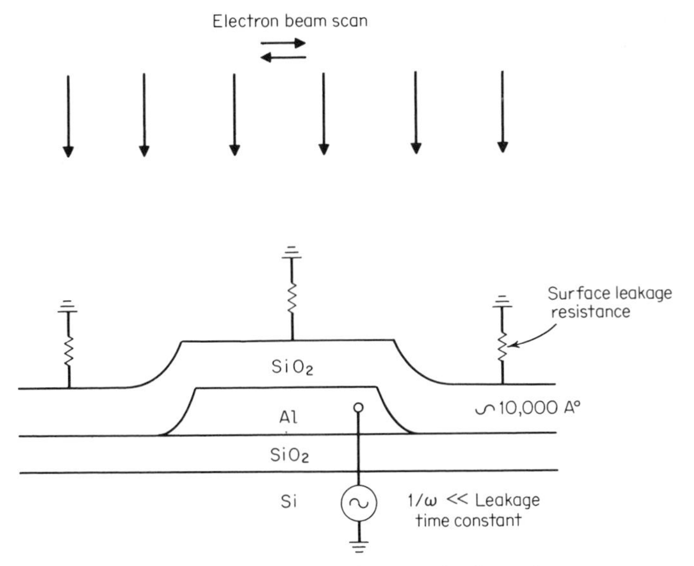

Fig. 9.6. Use of lead modulation to obtain voltage contrast through an insulating overcoat. (*From Crosthwait.*[22])

the backscattered high-energy electrons can be used to form an image of underlying metal.[21] An alternate approach that can be used in combination with voltage contrast to show which leads have voltage applied to them, even with overcoating, is shown in Fig. 9.6. If the voltage on the lead is modulated at a frequency high compared with the time for charge deposited on the insulator surface to leak away, there will be a modulation of the surface voltage above the lead because of capacitive coupling. This modulation is enough when operating a conventional TTL circuit at its rated voltages to enable lead breaks and other malfunctions to be observed even with a 10,000-Å overcoat of SiO_2. When insulating surfaces are to be examined for their own sake, they should first be coated very lightly with a metal film which can then be grounded.

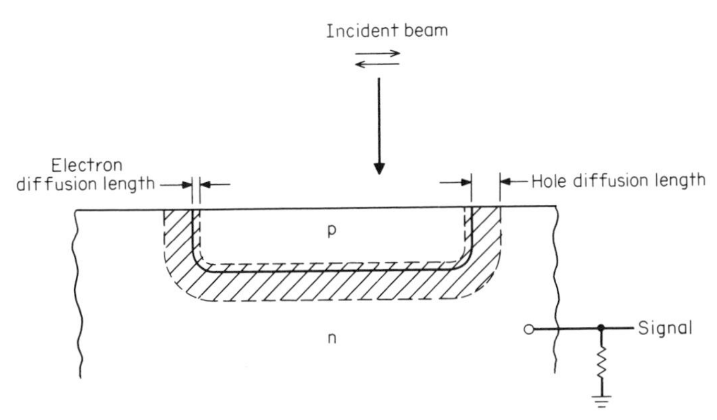

Fig. 9.7. Use of a p-n junction to produce an SEM signal. If the incident electron produces a hole-electron pair in the shaded region, current will flow.

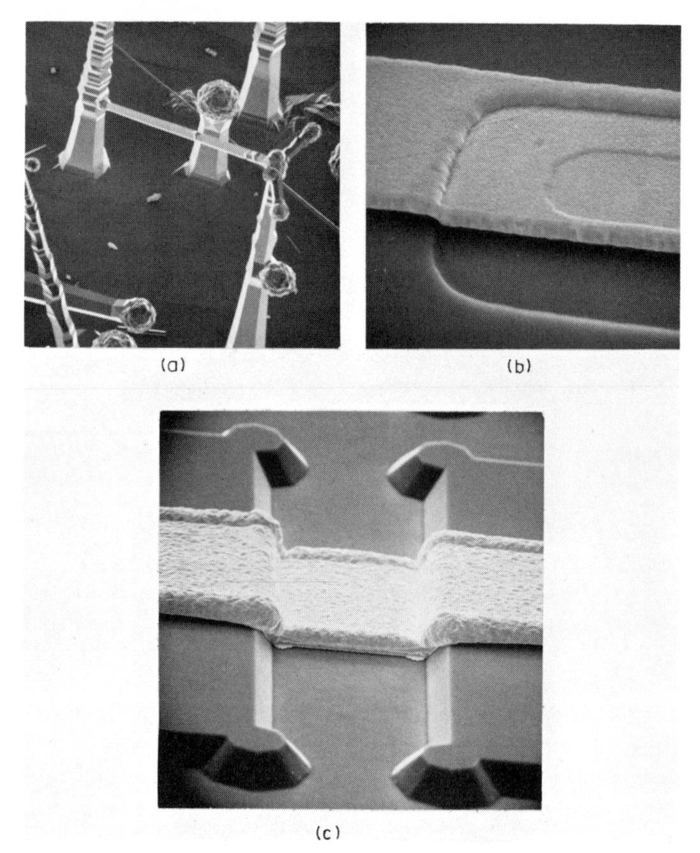

Fig. 9.8. Examples of SEM photographs. (*a*) Silicon needles several mils high growing up from a silicon substrate. (*b*) Ti:W/Au metallization contacting an integrated-circuit emitter. (*c*) A plated lead crossing mesas anisotropically etched into a silicon surface. (*Photographs courtesy of K. E. Bean, A. N. Akridge, and Farris D. Malone, Texas Instruments Incorporated.*)

A beam impinging close to a p-n junction will induce carriers which will contribute to a current flow if they reach the junction before recombining.[15] Thus, in Fig. 9.7, if the carriers are generated anywhere within the shaded region, they will produce a current which can be detected as was shown in Fig. 9.5. In materials without junctions, the beam can be used to modulate the conductivity and make conductivity profiles.

The mixing of signals alluded to earlier can sometimes be used to considerable advantage. For example, by combining the secondary electron signal with that from the p-n junctions, the junctions can be clearly defined, whereas otherwise some other contrast mechanism might only vaguely show them. Likewise the cathodoluminescence signal could be mixed with secondary emission.[15] There are also a variety of electronic manipulations that can be used to vary contrast. For example, the time derivative of the secondary electron signal can be used, either alone or mixed. To increase the dynamic range that can be recorded on the phosphor, fixed amounts of direct current can be removed in steps and a series of contours mapped out, each of which will have the full range of contrast allowed by the phosphor.[24]

SEM Applications. By far the greatest application of the SEM is in the inspection of structures and devices with high magnification and great depth of field. As an example, Fig. 9.8, which shows three representative SEM photographs, includes a photograph of some silicon crystallites growing up from a substrate that would have been impossible to produce any other way. Integrated-circuit metal coverage over oxide steps has been widely studied.[25-28] Some care must be exercised in looking at steps, since excess intensity and a shadowing can both occur because of the angles used. Large angles favor showing good detail when viewing a step in position *B* of Fig. 9.9, but in position *A* there is little difference between structures *a* and *b*. Therefore, whenever there is a question of interpretation, the surface should be viewed from several angles.* The fact that the beam does not hit normal to the surface will produce distortion which sometimes makes it difficult to visualize accurately the shape of the object being viewed and also makes it difficult to estimate dimensions even though the system magnification is approximately known. If the dimensions of some feature, e.g., an oxide thickness, are known, they can be used as a guide. If no features are known, polystyrene latex spheres are available in several sizes in the 0.1- to 1-μm range and can be added to the surface for a comparative calibration.

By using p-n-junction current as the signal, making charge collection maps, and combining with the outline of diffusions in planar structures, inversion layers can be located.[29-34] The electron beam can induce enough charge in the oxide to remove the inversion layer; so one should start with low enough intensity to prevent annealing before the opportunity to observe it. Should such premature annealing occur, however, it would be reflected in reduced leakage current, and this would not go unnoticed.

Electron-Beam Device Testing. Electron-beam probing can be used for some specialized electrical testing of semiconductor components and in principle has the advantage over mechanical probes because mechanical damage is minimized and the probe size can be made very small. Disadvantages are that the electron beam may crack the pump oil in the ambient above the sample and deposit a thin carbon layer over the semiconductor surface, that the interpretation of electron-beam signals

*Rather standard SEM microcircuit-viewing specifications (e.g., NASA) are available which when followed will allow reasonable correlation among different sets of observers.

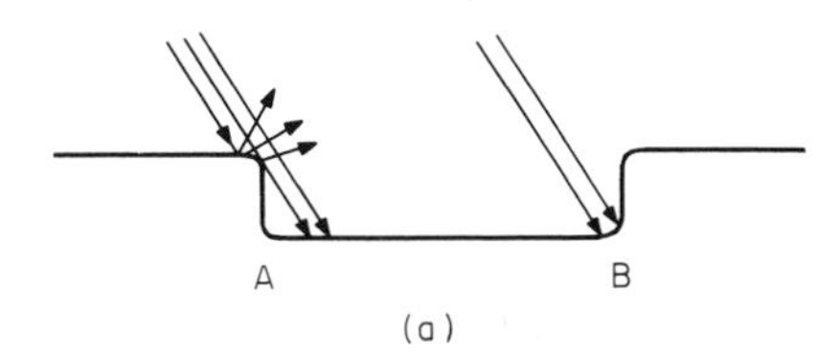

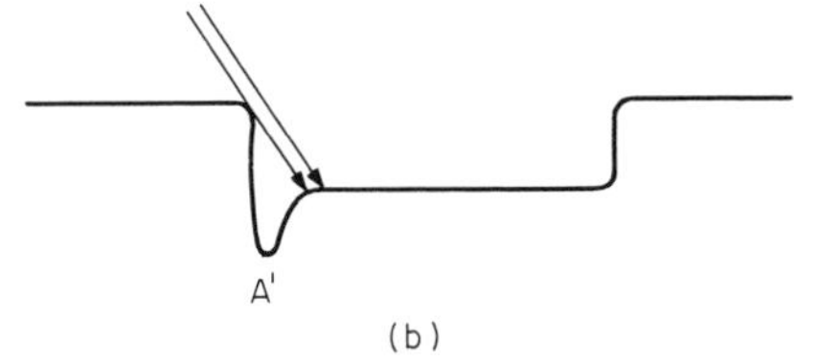

Fig. 9.9. Effect of angle on SEM detail. (*a*) The signal increases near the ledge because more secondary electrons can escape the solid. (*b*) Because of the viewing angle, the trench at *A′* is not seen.

may be difficult, and that it may be impossible to get enough numerical data to characterize the component completely.

Two modes of operation can be used. In one, the component under test is biased through the normal contacts and interconnects, and secondary emission at any selected point on the surface is used to estimate the relative voltage at that point. In the second mode, only one external connection is made to the component and the electron beam is used as the other one. Current from the fixed contact is then measured. The first method is particularly attractive when very complex circuitry is examined because it allows the point of failure to be seen as voltage contrast.[35-39]

9.4 ELECTRON DIFFRACTION[40-42]

An electron beam will be diffracted by the periodic lattice of any crystal it traverses, and if the optics of the TEM are slightly changed, a diffraction pattern rather than an image of the surface topography can be projected onto the screen. If the crystal is large with respect to the beam area, spots will be produced and can be interpreted in terms of crystal structure. For near perfect crystals, lines (Kikuchi lines) will also be seen; they are sometimes used to determine orientation. Samples with multiple crystallites considerably smaller than the beam size and randomly oriented will show rings. To a first approximation the ring diameter D (unless there is a stage of magnification between the sample and screen) is given by

$$D = \frac{2\lambda_e L}{d}$$

where λ_e is the electron wavelength, L the distance from sample to screen, and d the lattice spacing of the planes causing diffraction.

The arrangement as shown in Fig. 9.10*a* is well suited to studying small areas, and indeed the procedure often is to first examine the sample in the normal TEM mode and then change optics and look at the diffraction of the same region (selected-area electron diffraction). However, transmission electron diffraction (TED) suffers from the same disadvantage as the TEM in that a very thin sample is required. The diffracted beam and the normal transmitted beam can also be combined to give transmission electron diffraction contrast microscopy. In this case crystallographic defects in the thin section become visible and single dislocations, stacking faults, etc., can all be studied.* For pertinent references see Chap. 2.

Diffraction patterns can also be produced from backscattered electrons using the geometry of Fig. 9.10*b*. Now a sample of any reasonable thickness can be used, but the problem becomes one of signal strength. The backscattered signal goes down as the electron energy goes up; so operation is restricted to electron energies of tens or hundreds of electron volts rather than the tens or hundreds of kilo electron volts of a TED. This sort of instrument gives low-energy electron diffraction, or LEED. Because the energies are low, the beam penetration is very shallow and diffraction is from the first few atomic layers next to the surface. Such equipment is thus capable of studying surface contamination and/or surface reorientation effects.

Higher-energy beams can be used for reflection diffraction (RHEED) (Fig. 9.10*c*) by having a small grazing angle and depending on forward-scattered electrons, whose

*TED is almost always done with a conventional transmission electron microscope. The other methods described usually involve equipment built especially for the purpose.

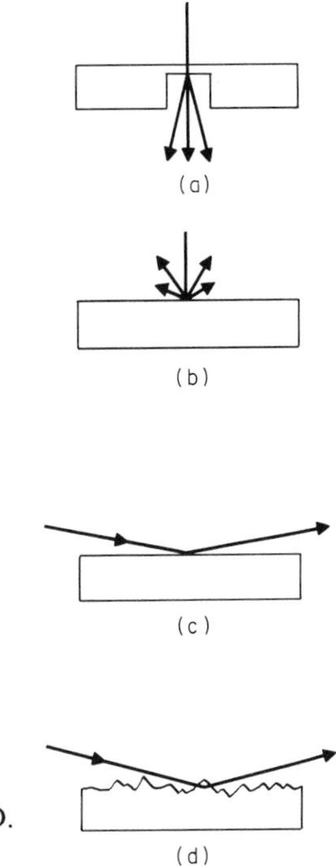

Fig. 9.10. Various electron-diffraction configurations. (*a*) TED. (*b*) LEED. (*c*) RHEED. (*d*) RHEED.

yield does not go down as energy increases. If the surface is reasonably rough as in Fig. 9.10*d*, the beam can penetrate the asperities and the system is basically the same as TED. However, if the surface is atomically quite smooth, at shallow angles, penetration is quite low, and RHEED will respond to essentially the same few surface layers as LEED. One point to remember is that even though the beam-spot size may be small, the low grazing angles will allow contribution from a ribbon of appreciable length, and localized measurements become difficult. RHEED is not as widely used as LEED for studying surfaces but has one advantage in that small local areas of differing crystal structure are more easily observed.[43]

Display of the diffracted beam may be by photographic film, direct viewing on a fluorescent screen, or collecting the electrons in a Faraday cup. In any case grids can be used to reject lower-energy electrons not associated with the elastically scattered beam. The Faraday cup can be used with electronic-scanning systems to produce a CRT display similar to that obtained from a densitometer profile of the film and is useful in observing time-dependent phenomena such as those which may develop during evaporation.[44] This application presupposes that the diffraction system is combined with other equipment and used in conjunction with it. If provisions are made only for heating the sample, annealing and surface contamination studies* can be made, but with evaporation or molecular-beam[45] sources, epitaxy and polycrystalline growth can be monitored *in situ*. Various gases can be admitted to the vacuum chamber and surface interactions studied.

*Surface contamination devoid of structure is more often studied by electron spectroscopy (e.g., Auger), discussed later. Diffraction is used if the contamination reacts with the surface and produces a change in structure or, occasionally, by observing an overall reduction in diffraction intensity because of masking by a disordered layer of contaminant.[46]

Table 9.2. X-Ray Testing Methods

Method	Parameter measured	End use
Diffractometer	Lattice spacing	Thermal-expansion coefficient, lattice strain, composition
	Linewidth	Crystal perfection
	Line intensity	Polarity through structure factor
Double-crystal spectrometer	Linewidth	Crystal perfection
Laue reflection	Lattice spacing	Orientation, estimate of amount of misorientation
Topography	Display of dislocations, stacking faults, twins	Same as parameters measured
X-ray microprobe	Characteristic line energy	Composition analysis

When really clean surfaces are to be studied, great care is necessary to prevent the instrument itself from contaminating the surface; so a vacuum of 10^{-9} torr or better is required and the clean surface is produced by cleaving the sample in the diffraction chamber or by extended heating at elevated temperatures.

One of the more interesting aspects of the LEED clean-surface studies on semiconductors has been the observation of a superlattice with spacing much larger than the normal d values, but always an integral number of d values. For example on (111) silicon surfaces a 7×7 spacing, as well as several others, may be found, depending on prior surface and heat treatments.[41]

9.5 X-RAYS

X-rays are generated when high-energy electrons strike some other material. Several sharp peaks will be superimposed on a background whose shape is similar to that of blackbody optical radiation. The wavelength of the peaks is characteristic of the target material, while the minimum wavelength and general shape of the continuous spectrum are primarily functions of the energy of the impinging electrons (applied accelerating voltage). For electron-acceleration voltages corresponding to minimum wavelengths longer than those of the characteristic peaks (lines), no lines will occur. For most purposes, the x-rays are first generated and then used in a variety of equipment. However, the procedure can be reversed; i.e., an unknown material can be bombarded with a beam of electrons and the wavelengths of the characteristic radiation measured. From this a composition analysis is possible, and because of the small spot size possible with an electron beam, very small regions can be examined. This procedure is usually referred to as *x-ray microprobe analysis.*

Table 9.2 summarizes the more common applications of x-ray technology to semiconductor measurements. Diffraction is the phenomenon most used. A simple diffractometer schematic is shown in Fig. 9.11. Constructive interference, i.e., a peak in x-ray intensity, will occur when

$$n\lambda = 2d \sin \theta$$

where n is the order, λ the x-ray wavelength, and d the spacing between two consecutive scattering planes. This is Bragg's law and is the basis for crystal diffractometers. Provisions are made to allow the θ and 2θ angles to be quite precisely set, and for many applications, a gear train is provided to rotate simultaneously the

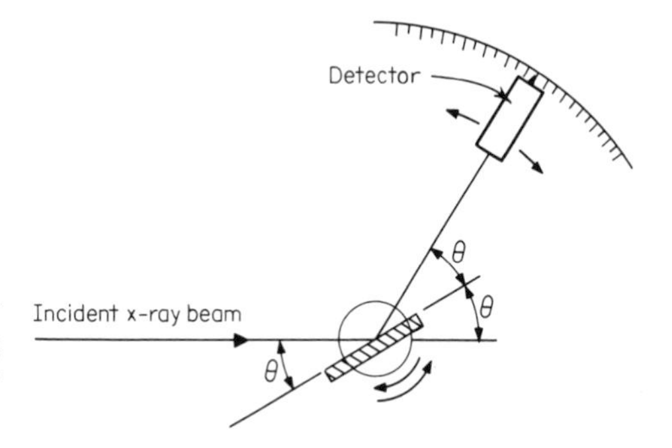

Fig. 9.11. Diffractometer schematic. Measured from the incident beam, the reflecting plane must be rotated to θ and the detector to 2θ for maximum intensity.

crystal by θ and the detector arm by 2θ. Beam-width control is by slits, and wavelength control is by using a combination of a characteristic line of the chosen target material and the absorption edge of a specific filter material. Perhaps the most obvious use of a diffractometer is to determine d from a measurement of θ. The value for d can then be used in a number of ways. If a material with known lattice spacings is being examined, the orientation can be determined by matching the d with previously calculated spacings for various orientations (see Chap. 1). If provisions can be made for observing the material from many directions and measuring the spacings for several orientations (as, for example, if a powder sample were used), they can be compared with tables of known spacings and the sample composition determined.

Double-Crystal Spectrometer. The primary purpose of this instrument is to provide as nearly a monochromatic and parallel beam as possible so that the line width observed is dependent on the sample and not on the instrument. One way of accomplishing this is by using the diffracted beam from one crystal as the x-ray

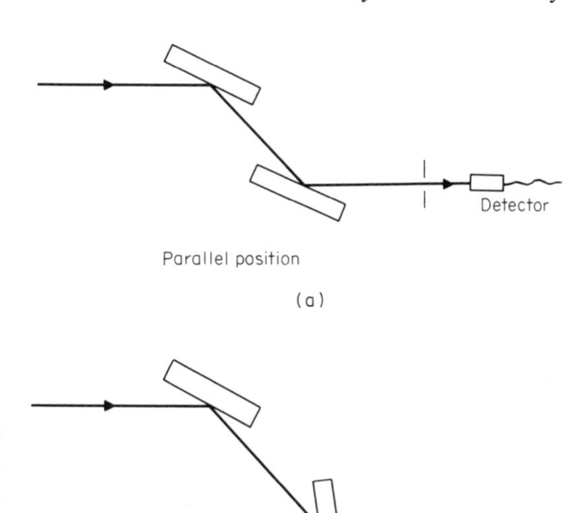

Fig. 9.12. Two geometries for double-crystal spectrometers. The conditions of the spectrometer are described by first listing the radiation used and then the reflection order for each crystal. $K_{\alpha 1}$ (11) means the first-order reflection is to be used on both crystals, and by convention, the fact that the second number is positive means the crystals are arranged in antiparallel geometry.

source for a second crystal as shown in Fig. 9.12. The first diffracted beam will still be somewhat divergent, because the crystal is not perfect and because slightly different x-ray wavelengths will be diffracted at different angles, but in general the beam is far more parallel than one defined only by slits. The deviation from parallelism is minimized by obtaining the most perfect first crystal possible, and by using as nearly monochromatic radiation as possible. Since less of the total energy is reflected as the beam width is narrowed, the intensity continually decreases so that even with a very intense source, the second diffracted beam will be quite weak, and some care must be exercised in the detection system. Alignment is also very important and, if not carefully done, will distort the line shape and shift the position of the peak.[47-49]

X-Ray Topography. The diffractometers just mentioned are not the only methods by which diffraction effects can be observed. For example, by the proper choice of x-ray, sample, and film geometry, the additional diffraction effects resulting from crystal defects such as dislocations and stacking faults will produce a photographic image of the defect (x-ray topography). The photograph is a 1:1 reproduction of the sample and all dislocations in it which have a component of their Burger's vector perpendicular to the diffracting plane. Any magnification must be achieved by subsequent enlargement of the film and is restricted to a few hundred $\times$ because of film grain. However, in semiconductor work magnifications of less than 100 are usually used. Because of the low magnification, and hence poor resolution when compared with electron-microscope observation, the method is most appropriate when the dislocation density is relatively low. Topography is widely used to study damage induced as a function of device-processing steps and has an advantage over other methods of being nondestructive.

Figures 9.13 and 9.14 show some of the geometries used.[50-60] Figure 9.13*a* is that of transmission Berg-Barrett and requires the absorption coefficient–thickness product μt of the sample to be less than 1 for good contrast. A disadvantage of this equipment is that it requires a well-collimated beam and the area of the crystal covered is quite small. Figure 9.13*b* shows the projection-topography procedure in which the crystal and film, but not the shield, are slowly moved perpendicular to the diffracted beam. This works well as long as the crystal lattice is quite flat (unstrained) over the whole sample area. However, if there is any bending or warping, only a portion of the crystal will satisfy diffraction conditions and a spotty photograph will result.

To improve contrast, it is usually desirable to keep the film as close to the crystal face as possible. However, as can be seen in Fig. 9.13*b*, proximity and large-distance scanning are not very compatible; so an alternate arrangement with the film and translation parallel to the crystal as in Fig. 9.13*c* is often used. The scan system of Fig. 9.13*b* creates considerable distortion between the vertical and horizontal scales. Figure 9.13*c* corrects that deficiency, but if it is desired to keep the original Lang geometry, individual scan controls for sample and film can be used and each allowed to move parallel to its face.

To minimize warping effects, scanning combined with oscillation (SOT), Fig. 9.13*d* can be used, and topographs of complete silicon slices after processing (which usually tends to produce considerable bowing) can be made.

Film contrast is dependent on having a very monochromatic x-ray source; so for some studies, e.g., stacking faults, a double-crystal spectrometer geometry as in Fig. 9.13*e* may be helpful.

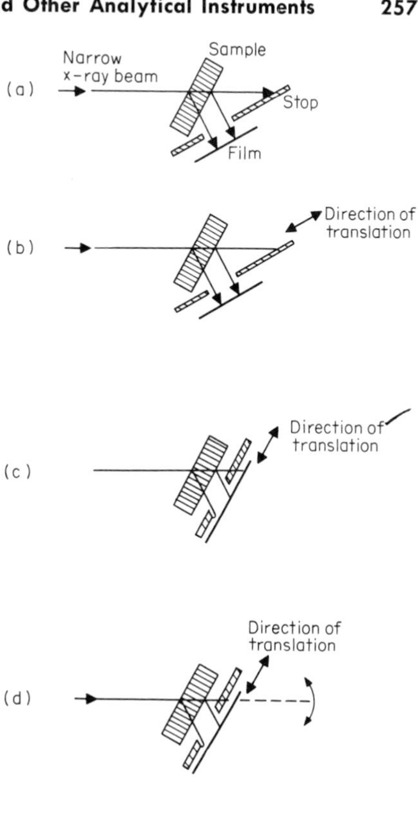

Fig. 9.13. Lang x-ray topography with variations. (*a*) Original (narrow x-ray beam). (*b*) With scanning capability (film and sample are moved in unison perpendicular to the diffracted beam). (*c*) Film parallel to sample surface (film and sample are moved in unison parallel to the sample face). (*d*) Scan combined with oscillation to observe warped surfaces (film and sample are moved in unison parallel to the sample face as well as rocked about the beam direction). (*e*) Combined with double-crystal monochrometer.

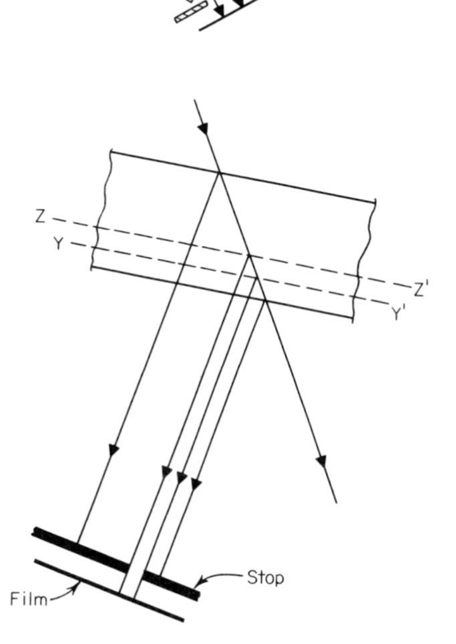

Fig. 9.14. By adjusting the stop as shown, only the diffraction contributions between planes *Y-Y'* and *Z-Z'* are recorded. (*Adapted from Lang.*[50])

If it is desired to look at an interior portion of the crystal, stops can be arranged so that only the contribution from the desired region is allowed to hit the film, as is shown in Fig. 9.14.[50] Two-layered structures with different lattice spacings (e.g., GaAs epitaxially grown on Ge) will give two topographs which can then be examined to see which defects originated in which layer without using any additional stops.[51] In examining wedge-shaped samples of high-perfection crystals, e.g., the rounded edge of a silicon chemically polished slice, banding may be observed which is due not to lattice strain but rather to interference effects.[61] However, distortion of these "Pendellösung" fringes can be used to estimate strain nonuniformities such as occur near dislocations.[62]

For a thick crystal, i.e., one in which $\mu t \gg 1$, little x-ray transmission would normally be expected, but by careful alignment of the crystal for Bragg reflection, enhanced transmission will occur if the crystal is highly ordered but not where there is a lattice defect.[63-65] This method has been used more often for Ge and GaAs than has Lang, since their higher atomic number gives higher absorption coefficients and makes the Lang method ($\mu t \ll 1$) less attractive because of the thinner samples or harder x-rays required.

Reflection geometry is also applicable to topography and is most appropriate if the surface or thin layers such as an epitaxial layer are to be examined. Modifications similar to some of those used for transmission, e.g., double-crystal monochrometer, and the use of selective stops to define interior sections, have been used.[60,66] Topographs can also be made using white radiation and a large-cross-section x-ray beam.[67] In this manner the spots will also be large and will show crystallographic detail. An application for which it is particularly suited is the rapid mapping of slices which may contain polycrystalline regions.[68]

Different diffraction planes can be used to generate the topographs, and by considering which ones show a particular dislocation, that dislocation can be characterized as to direction and Burger's vector.[69]

A variety of special-purpose cameras have been constructed. For example, a vidicon has been substituted for the film so that viewing can be in real time.[70] By providing a high-temperature sample chamber, dislocation motion can be detected, and in fact, the deleterious effects of high-temperature processing can be examined *in situ*.[71-73] Further details relating to equipment, precautions, interpretation, etc., are to be found in Refs. 74 to 81.

Microprobe Analysis.* By irradiating a material with a high-energy beam of electrons, it can be made to emit x-rays just as does the target of a conventional x-ray tube. The presence or absence of characteristic emission peaks of the various elements can then be used as a basis for compositional analysis.[82] Figure 9.15 is a schematic of such equipment. Since a controlled electron beam is an important part of this equipment, it is often combined with a scanning electron microscope. By scanning the incident electron beam across the surface, a map of the concentration of chosen elements can be prepared. As is normal with x-ray analysis, the sensitivity is rather poor; however, it is relatively nondestructive† and has good surface resolution. It has been used for semiconductor impurity analysis[83,84] but is much more

*For more extensive information relating to x-ray microprobing see, for example, the series "Advances in X-Ray Analysis," published by Plenum Press, and Ref. 87.

†Excluding radiation damage and the possibility of the deposition of pump-oil compounds on the surface.

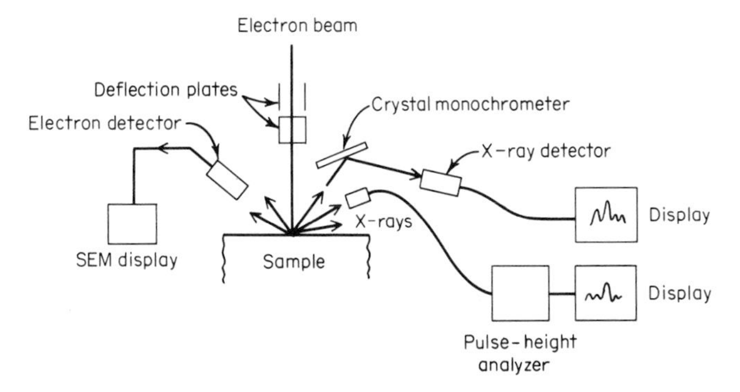

Fig. 9.15. X-ray microprobe. The surface to be studied can be scanned by deflecting the electron beam. A crystal detector combined with a pulse-height analyzer may be used or a monochrometer and conventional detector. For correlation purposes most equipment has one or more varieties of electron detectors so that it can be operated in a scanning-electron-microscope mode. Depending on the instrument, beam deflection may be either electrostatic, as shown, or magnetic.

useful in examining the metallization and dielectric systems used in semiconductor-device manufacture.[85] In order to assist in interpretation of the data if quantitative results are desired, various computer-based computations can be used.[86,87] With care a microprobe can be used for light-element analysis, but performance is considerably better for atomic weights above 12 to 15. By using charged particles rather than electrons to generate x-rays, the general techniques can be extended to much lighter elements.[88] However, the requirement for a particle accelerator is introduced, and lateral resolution becomes very poor.

9.6 ELECTRON SPECTROSCOPY[89-93]

Electron spectroscopy concerns itself with the energy spectrum of electrons emitted from atoms of the material being studied. Depending on the method of excitation, the electrons can come either from the valence shell or from an inner one, and may be either a one-step or a two-step process as shown in Fig. 9.16. They will have an energy characteristic of the kind of atom and the particular electron levels involved in the transition. Therefore, by determining the energy of the ejected electron, the element from which it came can be deduced. The electrons generally have a very low energy and are absorbed before reaching the surface if they are generated at depths of greater than perhaps 20 Å. This restricts electron spectroscopy to a study of the top few atomic layers of a material, but it also allows the surface layers to be studied independently of the remainder.

There are several kinds of spectroscopy (Table 9.3) based on the particular emission process used. Photoelectron emission is a one-step process in which an electron is directly emitted if the energy of the incident photon is high enough. If the electron is removed from one of the inner shells (either by photoionization or by an electron beam), the hole can be filled by either a valence electron falling into it, coupled by x-ray fluorescence, or by ejection of another (Auger) electron. The relative

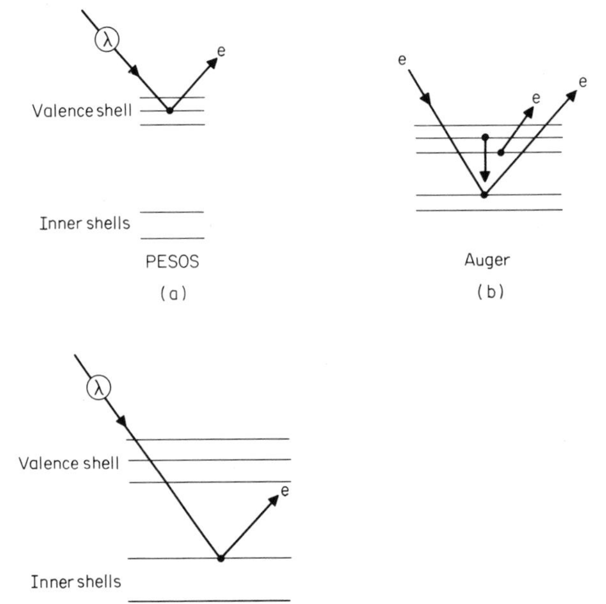

Fig. 9.16. Origin of electrons measured in electron spectroscopy. Auger electrons are the ones most used in semiconductor studies.

number of each is determined by the incident energy. Below 10 keV the Auger process is favored. Above it x-ray fluorescence is more probable. Auger spectroscopy (AES) is the one which has found most application in semiconductor studies and, when combined with sputter etching, can be used for depth profiling as well as surface investigations.

Figure 9.17*a* shows the general character of the electron energy spectrum produced by incident electrons. Those which leave with the same energy as the incident electrons were elastically scattered and are the ones used for electron diffraction. The peak near zero energy is from secondary electrons generated by multiple

Table 9.3. Electron-Beam Instrumentation Acronyms

ESCA	Electron spectroscopy for chemical analysis. Sometimes used as synonym for PESIS, often used to cover whole area of photoelectron spectroscopy
PESIS	Photoelectron spectroscopy of inner-shell electrons
PESOS	Photoelectron spectroscopy of outer-shell electrons
UPS	Ultraviolet photoelectron spectroscopy
XPS	X-ray photoelectron spectroscopy
INS	Ion-neutralization spectroscopy
FES	Field-emission spectroscopy
AES	Auger electron spectroscopy
SEM	Scanning electron microscopy
TEM	Transmission electron microscopy
TEDCM	Transmission electron diffraction contrast microscopy
TED	Transmission electron diffraction
RHEED	Reflected high-energy electron diffraction
RED	Same as RHEED
IED	Incident electron diffraction, same as RHEED
LEED	Low-energy electron diffraction

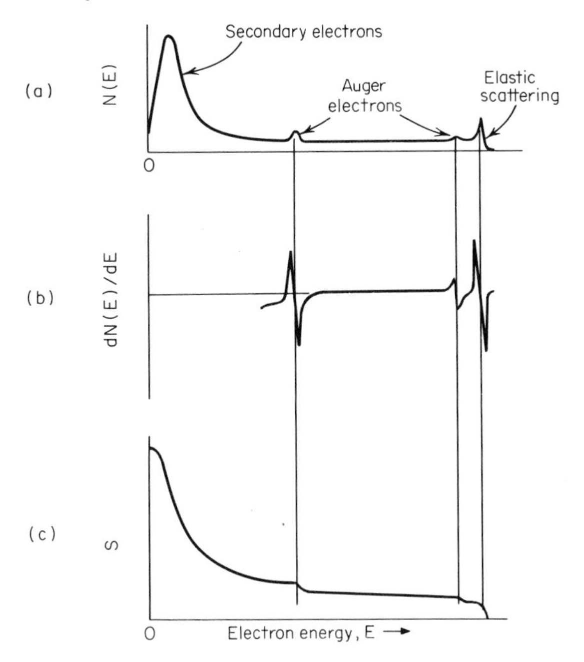

Fig. 9.17. Electron energy spectrum.

collisions within the material being examined. Between the secondary and the elastic peaks there is fine structure due to Auger transitions. The peaks are small; so the derivative $dN(E)/dE$ is used to enhance their detectability, and appears as in Fig. 9.17*b*. Experimentally, two kinds of energy analyzers are commonly used. One, with a retarding field, shields the actual detector by a grid or series of grids whose potential

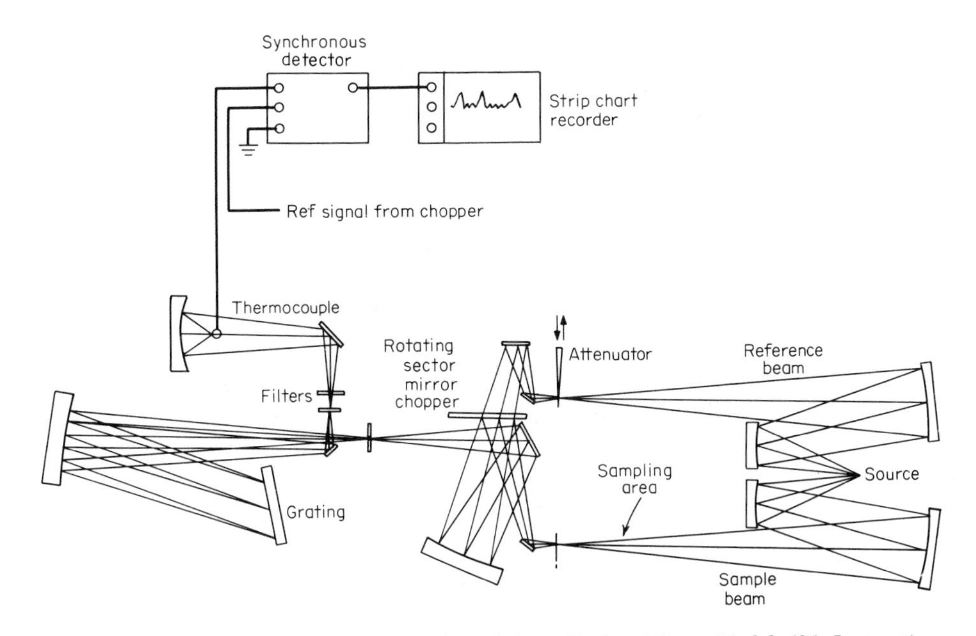

Fig. 9.18. Infrared spectrometer. (*Adapted from Perkin-Elmer Model 621 Instruction Manual.*)

can be varied. In that manner, only electrons above a given energy can penetrate to the detector. The signal S for a given setting is proportional to the total number of electrons above that energy E, or

$$S = \int_{E_X}^{\infty} N(E)\, dE$$

$S(E)$ appears as in Fig. 9.17*c*. When this sort of detector is used, two differentiations (both done electronically) are required to give the dN/dE energy spectrum of Fig. 9.17*b* that is normally cataloged. The other type of analyzer uses an electrostatic-lens system and allows only electrons in a small energy window of E to $E + dE$ to reach the detector. Its signal is then $N(E)$, or that of Fig. 9.17*a*.

A number of transitions will be probable for each element, so that the spectra can become quite complicated if there are many species on the surface. While calculations of the expected energies have been made, experimental spectra obtained from known materials are usually used.[93] To facilitate identification, the major peaks vs. elements are tabulated and can be used for initial screening. Fortunately there is little overlap between elements, and further, because of the extreme purities required in semiconductor operations, few species are to be observed on any given sample. Sensitivity is very good; approximately 10^{12} atoms per square centimeter can be detected.[92] This corresponds to 10^{18} to 10^{19} per cubic centimeter in the bulk, or to a few hundredths of a monolayer.

In common with LEED systems, Auger spectroscopy must be used in very high vacuums, i.e., $\sim 10^{-10}$ torr; otherwise surface contamination from the system may build up too rapidly for measurements to be made. Since the vacuum and electron-beam requirements are similar for LEED and AES, the two are often combined into one multipurpose instrument. In order to extend the capability to depth profiling, an ion sputter gun can also be combined with the other elements to erode away the surface.[94,95] The vertical depth scale is established from the sputter etch rate. By making sure that the area removed by sputtering is much larger than the electron beam, any chance of contribution from the sidewalls of the depression is eliminated.

AES can be used both for studying contaminants left on presumably clean surfaces and for the analysis of thin films, both metallic and dielectric. It has been used, for example, to look for metallic impurities such as Ni, Fe, and Au which might participate in the formation of the superlattices discussed earlier.[96,97] Diffusion data in very thin layers, e.g., Si into thin evaporated Au layers[98,99] and Ni/Cr thin-film resistor[91] interdiffusion, can be obtained more easily than from other types of measurements. Auger signals are generated for light elements from atomic weight 3 on and thus can provide more information than x-ray microprobe analysis. In particular, organic surface contaminants as a function of surface treatments such as vapor etching, aqueous etches and rinses, and asher photoresist removal have all been studied with AES.[100-106] Light-element dielectrics such as SiO_2, silicon nitride, and silicon oxynitride can also be analyzed in this manner.[107]

9.7 OPTICAL SPECTROSCOPY

Optical spectroscopy, which involves measuring the relative intensity of light as a function of wavelength, is categorized by both the wavelength of interest (infrared, visible, ultraviolet) and whether the light is being emitted, absorbed, or reflected by the sample.

Emission Spectroscopy. The sample to be analyzed is subjected to a high temperature such as an arc, flame, or spark which volatilizes it and breaks molecular bonds. The atoms will then absorb energy and re-emit with a line spectrum characteristic of each individual atomic species. In the uv and visible regions gratings are commonly used as dispersive elements, and the lines are recorded on film. Because of the extraordinarily low impurity levels required in semiconductors, it is not as helpful as some other methods in determining doping levels. However, for raw materials and the myriads of chemical supplies used in the semiconductor industry it probably is the most widely used of analytical tools.[108,109]

Absorption Spectroscopy. A high-intensity light source containing emission lines of the specific element to be analyzed for is passed through a flame of a gas burner into which the material to be analyzed has been injected. In this case, if any of the specific element is indeed present, it will absorb at the same frequency as the light-source emission and cause a reduction in the intensity. Such equipment has the disadvantage of requiring a separate light source for each element of interest and either a gaseous or liquid source, but for the routine analysis of one element it is very fast and inexpensive.

Flame Photometry. Flame photometry is really emission spectroscopy in which the sample is injected into a flame much as in absorption spectroscopy. Indeed, the burners are the same and commercial equipment usually combines both modes of operation. Unlike conventional emission spectroscopy, this one has extreme sensitivity (parts per billion to parts per trillion) for certain elements and can be used to study semiconductor wash-solution contamination by alkali metals such as potassium and sodium.[110-112]

Infrared-Absorption Spectroscopy. The high temperatures used to produce the emission bands used in the previously discussed methods break all molecular bonds and hence will give no information concerning molecular structure or bonding configurations. By keeping the sample near room temperature and measuring the infrared absorption, spectra are obtained which are dependent on particular bonds. Infrared spectroscopy is most used for analyzing organic compounds but is applicable to semiconductor analysis. For example,[113] the 9-μm band due to Si—O bonding will allow oxygen incorporated into the Si lattice to be determined down to about 10^{17} atoms per cubic centimeter. Similarly, the Si—C bond produces a 16-μm absorption band used for carbon analysis.[114] Other applications relating to the semiconductor industry involve the examination of surface films such as doped oxides[115] and the stains that sometimes are found on Si surfaces after etching.[116]

Infrared instruments may use either gratings or alkali-halide prisms for dispersion. The prisms are water-soluble; so they must be protected from high-moisture environments (many prisms are lost because nitrogen purges are inadvertently shut off). Photographic film is not sensitive for wavelengths much beyond 1 μm; so infrared spectrometers use detectors such as thermopiles or semiconductor photoconductive cells. The wavelength is then scanned by rotating the grating or prism, and the detector output is recorded on a strip chart. Figure 9.18 shows a typical spectrometer schematic.

In order to afford greater sensitivity and to minimize the effects of atmospheric attenuation, a double-beam instrument is ordinarily used. That is, the beam splits and goes through two similar paths, except that one of them has provisions for inserting the sample. The light is chopped (usually mechanically) so that signal processing can be ac. For high absorption, the signal-to-noise ratio may be very poor; so some variety of phase-sensitive detection is almost always used.

Most qualitative analyses will use only a strip-chart recording of relative transmittance vs. wavelength. However, for quantitative analysis (such as for O and C in Si and Ge) the actual absorption coefficient α must be determined for one or more of the bands. This is done through the equation

$$\frac{I}{I_0} = \frac{e^{-\alpha x}(1 - R)^2}{1 - R^2 e^{-2\alpha x}} \tag{9.1}$$

where I_0 is the intensity of the incident beam, I the intensity of the light transmitted, R the reflection coefficient, and x the sample thickness. This equation assumes normal incidence, parallel sides on the sample, and a surface finish adequate to prevent scattering (which usually means an optical finish). The solution of Eq. (9.1) is (from ASTM F 45-64T)

$$\alpha = \frac{B}{x} \tag{9.2}$$

where $$B = \log_e \frac{1 + \sqrt{1 + 4C^2R^2}}{2C}$$

$$C = \frac{T}{(1 - R)^2}$$

and $$T = \frac{I}{I_0}$$

The reflection coefficient R can be measured separately, or for well-characterized materials, the known value at the wavelength of interest may be inserted into the equation. n will ordinarily be the number found in tables, and for α small,

$$R = \left(\frac{n_1 - n_2}{n_1 + n_2}\right)^2 \tag{9.3}$$

where n_1 is the index for the semiconductor and n_2 equals 1 for operation in air or vacuum. If the complex dielectric constant $N = n - ik$ is used for optical characterization,* the absorption coefficient is given by

$$\alpha = \frac{n\pi k}{\lambda} \tag{9.4}$$

Emittance Spectroscopy.[117,118] Instead of measuring the absorption coefficient directly as in absorption spectroscopy, in regions where $\alpha x \ll 1$, the thermal emission may be measured. It is proportioned to αx and, because of the difficulty of determining α for the case of $\alpha x \ll 1$, may be easier to measure.

Reflectance Spectroscopy. The reflectance spectrum will also contain information similar to that of the other methods and is sometimes used to study the effect of adsorbed layers on surfaces. A more unusual semiconductor application is that of electroreflectance, in which the reflectance from a space-charge region generated either by immersion in a suitably biased electrolyte or by depositing a thin semitransparent metal (Schottky diode) over the surface is measured. From such measurements band-structure information and doping levels can sometimes be deduced.[119-121]

*For instance, an ellipsometer might be used to determine N at the wavelength of interest.

9.8 MASS SPECTROSCOPY

Mass analyzers convert a portion of the sample to be analyzed into ions and then, by a combination of curved condenser plates and a magnetic field, bring all ions with the same charge-to-mass ratio to a common focal point. These instruments in general have greater sensitivity than optical-emission spectroscopes and for that reason are very useful in semiconductor analysis.

Solid-source mass analyzers are most applicable and ordinarily make use of a high-frequency, high-voltage spark between two electrodes cut from the material to be examined. However, if the material is available only in powder form, it can be pressed into electrodes with a binder, although sensitivity is lost because a portion of the beam will come from the binder. If only thin layers are available, a counter-electrode of some other material can be used, but again, sensitivity suffers.[108,109]

Ion-Microprobe Mass Analyzer.[122,123] The mass spectrograph has adequate sensitivity for many useful semiconductor measurements but does lack in spatial resolution. In order to correct that deficiency, a beam of ions such as argon can be focused onto a small area on the surface to be analyzed and used to sputter it away locally. The secondary ions from the surface can then be collected and analyzed by a mass spectrometer. A schematic of such an instrument is shown in Fig. 9.19. By moving the sputtering ions across the surface in raster fashion and setting the spectrometer to detect a particular e/m value, topographic scans of a given element's concentration can be made. By stopping the beam and sputtering a hole into the surface, an impurity profile normal to the surface can be generated, although calibration is very difficult because of problems associated with determining and maintaining given sputtering rates. A further difficulty is that the walls of the sputtered hole will not be exactly perpendicular, so that at all times there will be some sputtered contribution

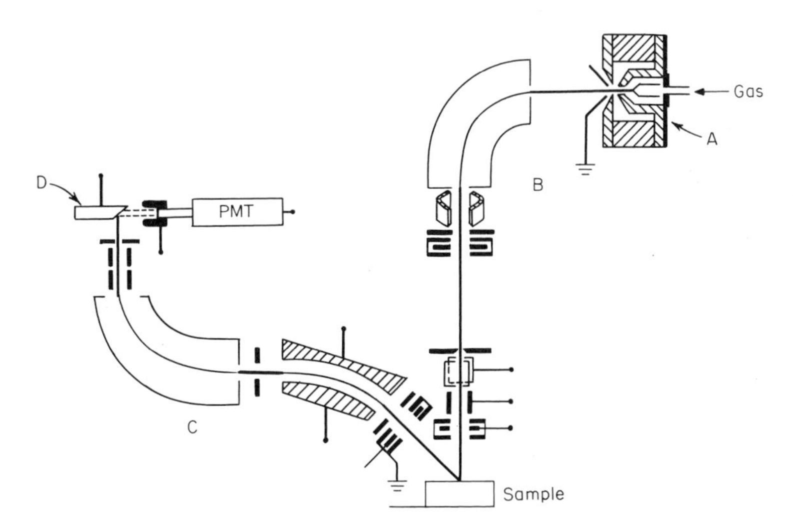

Fig. 9.19. Schematic of an ion-probe mass analyzer. A is the primary ion source. The magnetic and electrostatic optics of B are used to focus the beam. C is a mass spectrometer which analyzes the secondary ions which are ejected from the material studied. The secondary ions generate electrons at D which are detected by a scintillator-photomultiplier combination. Display can then be by a variety of methods.

over the whole depth of the hole. The measured profile will thus not be correct and furthermore will vary from instrument to instrument (particularly those of different manufacturers). In many instances the error is negligible, but it is also possible to change the character of the profile completely. Accordingly considerable care must be exercised in evaluating and interpreting such data.

9.9 ION BACKSCATTERING

Figure 9.20*a* is a schematic of the procedure used in ion-scattering experiments. A high-energy ion beam (up to several million electron volts) capable of penetrating into the material to be used is allowed to impinge on its surface, and the scattered beam energy is analyzed.[124] The scattered energy will depend both on the relative masses of the colliding particles (beam and sample constituents) and on the depth at which the collisions take place. A typical idealized spectrum will appear as the solid line in Fig. 9.20*b*. Should the sample be single-crystal and carefully aligned along a crystallographic axis, the beam can penetrate much more deeply by "channeling" between rows of atoms before a collision occurs. Thus the energy of the scattered beam will be substantially reduced, as indicated by the dotted line.

If there are misplaced atoms in the channel, i.e., interstitials, there will be additional scattering. For example, if the surface is damaged by ion bombardment so that interstitials of the sample are formed, an energy peak will occur as in Fig. 9.21*a*. Should the interstitials be foreign atoms, a peak will still occur, and if the mass of the foreign interstitial is different from the host, the energy of the scattered ions will be different and will be displaced. Figure 9.21*b* shows the appearance if the interstitial has a mass greater than that of the host atoms.

Sensitivity can be very good, and the technique can be used for analyzing films and contamination layers on the surface of semiconductors. Examples include composition of silicon nitride and aluminum oxide films deposited on Si;[125,126] the

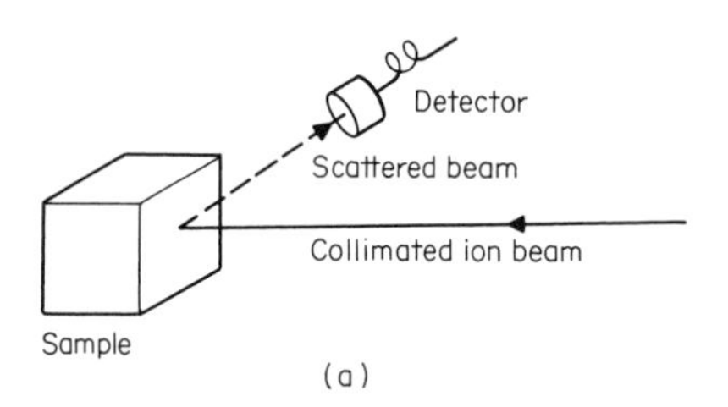

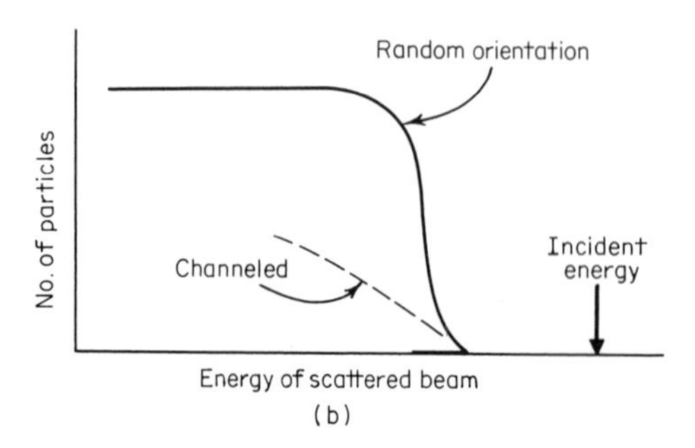

Fig. 9.20. Ion backscattering.

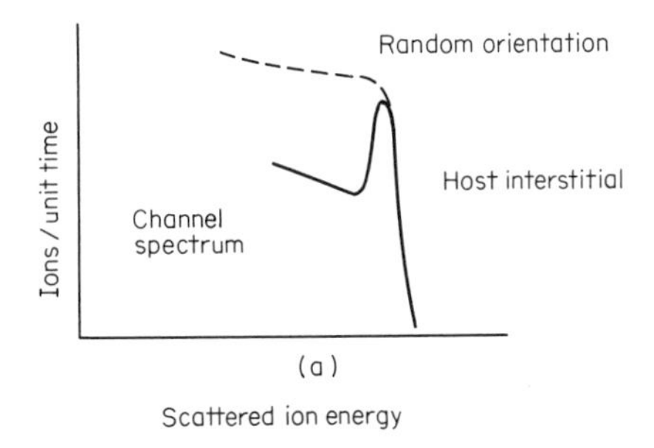

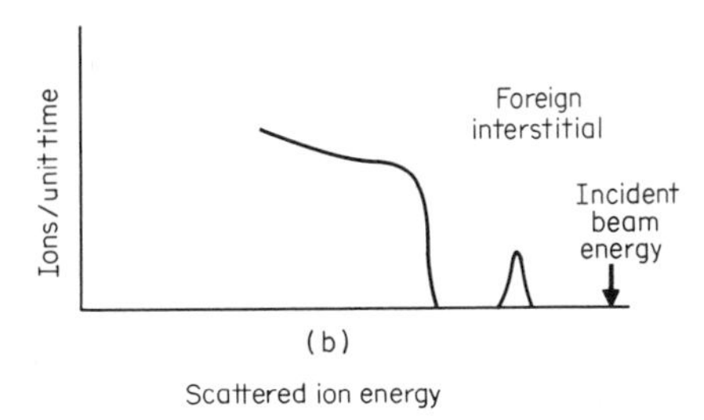

Fig. 9.21. Use of ion scattering to detect interstitials.

density of amorphous Si films;[127] the growth of secondary films, e.g., an SiO_2 layer over gold deposited on Si;[128] and miscellaneous surface contamination.[129,130]

9.10 RADIOACTIVE METHODS

Neutron-Activation Analysis.[108] Analysis is performed by irradiating the sample with neutrons, which converts a portion of the impurities into some radioactive species, and then monitoring the decay time of the activity. The activity A will be given by

$$A = \phi\sigma N(1 - e^{-\lambda t})$$

where ϕ is the integrated neutron flux, σ the absorption cross section of a given species, N their number, λ the decay constant of the newly generated radioactive species, and t the time after irradiation. In general, all the impurities as well as the matrix will become radioactive; so means of separating signals must be devised. The choice of detecting γ- or β-rays will remove some conflict. The short-lifetime components can often be allowed to decay so that only those with longer lifetimes are observed. Finally, some chemical separation may be required.

Because of the sensitivity required, the matrix activity can be of particular trouble, and indeed for most III–Vs is bad enough to virtually preclude its use. However, for Si and germanium it is a very useful and widely used method.

Charged-Particle-Activation Analysis.[131] Rather than neutron activation, a number of charged-particle reactions can be used. Some of these will allow greater sensitivity to elements such as boron, carbon, nitrogen, and oxygen.

Radioactive-Tracer Analysis. If a radioactive isotope of a particular element is used, its movement as a function of processing can be followed by detecting the radiation. The element whose motion is to be studied can be totally radioactive, but more likely will be considerably diluted with normal atoms. It is assumed that the radioactive atoms will be uniformly distributed among the others and will behave identically with them except for the radioactivity. Radiotracer studies have limited application but can be very useful in tracking the introduction of impurities if suitable radioactive elements are available in a form that can be used. Such studies are most commonly applied to diffusion[132] and segregation-coefficient determinations.[133] They have proved very helpful in establishing a direct correlation between etch-defined striations in melt-grown crystals and actual impurity-concentration nonuniformities.[134] Through autoradiograms, local variations of surface buildup of impurities can be studied, and in particular, redistributions after various washes and swabbing procedures can be noted.[108]

There are several ways of applying tracer data to diffusion-coefficient determination. They all depend on interpretations based on a combination of the diffusion-boundary conditions and properties of the emitted radiation. Such studies should always consider both facets of interpretation before conclusions are drawn. Probably the easiest method is to remove thin layers sequentially from the diffused surface (mechanically, by etching, or by oxidation and subsequent removal of the oxide) and count the activity of each layer. Alternately the activity remaining can be counted. In such cases the possibility of absorption of radioactivity originating well below the surface must be taken into account. One major source of error occurs in either case if there are unexpected pile-ups of the tracer. For example, if diffusion through a thick SiO_2 layer were being studied, pinholes could let the diffusant reach the silicon, diffuse into it, and remain despite innumerable oxide-etch cycles.

If the absorption coefficient of the radiation in the semiconductor material being studied is known, a measure of its intensity before and after diffusion combined with an assumed profile will allow the diffusion coefficient to be calculated. Should the diffusion be quite deep, a perpendicular section can be taken and the intensity profiled in a manner such as that described in Ref. 132. If it is suspected that a portion of a normally ionized impurity is not ionized and therefore not contributing to electrical conductivity (e.g., phosphorus in high concentrations in silicon), a heavy-concentration diffusion can be made using a radioactive source. Layers can then be sequentially stripped away and simultaneous counting and electrical-resistivity readings made. The impurity concentration as determined by radioactivity and the electrical concentration as determined by resistivity can then be compared.

REFERENCES

1. V. K. Zworykin et al., "Electron Optics and the Electron Microscope," John Wiley & Sons, Inc., New York, 1945.
2. Cecil E. Hall, "Introduction to Electron Microscopy," McGraw-Hill Book Company, New York, 1953.
3. Anon., TV Display Systems and Image Intensifier for Electron Microscopes, *Norelco Reptr.*, **20**:18–23 (April 1973).
4. Desmond H. Kay (ed.): "Techniques for Electron Microscopy," F. A. Davis Company, Philadelphia, 1965.
5. R. J. Jaccodine, Interaction of Diffusion and Stacking Faults in Si Epitaxial Material, *J. Appl. Phys.*, **36**:2811–2814 (1965).
6. M. V. Sullivan and R. M. Finne, Meeting of Electrochemical Society, Houston, October 1960.
7. G. R. Booker and R. Stickler, Method of Preparing Si and Ge Specimens for Examina-

tion by Transmission Electron Microscopy, *Brit. J. Appl. Phys.,* **13**:446–448 (1962); G. Das and N. A. O'Neil, Preparation of Large-Area Electron-Transparent Samples From Silicon Devices, *Solid-State Tech.,* **17**:29–31 (August 1974).
8. B. A. Irving, The Preparation of Thin Films of Germanium and Silicon, *Brit. J. Appl. Phys.,* **12**:92–93 (1961).
9. E. Biedermann and K. Brack, Preparation of GaAs Specimens for Transmission Electron Microscopy, *J. Electrochem. Soc.,* **113**:1088 (1966).
10. Eugene S. Meieran, Transmission Electron Microscope Study of Gallium Arsenide, *J. Appl. Phys.,* **36**:2544–2549 (1965).
11. B. D. Chase, D. B. Holt, and B. A. Unvala: Jet Polishing of Semiconductors, *J. Electrochem. Soc.,* **119**:310–313 (1972).
12. H. Abrams and R. N. Tauber, Transmission Electron Microscopy of As-grown PbSe Single Crystals, *J. Electrochem. Soc.,* **116**:103–104 (1969).
13. E. Levine and R. N. Tauber, The Preparation and Examination of PbTe by Transmission Electron Microscopy, *J. Electrochem. Soc.,* **115**:107–108 (1968).
14. T. G. R. Rawlins and L. E. Bresselard, Electron Microscope Replica Study of Epitaxial Silicon Nucleation on Silicon, *Trans. Met. Soc. AIME,* **236**:280–284 (1966).
15. P. R. Thornton, "Scanning Electron Microscopy," Chapman & Hall, Ltd., London, 1968; Oliver C. Wells (ed.), "Scanning Electron Microscopy," McGraw-Hill Book Company, New York, 1974.
16. Thomas E. Everhart, Deflection-Modulation CRT Display, *Proc. IEEE,* **54**:1480–1482 (1966).
17. Norman A. Foss, Deflection-Modulation Display of Depletion Movement in Epitaxial and Metal-Semiconductor Junctions, *J. Appl. Phys.,* **41**:823–825 (1970).
18. G. Y. Robinson, Scanning Electron Microscopy Study of the Avalanche Multiplication Factor, *Proc. IEEE,* **57**:2169–2170 (1969).
19. T. E. Everhart, O. C. Wells, and C. W. Oatley, Factors Affecting Contrast and Resolution in the Scanning Electron Microscope. *J. Electron. Control,* **7**:97–111 (1959).
20. R. P. Beaulieu, C. D. Cox, and T. M. Black, "An SEM Surface Voltage Measurement Technique," *10th Annual Proceedings Reliability Physics,* pp. 32–35, 1972 (and references contained therein).
21. A. J. Gonzales, "Factors Involved in the Scanning Electron Microscope Analysis of Glass Passivated Devices," *9th Annual Proceedings Reliability Physics,* pp. 142–148, 1971.
22. D. L. Crosthwait, Jr., "Synchronous Biasing of Integrated Circuits in the SEM," *Proceedings 6th Annual Stereoscan Colloquium,* Kent Cambridge Scientific, Chicago, 1973.
23. Kurt F. J. Heinrich, Charles Fiori, and Harvey Yakowitz, Image-Formation Technique for Scanning Electron Microscopy and Electron Probe Microanalysis, *Science,* **167**:1129–1131 (1970).
24. P. R. Thornton, D. V. Sulway, and D. A. Shaw, Scanning Electron Microscopy in Device Diagnostics and Reliability Physics, *IEEE Trans. Electron. Devices,* **ED-16**:360–371 (1969).
25. T. E. Everhart, O. C. Wells, and R. K. Matta, Evaluation of Passivated Integrated Circuits Using the Scanning Electron Microscope, *J. Electrochem. Soc.,* **111**:929–936 (1964).
26. M. J. Alter and B. A. McDonald, The SEM as a Defect Analysis Tool for Semiconductor Memories, *9th Annual Proceedings Reliability Physics,* pp. 149–154, 1971.
27. John W. Adolphsen and Robert J. Anstead, Use a SEM on a Production Line?!! *8th Annual Proceedings Reliability Physics,* pp. 238–243, 1970.
28. Robert J. Anstead, Failure Analysis Using a Scanning Electron Microscope, *Sixth Annual Reliability Physics Symposium Proceedings,* pp. 127–137, 1967.
29. P. R. Thornton et al., Failure Analysis of Microcircuitry by Scanning Electron Microscopy, *Microelectron. Reliability,* **6**:9–16 (1967).
30. P. R. Thornton et al., Device Failure Analysis by Scanning Electron Microscopy, *Microelectron. Reliability,* **8**:33–53 (1969).

31. I. G. Davies et al., The Direct Observation of Electrical Leakage Paths due to Crystal Defects by Use of the Scanning Electron Microscope, *Solid State Electron.,* **9**:275–279 (1966).
32. P. R. Thornton et al., Quantitative Measurements by Scanning Electron Microscopy, I. The Use of Conductivity Maps, *Microelectron. Reliability,* **5**:291–298 (1966).
33. J. W. Thornhill and I. M. Mackintosh, Application of the Scanning Electron Microscope to Semiconductor Device Structures, *Microelectron. Reliability,* **4**:97–100 (1965).
34. R. K. Matta, Application of the Scanning Electron Microscope to Semiconductors, *Solid State Tech.,* **12**:34–42 (1969).
35. G. S. Plows and W. C. Nixon, Operational Testing of LSI Arrays by Stroboscopic Scanning Electron Microscopy, *J. Sci. Instr.,* **1**:595–600 (1968).
36. G. V. Lukianoff and R. C. Mullaney, "Electrical Continuity Testing of Passivated Metallization by Scanning Electron Microscopy," *8th Annual Proceedings Reliability Physics,* pp. 244–246, 1970.
37. S. K. Behera and D. P. Speer, "A Procedure for the Evaluation of Failure Analysis of MOS Memory Circuits Using the Scanning Electron Microscope in Potential Contrast Mode," *10th Annual Proceedings Reliability Physics,* 1972.
38. J. P. Fleming, Electron Beam Testing, Gentle and Fast, *Electronics,* **42**:92–94 (1969).
39. J. F. Norton and H. L. Lester, "Integrated Circuit Diagnostics Using Electron Beam Probes," presented at the National Electronics Conference, Chicago, 1969.
40. P. B. Hirsch, R. B. Nicholson, A. Howie, and D. W. Pashley, "Electron Microscopy of Thin Crystals," Butterworth & Co. (Publishers), Ltd., Washington, 1965.
41. P. J. Estrup and E. G. McRae, Surface Studies by Electron Diffraction, *Surface Sci.,* **25**:1–52 (1971).
42. E. Bauer, Reflection Electron Diffraction, chap. 15; and Low Energy Electron Diffraction, chap. 16, in R. F. Bunshah (ed.), "Techniques of Metals Research," Interscience Publishers, Inc., New York, 1969.
43. R. C. Henderson, W. J. Polito, and J. Simpson, Observation of SiC with Si(111)-7 Surface Structure Using High-Energy Electron Diffraction, *Appl. Phys. Lett.,* **16**:15–17 (1970).
44. C. W. B. Grigson, Improved Scanning Electron Diffraction System, *Rev. Sci. Instr.,* **36**:1587–1593 (1965).
45. A. Y. Cho, GaAs Epitaxy by a Molecular Beam Method: Observations of Surface Structure on the (001) Face, *J. Appl. Phys.,* **42**:2074–2081 (1971).
46. G. J. Russell and D. Haneman, Oxygen Adsorption on Vacuum Cleaved GaAs at Liquid Nitrogen Temperature by LEED, *Surface Sci.,* **27**:362–366 (1971).
47. Herbert W. Schnopper, Spectral Measurements with Aligned and Misaligned Two-Crystal Spectrometers, I, Theory of the Geometrical Window, *J. Appl. Phys.,* **36**:1415–1423 (1965).
48. Herbert W. Schnopper, Spectral Measurements with Aligned and Misaligned Two-Crystal Spectrometers, II, Alignment, *J. Appl. Phys.,* **36**:1423–1430 (1965).
49. Herbert W. Schnopper, Erratum, *J. Appl. Phys.,* **36**:3692 (1965).
50. A. R. Lang, Crystal Growth and Crystal Perfection: X-Ray Topographic Studies, *Discussions Faraday Soc., (38)* 292–297 (1964).
51. Eugene S. Meieran, Reflection X-Ray Topography of GaAs Deposited on Ge, *J. Electrochem. Soc.,* **114**:292–295 (1967).
52. A. R. Lang, A Method for the Examination of Crystal Sections Using Penetrating Characteristic X-Radiation, *Acta Met.,* **5**:358–364 (1957).
53. A. R. Lang, Direct Observation of Individual Dislocations by X-Ray Diffraction, *J. Appl. Phys.,* **29**:597–598 (1958).
54. A. R. Lang, The Projection Topograph: A New Method in X-Ray Diffraction Microradiography, *Acta Cryst.,* **12**:249–250 (1959).
55. A. R. Lang, Studies of Individual Dislocations in Crystals by X-Ray Diffraction Microradiography, *J. Appl. Phys.,* **30**:1748–1755 (1959).

56. G. H. Schwuttke, X-Ray Diffraction Microscopy Study of Imperfections in Silicon Single Crystals, *J. Electrochem. Soc.,* **109**:27–32 (1962).
57. A. E. Jenkinson and A. R. Lang, X-Ray Diffraction Topographic Studies of Dislocations in Floating-Zone Grown Silicon, in J. B. Newkirk and J. H. Wernick (eds.), "Direct Observations of Imperfections in Crystals," Interscience Publishers, Inc., New York, 1962.
58. G. H. Schwuttke, New X-Ray Diffraction Microscopy Technique for the Study of Imperfections in Semiconductor Crystals, *J. Appl. Phys.,* **36**:2712–2721 (1965).
59. Kazutake Kohra, Mitsuru Yoshimatsu, and Ikuzo Shimizu, X-Ray Observation of Lattice Defects Using a Crystal Monochrometer, in J. B. Newkirk and J. H. Wernick (eds.), "Direct Observations of Imperfections in Crystals," Interscience Publishers, Inc., New York, 1962.
60. Mitsuru Yoshimatsu, Atsushi Shibata, and Kazutake Kohra, A Modification of the Scanning X-ray Topographic Camera (Lang's Method), in Gavin R. Mallett, Marie Fay, and William M. Mueller (eds.), "Advances in X-Ray Analysis," vol. 9, Plenum Press, Plenum Publishing Corporation, New York, 1966.
61. N. Kato and A. R. Lang, A Study of Pendellösung Fringes in X-Ray Diffraction, *Acta Cryst.,* **12**:787–794 (1959).
62. N. Kato, Dynamical Diffraction Theory of Waves in Distorted Crystals, II, Perturbation Theory, *Acta Cryst.,* **16**:282–290 (1963).
63. W. W. Webb, X-Ray Diffraction Topography, in J. B. Newkirk and J. H. Wernick (eds.), "Direct Observations of Imperfections in Crystals," Interscience Publishers, Inc., New York, 1962.
64. Volkmar Gerald, X-Ray Methods for Detection of Lattice Imperfections in Crystals, in William M. Mueller (ed.), "Advances in X-Ray Analysis," vol. 3, Plenum Press, Plenum Publishing Corporation, New York, 1960.
65. G. H. Schwuttke, Direct Observation of Imperfections in Semiconductor Crystals by Anomalous Transmission of X-Rays, *J. Appl. Phys.,* **33**:2760–2767 (1962).
66. J. K. Howard and R. D. Dobrott, Compositional X-Ray Topography, *J. Electrochem. Soc.,* **113**:567–573 (1966).
67. L. G. Schulz, Method of Using a Fine-Focus X-Ray Tube for Examining the Surface of Single Crystals, *J. Metals,* **6**:1082–1083 (1954).
68. Laurence N. Swink and Maurice J. Brau, Rapid Non-destructive Evaluation of Macroscopic Defects in Crystalline Materials: The Laue Topography of (Hg, Cd) Te, *Met. Trans.,* **1**:629–634 (1970).
69. Junji Matsui and Tsutomu Kawamura, Spotty Defects in Oxidized Floating-zoned Dislocation-free Silicon Crystals, *Japan. J. Appl. Phys.,* **11**:197–205 (1972).
70. Jun-ichi Chikawa, Isao Fujimoto, Sigeaki Endo, and Kiyoshi Mase, X-Ray Television Topography for Quick Inspection of Si Crystals, in Howard R. Huff and Ronald R. Burgess (eds.), "Semiconductor Silicon," The Electrochemical Society, Princeton, 1973.
71. Heinrich Grienauer, High Temperature X-Ray Topography of Silicon in Howard R. Huff and Ronald R. Burgess (eds.), "Semiconductor Silicon," The Electrochemical Society, Princeton, 1973.
72. I. Blech, J. Guyaux, and G. Cooper, High Temperature Camera for X-Ray Topography, *Rev. Sci. Instr.,* **38**:638–641 (1967).
73. Jun-ichi Chikawa, Isao Fujimoto, and Takao Abe, X-Ray Topography Observation of Moving Dislocations in Silicon Crystals, *Appl. Phys. Lett.,* **21**:295–298 (1972).
74. U. K. Bonse, M. Hart, and J. B. Newkirk, X-Ray Diffraction Topography, in J. B. Newkirk and G. R. Mallett (eds.), "Advances in X-Ray Analysis," vol. 10, pp. 1–8, Plenum Press, Plenum Publishing Corporation, New York, 1967.
75. E. M. Juleff, A. G. Lapierre III, and R. G. Wolfson, The Analysis of Berg-Barrett Skew Reflections and Their Applications in the Observation of Process-induced Imperfections in (111) Silicon Wafers, in J. B. Newkirk and G. R. Mallett (eds.), "Advances in X-Ray

Analysis," vol. 10, pp. 173–184, Plenum Press, Plenum Publishing Corporation, New York, 1967.
76. A. Authier, Contrast of Dislocation Images in X-Ray Transmission Topography, in J. B. Newkirk and G. R. Mallett (eds.), "Advances in X-Ray Analysis," vol. 10, pp. 9–31, Plenum Press, Plenum Publishing Corporation, New York, 1967.
77. S. B. Austerman and J. B. Newkirk, Experimental Procedures in X-Ray Diffraction Topography, in J. B. Newkirk and G. R. Mallett (eds.), "Advances in X-Ray Analysis," vol. 10, pp. 134–152, Plenum Press, Plenum Publishing Corporation, New York, 1967.
78. H. P. Layer and R. D. Deslattes, A Simple Nonscanning Camera for X-Ray Diffraction Contrast Topography, *J. Appl. Phys.,* **37**:3631–3632 (1966).
79. J. S. Makris and C. H. Ma, A Modified X-Ray Diffraction Microscope Technique for Study of Dislocations in Crystals, *Trans. Met. Soc. AIME,* **230**:1110–1112 (1964).
80. G. Dionne, High-Resolution X-Ray-Diffraction Topography Using K_β Radiation, *J. Appl. Phys.,* **38**:4094–4096 (1967).
81. G. J. Carron, X-Ray Topographic Camera, *Rev. Sci. Instr.,* **38**:628–631 (1967).
82. Wilhad Reuter, Electron Probe Microanalysis, *Surface Sci.,* **25**:80–119 (1971).
83. D. B. Wittry, J. M. Axelrod, and J. O. McCaldin, Use of the Electron Probe X-Ray Microanalyzer in the Study of Semiconductor Alloys, in Harry C. Gatos (ed.), "Properties of Elemental and Compound Semiconductors," Interscience Publishers, Inc., New York, 1960.
84. M. I. Nathan and S. H. Moll, Electron Beam Micro Analysis of Germanium Tunnel Diodes, *IBM J. Res. Develop.,* **6**:375–377 (1962).
85. Frank J. Cocca and Kenneth G. Carroll, Electron Microprobe Analysis of Impurity Heterogeneities in Thermally Grown Silicon Oxide, *IEEE Trans. Electron. Devices,* **ED-15**: 962–966 (1968).
86. S. S. So and H. R. Potts, Computer Programs for Quantitative and Semiquantitative Analysis with the Electron Microprobe Analyzer, *J. Electrochem. Soc.,* **115**:64–70 (1968).
87. Rolf Woldseth, "X-Ray Energy Spectrometry," Kevex Corp., Burlingame, Calif., 1973.
88. Wendland Beezhold, Ion-induced Characteristic X-Ray Micro-Analysis, in Howard R. Huff and Ronald R. Burgess (eds.), "Semiconductor Silicon," The Electrochemical Society, Princeton, 1973.
89. Thomas A. Carlson, Electron Spectroscopy for Chemical Analysis, *Physics Today,* **25**:30–39 (January 1972).
90. Homer D. Hagstrum, Electronic Characterization of Solid Surfaces, *Science,* **178**:275–282 (1972).
91. Roland E. Weber, Auger Electron Spectroscopy for Thin Film Analysis, *Research/Development,* **23**:22–28 (October 1972).
92. Chuan C. Chang, Auger Electron Spectroscopy, *Surface Sci.,* **25**:53–79 (1971).
93. Paul W. Palmberg et al., "Handbook of Auger Spectroscopy," Physical Electronic Industries, Edina, Minn., 1972.
94. J. M. Morabito and J. C. Tsai, In-Depth Profiles of Phosphorus Ion-implanted Silicon by Auger Spectroscopy and Secondary Ion Emission, *Surface Sci.,* **33**:422–426 (1972).
95. P. W. Palmberg, Use of Auger Electron Spectroscopy and Inert Gas Sputtering for Obtaining Chemical Profiles, *J. Vacuum Sci. Tech.,* **9**:160–163 (1972).
96. J. T. Grant and T. W. Haas, Auger Electron Spectroscopy of Si, *Surface Sci.,* **23**:347–362 (1970).
97. Norman J. Taylor, Thin Reaction Layers and the Surface Structure of Silicon (111), *Surface Sci.,* **15**:169–174 (1969).
98. Tadoshi Narusawa, Auger Electron Emission from Gold Deposited on Silicon (111) Surface, *Japan. J. Appl. Phys.,* **10**:280–281 (1971).
99. T. Narusawa, S. Komiya, and A. Hiraki, Auger Spectroscopic Observation of Si-Au Mixed Phase Formation at Low Temperatures, *Appl. Phys. Lett.,* **21**:272–273 (1972).

100. R. L. Moon and L. W. James, Auger Spectra of Hcl Vapor-etched n^+ GaAs 100 Substrates, *J. Electrochem. Soc.*, **120**:581–583 (1973).
101. John J. Uebbing and Norman J. Taylor, Auger Electron Spectroscopy of Clean Gallium Arsenide, *J. Appl. Phys.*, **41**:804–808 (1970).
102. B. A. Joyce and J. H. Neave, An Investigation of Silicon-Oxygen Interactions Using Auger Electron Spectroscopy, *Surface Sci.*, **27**:499–515 (1971).
103. R. C. Henderson, Silicon Cleaning with Hydrogen Peroxide Solutions: A High Energy Electron Diffraction and Auger Electron Spectroscopy Study, *J. Electrochem. Soc.*, **119**:772–775 (1972).
104. Chuan C. Chang, Contaminants on Chemically Etched Silicon Surfaces: LEED-Auger Method, *Surface Sci.*, **23**:283–298 (1970).
105. J. H. Affleck, Auger Spectroscopy and Silicon Surfaces, in Charles P. Marsden (ed.), "Silicon Device Processing," National Bureau of Standards, *Spec. Pub.* 337, 1970.
106. John J. Uebbing, Use of Auger Electron Spectroscopy in Determining the Effect of Carbon and Other Surface Contaminants on GaAs-Cs-O Photocathodes, *J. Appl. Phys.*, **41**:802–804 (1970).
107. H. G. Maguire and P. D. Augustus, The Detection of Silicon-Oxynitride Layers on the Surfaces of Silicon-Nitride Films by Auger Electron Emission, *J. Electrochem. Soc.*, **119**:791–793 (1972).
108. Philip F. Kane and Graydon B. Larrabee, "Characterization of Semiconductor Materials," McGraw-Hill Book Company, New York, 1970.
109. J. Paul Cali (ed.), "Trace Analysis of Semiconductor Materials," The Macmillan Company, New York, 1964.
110. William R. Knolle, Flame Emission Analysis of Potassium Contamination in Silicon Slice Processing, *J. Electrochem. Soc.*, **120**:987–991 (1973).
111. J. E. Barry, H. M. Donega, and T. E. Burgess, Flame Emission Analysis for Sodium in Silicon Oxide Films and on Silicon Surfaces, *J. Electrochem. Soc.*, **116**:257–259 (1969).
112. William R. Knolle, and Theodore F. Retajczyk, Jr., Monitoring Sodium Contamination in Silicon Devices and Processing Materials by Flame Emission Spectrometry, *J. Electrochem. Soc.*, **120**:1106–1111 (1973).
113. W. Kaiser and P. H. Keck, Oxygen Content of Silicon Single Crystals, *J. Appl. Phys.*, **28**:882–887 (1957).
114. A. R. Bean and R. C. Newman, The Solubility of Carbon in Pulled Silicon Crystals, *J. Phys. Chem. Solids*, **32**:1211–1219 (1971).
115. W. A. Pliskin, Use of Infrared Spectroscopy for the Characterization of Dielectric Films on Silicon, in Howard R. Huff and Ronald R. Burgess (eds.), "Semiconductor Silicon," The Electrochemical Society, Princeton, 1973.
116. K. H. Beckmann, Investigation of the Chemical Properties of Stain Films on Silicon by Means of Infrared Spectroscopy, *Surface Sci.*, **3**:314–332 (1965).
117. M. Blattë, Emittance Spectroscopy as a Diagnostic Tool for Semiconductors, in Rolf R. Haberecht and Edward L. Kern (eds.), "Semiconductor Silicon," The Electrochemical Society, New York, 1969.
118. D. L. Stierwalt and R. F. Potter, Lattice Absorption Bands Observed in Silicon by Means of Spectral Emissivity Measurements, *J. Phys. Chem. Solids*, **23**:99–102 (1962).
119. R. Sittig and W. Zimmerman, Doping Inhomogeneities in Semiconductors Measured by Electroreflectance, in Howard R. Huff and Ronald R. Burgess (eds.), "Semiconductor Silicon," The Electrochemical Society, Princeton, 1973.
120. Richard A. Forman, David E. Aspnes, and Manuel Cardona, Transverse Electroreflectance in Semi-insulating Silicon and Gallium Arsenide, *J. Phys. Chem. Solids*, **31**:227–246 (1970).
121. Manuel Cardona, Kerry L. Shaklee, and Fred H. Pollak, Electroreflectance at a Semiconductor-Electrolyte Interface, *Phys. Rev.*, **154**:696–720 (1967).
122. Helmut Liebl, Ion Microprobe Mass Analyzer, *J. Appl. Phys.*, **38**:5277–5283 (1967).

123. A. J. Socha, Analysis of Surfaces Utilizing Sputter Ion Source Instruments, *Surface Sci.*, **25**:147–170 (1971).
124. J. W. Mayer, L. Eriksson, and J. A. Davies, "Ion Implantation in Semiconductors," Academic Press, Inc., New York, 1970.
125. J. Gyulai et al., Analysis of Silicon Nitride Layers on Silicon by Backscattering and Channeling Effect Measurements, *Appl. Phys. Lett.*, **16**:232–234 (1970).
126. M. Kamoshida and J. W. Mayer, Backscattering Studies of Anodization of Aluminum Oxide and Silicon Nitride on Silicon, *J. Electrochem. Soc.*, **119**:1084–1090 (1972).
127. M. H. Brodsky, D. Kaplan, and J. F. Ziegler, Densities of Amorphous Si Films by Nuclear Backscattering, *Appl. Phys. Lett.*, **21**:305–307 (1972).
128. Akio Hiraki, Eriabu Lugujjo, and J. W. Mayer, Formation of Silicon Oxide over Gold Layers on Silicon Substrates, *J. Appl. Phys.*, **43**:3643–3649 (1973).
129. W. K. Chu, J. W. Mayer, M-A. Nicolet, T. M. Buck, G. Amsel, and F. Eisen, Microanalysis of Surface, Thin Films, and Layers Structures by Nuclear Backscattering and Reactions, in Howard R. Huff and Ronald R. Burgess (eds.), "Semiconductor Silicon," The Electrochemical Society, Princeton, 1973.
130. D. A. Thompson, H. D. Barber, and W. D. Mackintosh, The Determination of Surface Contamination on Silicon by Large Angle Ion Scattering, *Appl. Phys. Lett.*, **14**:102–103 (1969).
131. E. A. Schweikert and R. E. Wainerdi, Ultratrace Analysis of Light Elements Using Charged Particle Activation, Application to the Determination of Oxygen in Silicon, *J. Electrochem. Soc.*, **115**:249C (1968).
132. B. I. Boltaks, "Diffusion in Semiconductors," Academic Press, Inc., New York, 1963.
133. S. Nakanuma, Radiotracer Studies on the Incorporation of Phosphorus in Epitaxially Grown Silicon, *J. Electrochem. Soc.*, **111**:1199–1200 (1964).
134. G. R. Cronin, G. B. Larrabee, and J. F. Osborne, Annular Facets and Impurity Striations in Tellurium Doped Gallium Arsenide, *J. Electrochem. Soc.*, **113**:293–294 (1966).

Index